RESPONSIBLE MARINE AQUACULTURE

Responsible Marine Aquaculture

Edited by

Robert R. Stickney
Director
Texas Sea Grant College Program
College Station, Texas, USA

and

James P. McVey
Aquaculture Program Director
National Sea Grant College Program
Silver Spring, Maryland, USA

CABI *Publishing*

CABI *Publishing* is a division of CAB *International*

CABI Publishing
CAB International
Wallingford
Oxon OX10 8DE
UK

Tel: +44 (0)1491 832111
Fax: +44 (0)1491 833508
Email: cabi@cabi.org
Web site: www.cabi-publishing.org

CABI Publishing
10 E 40th Street
Suite 3203
New York, NY 10016
USA

Tel: +1 212 481 7018
Fax: +1 212 686 7993
Email: cabi-nao@cabi.org

A catalogue record for this book is available from the British Library, London, UK.

Library of Congress Cataloging-in-Publication Data

Responsible marine aquaculture / edited by Robert R. Stickney and James P. McVey.
p. cm.
Includes bibliographical references.
ISBN 978-0-85199-604-2
1. Mariculture--Management. 2. Mariculture--Environmental aspects.
I. Stickney, Robert R. II. McVey, James P.
SH138.R47 2002
639.8--dc21 2002000585

ISBN 978-0-85199-604-2

Typeset by AMA DataSet Ltd, UK.
Printed and bound by CPI Group (UK) Ltd, Croydon, CR0 4YY

Contents

Contributors ix

Foreword xiii

1. Introduction 1
Robert R. Stickney

2. Management of Marine Aquaculture: the Sustainability Challenge 21
M. Richard DeVoe and Catherine E. Hodges

3. Marine Mammals and Aquaculture: Conflicts and Potential Resolutions 45
Bernd Würsig and Glenn A. Gailey

4. Recreational Fishing and Aquaculture: Throwing a Line into the Pond 61
William D. Harvey and Larry D. McKinney

5. Aquaculture: Opportunity or Threat to Traditional Capture Fishermen? 71
Rollie Barnaby and Steve Adams

6. Advances in Marine Stock Enhancement: Shifting Emphasis to Theory and Accountability 79
Kenneth M. Leber

7. Aquatic Polyculture and Balanced Ecosystem Management: New Paradigms for Seafood Production 91
James P. McVey, Robert R. Stickney, Charles Yarish and Thierry Chopin

8. The Role of Marine Aquaculture Facilities as Habitats and Ecosystems 105
Barry A. Costa-Pierce and Christopher J. Bridger

9. Mangroves and Coastal Aquaculture 145
Claude E. Boyd

10. Environmental Effects Associated with Marine Netpen Waste with Emphasis on Salmon Farming in the Pacific Northwest 159
Kenneth M. Brooks, Conrad Mahnken and Colin Nash

11. Issues Associated with Non-indigenous Species in Marine Aquaculture 205
Robert R. Stickney

12. Genetic Changes in Marine Aquaculture Species and the Potential for Impacts on Natural Populations 221
William K. Hershberger

13. What Role Does Genetics Play in Responsible Aquaculture? 233
L. James Lester

14. Understanding the Interaction of Extractive and Fed Aquaculture Using Ecosystem Modelling 263
Mac V. Rawson, Jr, Changsheng Chen, Rubao Ji, Mingyuan Zhu, Daoru Wang, Lu Wang, Charles Yarish, James B. Sullivan, Thierry Chopin and Raquel Carmona

15. Shrimp Farm Effluents 297
Granvil D. Treece

16. Fish Meal: Historical Uses, Production Trends and Future Outlook for Sustainable Supplies 311
Ronald W. Hardy and Albert G.J. Tacon

17. The Use of Wild-caught Juveniles in Coastal Aquaculture and its Application to Coral Reef Fishes 327
Cathy Hair, Johann Bell and Peter Doherty

18. Contending with Criticism: Sensible Responses in an Age of Advocacy 355
Terrence R. Tiersch and John A. Hargreaves

Index 373

Contributors

S. Adams, New Hampshire Sea Grant, Kingman Farm, Durham, NH 03824, USA.

R. Barnaby, University of New Hampshire, Cooperative Extension/Sea Grant, 113 North Road, Brentwood, NY 03833, USA.

J. Bell, ICLARM – The World Fish Center, PO Box 500, GPO 10670, Penang, Malaysia.

C.E. Boyd, Department of Fisheries and Allied Aquacultures, Swingle Hall, Auburn University, Auburn, AL 36849, USA.

C.J. Bridger, College of Marine Sciences, Gulf Coast Research Laboratory, The University of Southern Mississippi, 703 East Beach Drive, Ocean Springs, MS 39564, USA.

K.M. Brooks, Aquatic Environmental Sciences, 644 Old Eaglemount Road, Port Townsend, WA 98368, USA.

R. Carmona, University of Connecticut, Department of Ecology and Evolutionary Biology, 1 University Place, Stamford, CT 06901–2315, USA

C. Chen, Georgia Sea Grant College Program, The University of Georgia, 220 Marine Sciences Building, Athens, GA 30602–3636, USA.

T. Chopin, University of New Brunswick, Centre for Coastal Studies and Aquaculture & Centre for Environmental and Molecular Algal Research, PO Box 5050, Saint John, New Brunswick E2L 475, Canada.

B.A. Costa-Pierce, Mississippi–Alabama Sea Grant Consortium, 703 East Beach Drive, Ocean Springs, MS 39564, USA. (Current address: Rhode Island Sea Grant College Program, University of Rhode Island Narragansett Bay Campus, Narragansett, RI 02882, USA.)

M.R. De Voe, South Carolina Sea Grant Consortium, 287 Meeting Street, Charleston, SC 29401, USA.

P. Doherty, Australian Institute of Marine Science, PMB No.3, Townsville, Queensland, Australia.

G.A. Gailey, Department of Marine Biology, Texas A&M University at Galveston, 4700 Avenue U, Building 303, Galveston, TX 77551, USA.

C. Hair, ICLARM – The World Fish Center, PO Box 500, GPO 10670, Penang, Malaysia.

R.W. Hardy, University of Idaho, Hagerman Fish Culture Experiment Station, 3059 National Fish Hatchery Road, Hagerman, ID 83332, USA.

J.A. Hargreaves, Department of Wildlife and Fisheries, Mississippi State University, Mississippi State, MS 39762, USA.

W.D. Harvey, Resource Protection Division, Texas Parks and Wildlife Department, 4200 Smith School Road, Austin, TX 78744, USA.

W.K. Hershberger, USDA/ARS/NCCCWA, 11876 Leetown Road, Kearneysville, WV 25430, USA.

C.E. Hodges, South Carolina Sea Grant Consortium, 287 Meeting Street, Charleston, SC 29401, USA.

R. Ji, Georgia Sea Grant College Program, The University of Georgia, 220 Marine Sciences Building, Athens, GA 30602–3636, USA.

K.M. Leber, Center for Fisheries Enhancement, Mote Marine Laboratory, 1600 Ken Thompson Parkway, Sarasota, 34236, USA.

L.J. Lester, Environmental Institute of Houston, University of Houston – Clear Lake, 2700 Bay Area Boulevard, Houston, TX 77058, USA.

C. Mahnken, National Marine Fisheries Service, Northwest Fisheries Science Center, Manchester Laboratory, 7305 Beach Drive East, Manchester, WA 98366, USA.

L.D. McKinney, Resource Protection Division, Texas Parks and Wildlife Department, 4200 Smith School Road, Austin, TX 78744, USA

J.P. McVey, National Sea Grant College Program, SSMC-3, Room 11829, 1335 East-West Highway, Silver Spring, MD 20910–3226, USA.

C. Nash, National Marine Fisheries Service, Northwest Fisheries Science Center, Manchester Laboratory, 7305 Beach Drive East, Manchester, WA 98366, USA.

M.V. Rawson, Jr, Georgia Sea Grant College Program, The University of Georgia, 220 Marine Sciences Building, Athens, GA 30602–3636, USA.

R.R. Stickney, Texas Sea Grant College Program, 2700 Earl Rudder Freeway South, Suite 1800, College Station, TX 77845, USA.

J.B. Sullivan, Department of Marine Sciences, The University of Georgia, 220 Marine Sciences Building, Athens, GA 30602–3636, USA.

A.G.J. Tacon, Aquatic Farms Ltd, 49–139 Kamehamehja Highway, Keneohe, HI 96744, USA.

T.R. Tiersch, Aquaculture Research Station, Louisiana State University Agricultural Center, Louisiana Agricultural Experiment Station, Baton Rouge, LA 70820, USA.

G.D. Treece, Texas Sea Grant College Program, 2700 Earl Rudder Highway South, Suite 1800, College Station, TX 77845, USA.

D. Wang, Hainan Marine Development, Planning and Design Institute, 10th Yiyuan Building No. 69, Haifu Road, Haikou, Hainan 570203, People's Republic of China.

L. Wang, Marine and Fishery Department of Hainan Province, Haikou, Hainan, People's Republic of China.

B. Würsig, Department of Marine Biology, Texas A&M University at Galveston, 4700 Avenue U, Building 303, Galveston, TX 77551, USA.

C. Yarish, University of Connecticut, Department of Ecology and Evolutionary Biology, 1 University Place, Stamford, CT 06901–2315, USA.

M. Zhu, The First Institute of Oceanography, State Oceanic Administration, PO Box 98, Qingdao 26603, People's Republic of China.

Foreword

As the fastest-growing food production sector in the world, modern aquaculture is maturing, with its emphasis shifting from the development of production technologies to assurance of longer-term economic and environmental sustainability. As in any rapidly growing industry, relatively infrequent examples can be cited of environmental or social abuses. As environmental groups, and even prestigious scientific journals, have increasingly targeted aquaculture as an unsustainable activity, the need for balanced and scientifically rigorous reviews of issues related to aquaculture sustainability has increased. The present volume (based on the special session on aquaculture sustainability at Aquaculture 2001 in Orlando, Florida) is particularly timely in this regard, providing information for continuing improvements in aquaculture sustainability while providing research-based data to respond to often exaggerated or inaccurate external criticisms.

Fish is an essential food source for mankind. It is mankind's largest source of animal protein (16%), and makes up an even higher proportion in areas where high quality protein is relatively scarce (21% overall in low-income food-deficit countries). Almost 1 billion people rely upon fish as their primary source of animal protein. According to the United Nations Food and Agriculture Organization (FAO), about three-quarters of major marine fish stocks are presently harvested to their maximum limits and there is little or no room for growth in international wild harvests. Aquaculture has provided a vitally important role in expanding supplies of limited aquatic food resources. Aquaculture production expanded from 7 million to 33 million metric tons from 1984 to 1999, and now provides over 30% of the world's total food fish. At a growth rate of 11% per year, aquaculture could surpass beef production by 2010 and by 2030 aquaculture will provide over half of the fish consumed by the world's people. Aquaculture has been growing almost six times faster in developing countries than in developed countries and FAO has stated that as an inexpensive source of a highly nutritious animal protein, aquaculture has become an important factor for improving food security, raising nutritional standards and alleviating poverty in the world's poorest countries.

With its extremely rapid development aquaculture has been associated with impacts that, in some cases, warrant criticism. The nature of this criticism ranges from valid targeted responses to specific problems, to grossly inaccurate generalizations and scientifically dubious literature and reports. These are then repeatedly cited and magnified, generating an overall negative image of a widely diverse industry. Aquaculture is in most cases a relatively environmentally responsible alternative for coastal economic development, which depends upon a healthy environment around farms and a strong community base. Many of the early unsustainable activities are not being repeated and enlightened producers have been quick to respond to, and improve, environmentally unsustainable practices. Researchers have made great strides in improving diet formulations and production technologies, significantly enhancing the outlook for long-term technical sustainability. Increasing scrutiny by investors, lending agencies, governments and consumers has led to policies which support improved site selection, regional planning and social sustainability.

With improving management practices and technologies, continuing long-term ecological and social assessments, good planning and effective regulatory policies, responsible aquaculture will continue to grow and play an increasingly important role in global food security and economic development. The often unfair assignment of a negative environmental label to aquaculture in general (through unbalanced ecological assessments and exaggerated generalizations) could have serious long-term negative consequences. With world population continuing to increase, the deficit between the supply from capture fisheries and increasing demand will continue to increase. The consequences of this could be devastating for global fisheries and for human populations in regions where population growth and protein deficits are highest.

The present volume provides a series of reviews focusing on technical advances, ecological issues, policy development and management practices for aquaculture sustainability. Issues including fish meal usage, effluents, stock enhancement, non-indigenous species, genetics, mangroves, polyculture and more are reviewed with the aim of separating fact from fiction and focusing on real alternatives for defining and resolving potential problems. The World Aquaculture Society (WAS) has supported this and other workshops and publications on responsible aquaculture in accordance with its mission to support the educational, scientific and technological development, and advancement of aquaculture throughout the world. By providing venues for the presentation, discussion and exchange of science-based information and the dissemination of research-based publications around the world, WAS will continue to play an important role in fostering balanced information exchange.

Craig L. Browdy
President World Aquaculture Society
Marine Resources Research Institute
South Carolina Department of Natural Resources
217 Ft Johnson Rd (PO Box 12559)
Charleston, SC 29422, USA
browdycl@musc.edu

Introduction

1

Robert R. Stickney

Director, Texas Sea Grant College Program, 2700 Earl Rudder Freeway South, Suite 1800, College Station, TX 77845, USA

Background

Buzzwords of the 1990s that have carried over into the 21st century are sustainable and sustainability. The meaning of them varies significantly, but two often heard definitions used in association with aquaculture are production of aquatic organisms without a net utilization of natural resources and production of aquatic organisms with a minimal net utilization of natural resources. Developing a production system that fits within the first definition may well be impossible so, from a practical standpoint, the latter definition seems more reasonable.

Rather than become bogged down with semantics (though the terms sustainable and sustainability can be found in this volume), we elected to use the term responsible in our symposium title. That term implies that the aquaculturist will not only be conscious of the utilization of natural resources but that he or she will also be dedicated to avoiding environmental impacts, attempt to produce healthful products in the case of species used for human food, and will follow the regulations and policies that have been established.

This volume incorporates manuscripts that were prepared based on presentations made in the Responsible Marine Aquaculture symposium held on 22 and 23 January 2001 in conjunction with Aquaculture 2001. The venue was a meeting of the World Aquaculture Society, Fish Culture Section of the American Fisheries Society, and the National Shellfisheries Association held in Orlando, Florida, USA. A total of 16 presentations, plus a brief introductory session, were scheduled over 2 days. Only one presenter was unable to attend.

Manuscripts for this volume were developed from those presentations. The first 16 chapters were developed from the presentations made at the symposium. Additional material was solicited to fill gaps. A summary of each chapter appears below.

The Sustainability Challenge

Chapter 2, by M. Richard DeVoe and Catherine E. Hodges, sets the tone for the detailed information that makes up the rest of the book. That chapter, entitled 'Management of Marine Aquaculture: the Sustainability Challenge', summarizes the basic requirements of marine aquaculture. Those are:

- Good water quality.
- Convenient access to the site.
- Exclusive rights to the site.
- Financial commitment.
- Commitment by government to the industry.

The complex requirements of marine aquaculture present serious management challenges, not only to the culturist but also to various other entities. Included are government, resource managers, scientists, extension personnel, coastal communities, and the public-at-large. The challenges imposed by each organization or group have often constrained the growth of the industry. Dealing with those entities must be done in a comprehensive fashion, a reality that was also stressed by McKinney and Harvey in Chapter 4. Some advocacy groups have been strongly opposed to aquacultural development. Chapter 18 by Tiersch and Hargreaves focuses on that issue, so that chapter is also closely related to Chaper 2.

DeVoe and Hodges note that marine aquaculture is a relatively new enterprise. Those who become involved can often be characterized as lacking in experience. They may not recognize the complexity of the industry. They may also be given inappropriate advice on site selection and they may fall short in their evaluations of demand for the product and the availability of marketing outlets. In many countries there are numerous legal and institutional constraints that must be faced. Another significant problem is a lack of understanding of the place of coastal and marine aquaculture relative to other coastal uses.

Among the user conflicts that arise, particularly in developed nations, are those that develop between marine aquaculturists and such users as recreational fishermen and boaters, commercial fishermen, environmentalists, and those who object to marine aquaculture on the basis of aesthetics. There can be disruption of rural/coastal communities when a new industry comes on the scene that does not appreciate the local culture and traditions. Each of these groups and their issues need to be addressed by the marine aquaculturist. Gaining acceptance and support will often be very difficult.

Many of the environmental impacts of marine aquaculture are discussed, and nearly all of those mentioned are considered in detail in other chapters within this volume. Included are impacts associated with water quality, the benthos, genetics, non-indigenous species, diseases, chemical use and others such as wetland destruction and sedimentation.

With respect to the institutional and legal environment, DeVoe and Hodges discuss the present state of the Food and Agriculture Organization (FAO) of the United Nations' aquaculture policy, but concentrate their remarks on the regulatory situation in the United States. They point out the lack of coordination among state and federal agencies, the uneven application of existing laws and regulations across the nation and the imposition of regulations that were not promulgated with the intent of applying them to aquaculture. The lack of blanket decisions by state regulators is also mentioned.

After describing the many problems that exist, DeVoe and Hodges describe what an effective aquaculture framework should consist of and the benefits of such a system. Basically, an effective policy framework should do the following:

- Designate zones for aquaculture.
- Provide exclusive property rights.
- Provide regulatory coordination and streamlining.
- Provide a risk safety net.
- Provide mechanisms that address issues associated with feeding, stocking, diseases and species selection.

DeVoe and Hodges indicate that there are various challenges that the marine aquaculture industry will have to face if it is to realize its potential. Among them is the fact that perception is reality. Aquaculture, particularly in coastal regions, has an image problem. While much progress has been made in addressing the objections of opponents, those contributions are not being recognized or widely reported. People must be engaged in aquaculture's development to obtain the necessary support for this relatively new resource use. Also, government has dual and sometimes conflicting roles regarding aquaculture. Government is a supporter and promoter of aquaculture, but also has a regulatory and enforcement role. With respect to the United States, DeVoe and Hodges suggest that the government should re-evaluate and reaffirm the nation's aquaculture policy, support sustainable marine aquaculture development, and strengthen policy development through improved coordination among the agencies involved.

To achieve sustainability in the marine aquaculture arena, government will have to take a leadership role. Planning is essential, but not an end in itself. Community involvement and conflict resolution are necessary components of a responsible development strategy if marine aquaculture is to gain acceptance and reach its potential. Without a commitment by government to address these activities, the industry may be imperiled.

Marine Mammals

Interactions between marine mammals and marine aquaculture can have negative impacts both on the mammals and on the species being cultured. The issues surrounding these types of interactions, which can occur in conjunction with extensive shellfish production as well as intensive aquaculture, are addressed in Chapter 3 by Bernd Würsig and Glenn A. Gailey.

With regard to marine mammal interactions with shellfish culture operations, the involvement is relatively minor. The most commonly reported problems revolve around sea otters that are drawn to shellfish beds, which represent readily available food supplies. More significant impacts on both aquaculture operations and marine mammals have been reported in association with netpen culture.

Marine mammals impact aquaculture by damaging the gear to gain access (often resulting in the escape of fish), preying upon the fish to which the mammals have gained access, and contaminating the water with their waste products. The highest percentages of reported problems from the Pacific Northwest region of North America have been in association with seals and sea lions. River otters and mink have also been reported as problems, and while these species are not marine mammals, they can enter marine waters and prey upon salmonids in netpens. Predation problems have also been reported in conjunction with interactions with various types of fish-eating birds.

Impacts on marine mammals are most often associated with entanglement in the gear associated with aquaculture production operations. Mammals that become entangled may be injured or they may drown.

To discourage marine mammals from interacting with cultured fishes, and in some cases to discourage them from decimating wild populations, a number of harassment techniques have been used. Included are firecrackers and cracker shells, acoustic harassment and deterrent devices, the playing of predator sounds, chasing with vessels, and the use of rubber bullets and blunt arrows. Most methods are effective in the short term, but the mammals quickly learn to ignore them. Adversive conditioning has also been used. This involves lacing the food – in this case a carcass of the prey species of interest – with lithium chloride, which imparts a foul taste. The mammals may ignore non-living prey once they determine that live fish do not contain the chemical.

Perhaps the most effective method for keeping marine mammals away from cultured fish is through exclusion. Physical barriers, such as predator nets, have been effective.

The last resort in dealing with marine mammals involves removal. This can take the form of lethal or non-lethal methods. Lethal removal may not be an option in nations, such as the United States, where laws fully protect marine mammals from being lethally taken. Non-lethal removal may involve capture and relocation or capture and maintenance in captivity. Permits may be required from the appropriate regulatory agency before removal (or harassment for that matter) can be employed.

Recreational Fishing

While not necessarily a global issue, interactions, including confrontations between marine aquaculturists and recreational fishing interests, have become a problem in some nations. The United States is a good example because there are millions of marine recreational fishermen, many of whom see aquaculture development in the coastal zone as having a negative impact on their pastime. The issue is addressed here in Chapter 4 by William D. Harvey and Larry D. McKinney entitled: 'Recreational Fishing and Aquaculture: Throwing a Line in the Pond'.

The authors point out that recreational fishing in the United States is a nearly $40 billion dollar industry. In Texas, the state in which the authors live and work, saltwater angling is the fastest-growing segment of the recreational fishing industry. Recreational anglers have begun to express concern about the environmental impacts of aquaculture, though the angling community has similar concerns when it comes to commercial fishing. Included are effects on the quality of recreational fishing, competition for resources and ecological disruption. Chapter 4 examines some of the perceptions of recreational anglers, looks at how those perceptions lead to conflict, and addresses how the conflicts can be mitigated or avoided.

Recreational fishermen see impacts on water quality due to the release of effluents from aquaculture facilities as having negative environmental impact and thus negative impacts on the species many of those anglers target. Recreational fishermen are also concerned about disease transmission from farmed to wild stocks and fear that there will be a loss of fitness of wild stocks when they interbreed with escapees from aquaculture facilities.

An example of a mariculture industry that came into conflict with the recreational fishing community is the Texas shrimp mariculture industry. Anglers were opposed to the construction of large pond complexes because of perceived negative impacts on habitat. There was also opposition to the prodigious amounts of water required for the pond systems, lack of treatment of discharges from the ponds, and the consequent enrichment of receiving waters. Those concerns and objections were reinforced by a fragmented regulatory framework, the use of exotic shrimp in the culture ponds and, during the mid-1990s, outbreaks of Taura syndrome virus in the cultured shrimp. While there has been a revamping of the regulatory situation in Texas, there has also been proactive addressing of many of the problems by the shrimp mariculture industry. Shrimp farmers now recirculate water and have, in some instances, developed constructed wetlands. The result has been a great reduction in water use and effluent volume, plus the effluent that is released is often treated. Effluent water is also heavily screened to prevent escape of exotic shrimp into the environment.

Harvey and McKinney stress the need for stakeholders in such disputes to be proactive and place the burden for assuming leadership on the mariculturists. Not only should the mariculture community shoulder responsibility for educating the public (in this case recreational fishermen, but certainly many others as well), but it must also educate the regulators. By doing so, the result can be significant savings in time and money.

Commercial Fishing

The relationships between aquaculture and commercial fishing are addressed in Chapter 5 by Rollie Barnaby and Steve Adams. The authors indicate that while many aquaculturists support the concept that their activities can potentially assist segments of the commercial fishing industry, commercial fishermen often do not agree. Instead, many see aquaculture as a threat. The perception often differs geographically. With the exception of not-for-profit salmon hatcheries, the state of Alaska bans commercial finfish aquaculture, though shellfish culture is allowed. With the collapse of the cod fishery off New England, many former commercial fishermen were retrained as salmon aquaculturists, so there is better support for aquaculture in that part of the United States, at least among commercial fishermen, than in other locales. Similarly, when a commercial fishing net ban was imposed in Florida, many commercial fishermen went into clam farming.

Controversy is also associated with marine fishery enhancement (see also Chapter 6). Opposition to the idea exists within the commercial fishing industry in some areas. The scientific community is also split on whether the practice should be employed as a means of aiding in the recovery of overfished populations. Comments range from 'it didn't work a century ago so it won't work now', to concerns about the genetic integrity of stocks and potential negative impacts on native populations by fish stocked in enhancement programmes.

Salmon farming has been responsible, in part, for falling salmon prices in the marketplace. While farming is not allowed in Alaskan waters, enhancement hatcheries have had the effect of increasing some runs and creating others, with the result that commercial fishermen who target those runs have maintained their landings and not suffered as much economically as fishermen who depend entirely on wild runs. Similarly, there is some indication that clam prices have fallen in parts of the United States as production of cultured clams has grown. Gulf of Mexico commercial shrimpers have also complained that cultured shrimp have had a negative impact on the price of the wild product (though in this case the vast majority of the cultured shrimp that reach North American markets are imported from Asia and Latin America).

For some species, declines in wild populations have virtually eliminated their availability in the marketplace, leading to erosion of the infrastructure associated with processing and marketing fishery products in general. Wholesalers would purchase farmed cod, haddock and flounder, for example, if those products were available.

Other than in conjunction with shellfish, where traditional fishing grounds may be placed off limits when aquaculture leases are granted, it is difficult to find a situation where aquaculture has seriously impacted commercial fishermen. Some commercial fishermen see a threat to their livelihoods from the trend towards leasing marine areas for aquaculture. One example is complaints about netpens in areas traditionally fished by lobstermen (an indication that retraining cod fishermen is not seen as a positive move by everyone involved in New England commercial fisheries). A basic argument is that leasing areas of the marine environment for aquaculture eliminates access to what have traditionally been the commons.

With controversy and limited areas that are highly suitable for finfish production in protected coastal waters, the industry is looking to develop open ocean sites. As open ocean aquaculture develops, the authors see opportunities for commercial fishermen. Their vessels, knowledge of ocean conditions, skills with various types of gear, and work ethic can be used both to their advantage and to that of aquaculturists. There is also a belief that some commercial fishermen will become aquaculturists (as has occurred to some extent in New England, Florida and elsewhere), though there is peer pressure against such conversions in many communities.

The authors of Chapter 5 caution that if commercial fishermen are to become major players in marine aquaculture, that will only happen if large corporate entities do not enter the business and become dominant. That result has been avoided in Japan, for example, where fishermen's cooperatives have been formed to provide the capital and infrastructure needed to support marine aquaculture. A similar approach may work in the United States. However, in Japan, there is no real distinction between commercial fishing and aquaculture. That same dynamic does not currently exist in the United States but it would have to be developed if the collaboration required to form and operate aquaculture cooperatives is to be realized.

Stock Enhancement

Marine fish stock enhancement began in the United States in the 1870s, largely without any apparent benefit to the fisheries, primarily because releases were largely restricted to eggs and larvae. Anadromous salmonids were a rare exception. With advances in production technology, it is now possible to rear many marine species to sizes at which their chances for survival in nature increase dramatically. In recent years, interest in enhancement has been renewed. Japan has been involved in the practice for over 30 years. Norway has an active programme and there has been some activity in the United States. Not everyone is convinced that marine stock enhancement can be successful, nor that it is an ecologically responsible activity. In Chapter 6, 'Advances in Marine Stock Enhancement: Shifting Emphasis to Theory and

Accountability', Kenneth M. Leber discusses the concept and the shifting paradigm under which enhancement is beginning to operate.

After providing a brief overview of United States activities with fish stock enhancement by the federal government, Leber points out that during almost the first century of the activity there was little or no attention paid to evaluating its impact. It has only been about a decade since research has begun to focus on the effectiveness of stocking marine and catadromous species. For a century beginning in the 1880s, marine research focused on hatchery technology. That topic continues to be of interest today, but since the 1990s production efficiency, tagging technology and release strategies have also become subjects of interest. In the current decade, competitive displacement of wild fish by stocked fish, economics associated with enhancement stocking, and genetic risks have been added to the list of topics of interest to researchers.

The practice of marine stock enhancement has become highly politicized and has polarized groups within the scientific and regulatory communities. At the same time response to enhancement stocking by the general public has been largely favourable, regardless of whether impact assessments are a part of the process. However, lack of assessment has resulted in a lack of scientific information and a general lack of control over the effects of the process.

Leber presents lists of pros and cons associated with marine stock enhancement. Reasons for developing a stock assessment programme include:

1. Replenishment of depleted fisheries.
2. Supplementation of weak year classes.
3. Assistance in the recovery of endangered species.
4. Establishment of urban fisheries.
5. Great potential as another fishery management strategy.
6. Provision of a mechanism to increase our knowledge about wild stocks.

On the other side are a number of objections that have been expressed. Among the concerns that opponents have raised are:

1. Stock assessment represents an expensive, arrogant use of technology.
2. The scientific background for stock enhancement is undeveloped and, thus, has many critical uncertainties.
   - Once a programme is initiated, it will be difficult to stop.
   - Stock enhancement is potentially harmful to wild stocks.
   - There are few good examples to build upon.
   - The practice can postpone or even displace the inevitable use of tough restrictions on fisheries.

Leber indicates that rigorous science is beginning to be applied to stock assessment. However, he also expresses the need to develop and test enhancement theory as well as to prioritize the key questions that need to be addressed. A responsible approach to marine stock enhancement would be to: (i) develop species management plans that prioritize species to be used in enhancement activities and identify the harvest and genetic objectives; and (ii) develop a sound enhancement strategy.

More of the findings from research in this arena should be published in the peer-reviewed literature. In addition, agency oversight of enhancement stocking is needed. Agencies need to: (i) determine how stock enhancement fits within the context of their overall management plans; (ii) consider the issues of genetic conservation and health management; and (iii) address the impact caused by releases.

Polyculture

'Aquatic Polyculture and Balanced Ecosystem Management: New Paradigms for Seafood Production' is the title of Chapter 7, the authors of which are James P. McVey, Robert R. Stickney, Charles Yarish and Thierry Chopin. The chapter begins with a discussion of the levels of nutrient inputs to coastal waters from a variety of sources, including natural upwelling of nutrient-rich deep water in such highly productive areas as the Pacific Ocean off Peru. Whatever the source, high nutrient levels can be reduced through biological activity, including the rearing of plants and animals in aquaculture. Polyculture of selected organisms could provide a better balance of ecosystem function than what is presently seen in some regions.

Marine fish and shellfish culture in ponds or cages, on the bottom, and in suspended culture (e.g. shellfish production employing rafts or longlines) can lead to degradation of water and sediment quality. Waste feed and metabolites are sources of phosphorus, nitrogen, and other nutrients required by primary producers and decomposers. Plants, including phytoplanktonic as well as macroalgal and rooted species, can remove these nutrients. In addition, selected microorganisms, such as denitrifying bacteria, that would help detoxify the water column and sediments, by converting ammonia to nitrate, could be encouraged in conjunction with an aquaculture system. Molluscs in the system could feed upon the phytoplankton and bacteria, while higher plants may be used in habitat restoration or, depending upon species, be harvested for food.

Developing balanced ecosystems in an aquaculture mode will require a significant amount of research and the development of sophisticated ecosystem models. The end result could be not only responsible marine aquaculture, but also a form of marine aquaculture that increases the level of sustainability above what is currently practised in most instances.

Aquaculture and Habitats

The concerns that have been expressed by recreational and commercial fishermen with respect to aquaculture were mentioned earlier in this chapter and are covered in detail in Chapters 4 and 5. Barry Costa-Pierce and Christopher Bridger delve much more deeply into the issue of public concern about marine aquaculture in Chapter 8: 'The Role of Marine Aquaculture Facilities as Habitat and Ecosystems'.

Altered habitats as a result of the construction of aquaculture facilities on land have caused a number of problems. Environmental issues include mangrove and wetland destruction (the mangrove issue is also discussed in Chapter 9), disruption of food webs as a result of the release of nutrient-rich, sediment-laden water, discharge of toxicants and, in the case of shrimp farms, abandonment (after a few years productivity may decline and farmers may build new facilities in a different location). Social issues include loss of traditional livelihoods and salt-water intrusion.

In recent years, various nations and such global organizations as the Food and Agriculture Organization of the United Nations (FAO) have begun to develop codes of practice aimed at the development of sustainable, or at least responsible, aquaculture. FAO defines it this way: 'Sustainable development is the management and conservation of the natural resource base and the orientation of technological and institutional change in such a manner as to ensure the attainment and continued satisfaction of human needs for present and future generations'.

While marine aquaculture is often viewed by the public as extractive or worse, destructive with respect to environmental resources, there are a number of forms of aquaculture that are being practised to produce products that are being used to actually create, rehabilitate, or enhance coastal ecosystems. Included are the production of marsh grasses used for planting coastal wetlands, culture of mangroves for outplanting, seagrass culture for recolonizing submerged grassland areas, and live rock aquaculture to reduce the collection of material from coral reefs. Chapter 8 provides details on some restoration efforts with mangroves and salt marshes. Management of oyster reefs is also discussed.

Oil and gas platforms, which are currently being evaluated as sites for aquaculture in cages as well as for hanging culture of molluscs, provide substrate for the attachment of wild marine organisms and also serve as fish-attracting devices. They may enhance the productivity in areas where they exist. The presence of aquaculture cages, which provide additional substrate, and the nutrients released from the caged fish are additional attractants for native populations.

Extensive pond aquaculture facilities are, in reality, managed wetlands. A primary example, though not marine, is ponds used for rice and crayfish production. Constructed wetlands, which are sometimes used to treat aquaculture effluents (see Chapter 15), can also attract large numbers of birds. These ecosystems attract a variety of bird species and often serve as locations for rookeries. Creating desirable bird habitat can be a two-edged sword, however, as many of the birds that are attracted are piscivorous and can devour a significant proportion of the aquaculture crop. In addition, bird droppings add to the nutrient levels in the water.

Aquaculturists are learning to work with the natural environments in which they establish their facilities. By taking that approach, both stakeholders and coastal habitats benefit.

Aquaculture in Mangroves

Claude E. Boyd focuses Chapter 9 on the topic 'Mangroves and Coastal Aquaculture', which builds upon the information presented in Chapter 8. Boyd indicates that mangroves are important in erosion control and act as windbreaks, are spawning locations for various marine species, are utilized by migratory birds and have other features that make them important components of the coastal environments in which they occur. Removal of mangroves can result in saline intrusion into rice fields, a decline in shrimp postlarval abundance, a decline in crab abundance and the acidification of the shrimp ponds. Acid soils lead to reduced shrimp yields.

While many shrimp farms in Latin America and Asia were constructed in mangrove forests in the past, a new ethic has spread through the industry and has been adopted by regulators in nations that now understand the importance of the mangroves along with the drawbacks of constructing ponds in mangrove areas. Drawbacks include low elevation, poor drainage characteristics and the previously mentioned acid soils.

Shrimp farms should be located outside of not only mangroves but also wetlands that have significant ecological value. Roads, canals and other infrastructure should not block the flow of water into and out of mangroves or other wetlands. Further, where mangroves must be removed to provide access, mitigation plantings in other areas should be conducted. As a result, there should be no net loss of mangrove habitat.

Avoiding damage to mangroves during construction is not the end of the story. Once a shrimp farm is in operation, employees should avoid disposing of such things as waste materials, treatment chemicals and piscicides in mangrove areas. Caution should be used to avoid polluting mangroves with oil and fuels. Water exchange in ponds should be kept to a minimum, and pond effluent should not be discharged into stagnant areas within mangroves.

Aquaculture is not the only activity associated with mangrove destruction. Boyd presents information on a number of other activities that have resulted in removal of mangroves. Some of these activities have resulted in much greater loss of mangrove habitat than has occurred as the result of aquaculture. In Africa, for example, where little coastal aquaculture has occurred, some 70% of the mangrove area has been destroyed to provide space for human activities.

Impacts on the Benthos and Water Column

Marine netpens can have significant local impacts, particularly with respect to the sedimentary environment immediately under and adjacent to them. There can also be observable effects in the water column. The topic has been studied in detail and is summarized by Kenneth Brooks, Conrad Mahnken and Colin Nash in Chapter 10.

The focus of Chapter 10 is on temperate-water marine culture, with particular emphasis on salmon netpen culture because that industry is large and well established. It is also the most widely studied cage or netpen culture industry with respect to impacts on benthic communities.

Most people think of losses of material from cage or netpen culture operations in terms of organic deposition in the forms of waste feed and fish faeces. However, there are other sources of waste. Included are mortalities, pharmaceuticals, biocides and trace metals, such as copper, that are associated with anti-fouling substances.

Studies have found that the accumulation of organic wastes immediately under netpens can accrete to the point that the rate of microbial degradation is exceeded and the microbial community will change from aerobic to anaerobic species. This leads to the production of hydrogen sulphide and can lead to the elimination of most benthic organisms. A few species, such as the polychaete *Capitella capitata*, are able to survive by remaining in contact with the overlying oxygenated water. Ultimately, the anaerobic sediments will also produce methane and a truly azoic environment may develop.

Azoic zones are not common to all netpen or cage farms. Hydrographic conditions lead to a great deal of variability, and well-sited farms should have sufficient flushing so the condition does not occur. Using best management practices, impacts on the benthic community are typically undetectable only a few metres outside the shadow of the netpens.

Chapter 10 provides an in-depth literature review concerning the topics mentioned above and a number of others associated with impacts of netpen salmon farming on the environment and on animal health, including human health. Problems can generally be avoided through proper siting and sizing of facilities and by employing proper management of the facilities. Governments typically require extensive monitoring to ensure that netpen facilities are in compliance with local and national environmental quality standards.

Non-indigenous Species Issues

Non-indigenous species are those organisms that have become established in a location outside of their native range. The introduction may be intentional (e.g. bringing horses and cattle to North America from Europe) or accidental (e.g. numerous species translocated in the ballast water of ships). With respect to marine aquaculture, various industries around the world are based on non-indigenous species. A prime example is the rearing of Atlantic salmon in Chile and on the west coast of North America. This topic is addressed by Robert R. Stickney in Chapter 11: 'Issues Associated with Non-indigenous Species in Marine Aquaculture'.

The use of a variety of marine aquaculture species, in addition to Atlantic salmon, is discussed. Included are species produced in aquaculture but which have also been used for enhancement stocking. Among the latter are striped bass (*Morone saxatilis*) in California and Pacific oysters (*Crassostrea gigas*) on the west coast of North America. The non-indigenous freshwater shrimp, *Macrobrachium rosenbergii*, which requires a moderately saline environment during the larval stages, received a great deal of attention during the 1970s, but interest in shrimp culture ultimately centred on non-indigenous marine species such as *Litopenaeus vannamei*, *L. stylirostris* and *Penaeus monodon*.

Two major concerns surround the use of non-indigenous species in marine aquaculture. The first is associated with the possibility of introducing diseases that will spread to wild populations. This is a particular concern with respect to shrimp in the United States where the industry is based on non-indigenous species. Several viruses have caused outbreaks that resulted in high levels of mortality in shrimp ponds and fuelled fears that the viruses might also infect native wild populations.

Disease issues are best addressed by controlling the importation of non-indigenous species, but some nations have taken the position that the only way they can build an aquaculture industry is to bring in species from other regions. Strict quarantine procedures are sometimes practised and there have been attempts to develop specific pathogen-free stocks.

The second issue with respect to non-indigenous species is concern that escapees from mariculture operations may become established and lead to the displacement of native species. It is a fact that in nearly any type of aquaculture facility, escape is likely to occur. There does not seem to be a foolproof way to totally eliminate the possibility of escape from a mariculture facility.

There are not many examples of escapees from mariculture operations having become established, let alone proliferating to the extent that they threaten native stocks. The possibility exists, however, and opponents to non-indigenous species remain strong in their resolve to eliminate these species from culture operations.

Sometimes even the culture of native species is objected to, particularly if there has been selective breeding involved. The fear is that escapees that are selectively bred for aquaculture will interbreed with their wild congeners and produce inferior offspring. Looming on the horizon is the possibility that genetically modified organisms (GMOs) will be utilized in aquaculture. Some GMOs have been produced in the laboratory, but there does not appear to be any significant commercial production of such animals at present. The issue will intensify as interest by producers to mass produce GMOs in culture facilities develops.

Genetic Changes and Potential Impacts

Although the face of aquaculture has changed very dramatically over the last 50 years and the changes are apparently accelerating, there is one feature that has not changed very much. This feature is the use of natural aquatic resources as the primary supply of genetic material for raising in production facilities. Currently, most of the industry relies directly or indirectly on the genetic resources provided from natural populations of aquatic organisms. Since a large amount of time and financial investment is needed to develop useful domesticated stocks, this will undoubtedly remain the primary source of genetic variability for some time into the future. Consequently, as William Hershberger points out in Chapter 12, some serious attention should be paid to the factors that affect natural aquatic genetic resources and their conservation.

Three major factors are identified that can have an impact on the natural genetic resources. The first involves environmental changes that can impact the genetics of these resources through direct effects on survival and through more subtle selective factors. The second factor is the capture fishery and the resulting changes brought about by overexploitation and increasingly selective fishing practices. Finally, the utilization of aquaculture in the restoration or enhancement of natural populations and in intensive cultivation can alter the natural genetic resources owing to differences in genetic make-up. With this diversity in potential effects, it is not too difficult to deduce that there is some genetic risk associated with introducing additional components to the natural genetic system. Minimization of genetic risk can be accomplished by maintaining genetic variability within the natural populations and a model is discussed to evaluate the characteristics of an aquaculture species that can be utilized to minimize loss.

Genetics in Responsible Aquaculture

Critics of aquaculture, according to L. James Lester, author of Chapter 13, mention such things as environmental damage, detriments to human nutrition and contributions to social inequity as being significant problems. To be responsible and responsive to critics, aquaculturists need to develop ethical codes of conduct and should be environmental stewards.

Examples of ethical codes for aquaculture include one developed by the Food and Agriculture Organization of the United Nations and one that is currently being developed by the US National Marine Fisheries Service. With respect to genetics in conjunction with aquaculture, any code of conduct should:

1. Conserve genetic resources.
2. Promote the contribution of genetics to food security.

3. Protect genetic diversity.
4. Establish standards of conduct for all persons involved in managing genetic resources.

Genetics can be used for monitoring the genetic composition of aquacultured species and for modifying the genetics of cultured species. Genetic technologies include biochemical monitoring, selective breeding, producing inbred lines, gynogenesis, hybridization, chromosome manipulation, sex ratio manipulation, cloning and the production of transgenics. Different approaches should be applied to the management of wild as opposed to domestic stocks. Lester discusses each approach. Some of the conclusions reached are:

1. Selection of outbreeding populations has been emphasized.
2. The use of inbreeding, hybridization and other technologies has been limited.
3. Breeding of aquatic species is not currently being controlled by large corporations.
4. Human health and equity are not harmed by the traits currently being manipulated.
5. Genetic management has caused few stewardship problems.
6. Genetic modification has raised few ethical questions.
7. Future issues may arise over proprietary rights to genes.

Ecosystem Modelling and Aquaculture

Modelling of marine ecosystems has become an important item in the toolbox of ecologists in recent years. In Chapter 14, 'Understanding the Interaction of Extractive and Fed Aquaculture Using Ecosystem Modelling', Mac Rawson and his co-authors indicate that one objective is to integrate various types of aquaculture into coastal management programmes.

Fed aquaculture, such as is associated with netpen fish culture and pond shrimp culture, adds organic matter to the system, leads to increases in the levels of microflora and plankton, and places additional oxygen demand on the system. Extractive aquaculture, for example, mollusc and seaweed production, leads to decreases in plankton and nutrient levels and, depending on the organism, may increase or decrease the levels of oxygen and carbon dioxide. Polyculture can be employed to integrate the two types of aquaculture and reduce the potential for eutrophication.

The authors have been conducting ecosystem modelling efforts in China and have attempted to predict how aquaculture practices may affect local conditions. The study area is Xincun Lagoon, a 20 km^2 water body associated with Hainan Island in Jiaozhou Bay. Fish, pearl oysters, shrimp and seaweeds are cultured in the lagoon. In addition to wastes from the fish cages, there are additional nutrient inputs from human sewage discharges and a river. The model takes into account the local hydrography, climate, water chemistry and other factors, including aquaculture, that affect nutrient and plankton levels. The model evaluates the influences of tide, river inflow and wind on the distributions of temperature, salinity, phytoplankton and phosphate in the lagoon. The model should be useful in helping decision-makers determine the types of aquaculture and locations where aquaculture can be practised which will have the least negative environmental impact.

Shrimp Farm Effluents

Granvil Treece addresses the topic of effluents from shrimp farms in Chapter 15. The topic has been a major issue in the United States, and particularly in Texas, which has the largest shrimp pond area in the country. The number of farms changes periodically, but averages about a dozen in the state. Farms range in size from a few hectares to a few hundred hectares.

During the 1980s and early 1990s, the majority of the farms were operated on a flow-through basis and continuously discharged enormous volumes of water. Objections to nutrient and sediment loading of receiving waters led to changes in the regulations concerning the amount of water that could be pumped. Those changes resulted in modifications in the operational procedures on the farms. Most farmers drastically reduced their water consumption rates by going into a recirculation mode.

Modifications included widening and deepening drainage ditches and installing weirs to prevent escape of shrimp when discharges did occur. Mechanical aerators and recirculation pumps were installed in the ditches so that the water could be recycled. In addition, stocking rates were reduced and the number of feedings per day was increased. Protein levels in the feed were reduced, aeration in the ponds increased and there was increased attention to water quality monitoring. All these activities were undertaken with the goal of cutting discharges to as close to zero as possible.

Currently, the ponds are stocked in April and there is no discharge from them until July, at which time recirculation of the water begins. The result has been a slight increase in salinity; small drops in pH, BOD and volatile suspended solids; and a considerable increase in total suspended solids in the ponds as compared with the intake water. Nitrate nitrogen does not differ between recirculated ponds and the intake water. Nitrite and ammonia levels have actually been reduced, as have total and reactive phosphorus levels.

At one carefully monitored farm, production levels on a weight of shrimp per unit surface area of ponds after the conversion to recirculation have been slightly higher in recent years than they were when the flow-through technique was being employed. At the same time water quality has been maintained within the limits set by the regulatory agency.

The Fish Meal Story

The history of fish meal use, production trends and the future with respect to sustainable supplies are the topics included in Chapter 16 by Ronald W. Hardy and Albert G.J. Tacon. In the United States, menhaden were being captured for their oil in the early 1800s. The residue after the oil was pressed from the fish – fish meal – was used primarily as fertilizer. Fish meal became an important component of animal feeds when swine and poultry production expanded greatly during the first half of the 20th century.

Prior to World War II, global fish meal production was about 1 million tonnes annually. That doubled by 1960 and gradually expanded to 6–7 million tonnes by the mid-1980s when production levelled off. There have been significant fluctuations associated with El Niño events as Peru is now a major producer of fish meal.

Of the 84–94 million tonnes of fisheries landings in the world, the amount of fish used to produce fish meal and fish oil ranges from 18.8 to 28.3 million tonnes. The fish meal obtained from these landings amounts to 6 to 7 million tonnes.

Major fish meal exporting countries are Peru, Chile, Denmark and Iceland. Japan, Norway, Thailand, the United States and Russia are major producers, but they utilize most of their production domestically. Peru and Chile produce some 75% of the fish meal traded in international commerce each year. Because that source dominates the market, fish meal prices can as much as double when production declines during El Niño years.

Fish meal is an important dietary ingredient because of its excellent balance of essential amino acids, and its relatively low cost on a unit protein basis. When the price of fish meal is reasonable, there are economic benefits for incorporating it in animal, including aquatic animal, feeds. When fish meal prices are elevated, other protein sources become more economical in least-cost formulations.

World capture fisheries landings have been static for several years and there is no expectation that they will increase in the future. Aquaculture is seen as the only way to significantly increase the availability of seafood in the marketplace. However, increasing aquaculture production will require using more feed, which means more protein-rich ingredients, such as fish meal, will be required. Currently, aquaculture utilizes only about 3% of the animal feed manufactured in the world. Poultry, swine and cattle account for some 94%. In terms of fish meal utilization, aquaculture feeds account for 34%, compared with 46% in poultry feeds and 9% in swine feeds.

As aquaculture grows, either there will be an increasing demand for fish meal for aquatic animal feeds or alternative protein sources will need to be employed. The latter may well be the case. Fish nutritionists have been developing practical diets that perform very well with reduced levels of fish meal. For example, catfish feeds contained about 3% fish meal in 2000 and are expected to contain no fish meal by 2010. During the same decade, the fish meal used in feeds for salmon, trout, shrimp and other species can be expected to drop significantly. However, because aquaculture production will increase, net utilization of fish meal in aquatic animals feeds can also be expected to increase.

To be viable in fish feeds, alternative protein sources must be economically competitive, must not negatively affect product quality, must not increase pollution associated with increased fibre and nutrient excretion, and must be readily available in the commodity marketplace. Some examples of materials that are being evaluated and utilized increasingly are wheat and maize gluten, oilseed meals and protein concentrates, pea and bean protein concentrates, rendered products such as meat and bone meal and poultry meal, and fisheries bycatch and processing wastes. Each of those products has limitations that make them less desirable than fish meal, but that does not mean that their use will not be increased in the future.

Hardy and Tacon indicate that the challenges for this decade include expanded recovery and utilization of seafood processing waste and bycatch, expanded research and development on grain and oilseed protein sources, continued development of feeds that produce low environmental impact, and ensuring that the available fish meal is suitable for use in fish feeds. Recent reports of polychlorinated biphenyls (PCBs) in salmon feeds have raised fears about the suitability of fish meal in feeds. Adequate sampling programmes, research on the distribution of PCBs in fish and consumer education are required to address that issue.

Ornamentals

The collection of ornamental fish and invertebrate species from wild populations has become a contentious issue as some populations have been decimated by the practice. This issue is addressed in Chapter 17 by Cathy Hair, Johann Bell and Peter Doherty, who present an alternative that incorporates both capture and culture. The authors discuss the fact that high percentages of larval stages of coral reef species will not survive in nature. By collecting animals in their early life stage, it is possible to gather large numbers without significantly affecting the numbers of individuals that successfully recruit to the reef populations. The collected animals can be put into a culture environment for a period of time until they are sufficiently developed to be marketed in the ornamental trade or as food fish. Trapping methods are described along with how those methods can be adapted to artisanal fisheries. While the technique has the potential to significantly increase the numbers of ornamentals available in the marketplace without causing further declines in natural populations, the authors readily admit that it is unlikely to meet the demand that exists for tropical marine fishes.

Dealing With Advocacy Groups

Chapter 18 by Terrence Tiersch and John Hargreaves stems from a presentation made at the 2001 World Aquaculture Society (WAS) meeting in Orlando, Florida, USA, in a session other than the one from which most of the chapters of this book were developed. The title, 'Contending with Criticism: Sensible Responses in an Age of Advocacy', is highly appropriate. Throughout this book considerable mention is made of the opposition that exists toward aquaculture and several of the topics discussed address topics that became issues only after advocacy groups raised them. In their chapter, Tiersch and Hargreaves describe tactics employed by the advocacy groups, including their use of the print and broadcast media to keep their agenda before the people. The authors then present a series of methods that aquaculturists can employ to respond to the criticisms.

The authors point out that valid criticisms have been lodged at the aquaculture community and that those criticisms have been or are being addressed. However, when an advocacy organization has an agenda, such as eliminating aquaculture so as to stop the consumption of animals as human food, the fact that an issue has been appropriately addressed does not mean that the opponents will accept that fact, nor that they will not continue to come up with new issues. Often, the approach is to predict an environmental tragedy which, on its own merits, is reason enough for some type of aquaculture to be eliminated. Interestingly, predictions of gloom and doom can backfire when the public, not seeing the horrors that have been predicted, tire of hearing about them and begin to doubt the veracity of the groups that are making the predictions.

Chapter 18 is a fitting way to end this volume. It contains very interesting and timely material that should be of broad interest within the much maligned components of the aquaculture community of practitioners, suppliers, educators and scientists.

Management of Marine Aquaculture: the Sustainability Challenge

2

M. Richard DeVoe and Catherine E. Hodges

South Carolina Sea Grant Consortium, 287 Meeting Street, Charleston, SC 29401, USA

Abstract

Marine aquaculture holds great promise throughout the world. However, a host of physical, environmental, social, economic, and legal and institutional factors must be addressed if marine aquaculture is to realize this potential. Applying the principles of sustainable development to aquaculture projects presents many challenges to the practitioner. Social–cultural considerations, conflicts with other, more traditional uses of coastal waters, environmental concerns, and existing legal and regulatory structures present obstacles to development of the industry. Diversity within the industry and the resources it requires, the locations in which it is practised, and its need for private and public support add to the complexity of the situation.

One of the most important issues facing the marine aquaculture industry is the negative image that it has today. While great strides have been made in designing environmentally friendly technologies, improving culture practices, developing codes of conduct and best management practices, and enhancing education and outreach programmes, recent negative publicity has set the aquaculture community back on its heels. While we must continue to deal with the realities of improving aquaculture practices, the issue now is one of perception, not reality. And it involves people, not fish.

Effective management of aquaculture in a sustainable fashion requires the involvement and cooperation of government, academia, the private sector, investors, communities and the public at large. Marine aquaculture should be looked at as an important component of expanding protein supplies, providing economic development opportunities and jobs, and offering communities the ability to diversify. But aquaculture expansion must also accommodate the need to maintain and improve the environment, social norms and overall quality of life. It is becoming increasingly clear that marine aquaculture must be planned, developed, managed and monitored with long-term sustainability in mind.

There are no easy solutions to the issues related to the sustainable development of marine aquaculture. Progress towards the future will depend in part on having learned from past mistakes and profiting from positive experiences. Marine aquaculture may still prove to provide enduring societal benefits, but only if we are very careful to ask and address the right questions. How we address these challenges – through integrated planning and management, community and public involvement, and education and outreach – will determine the future for sustainable aquaculture in the world.

Introduction

Marine aquaculture holds great promise throughout the world. However, a host of physical, environmental, social, economic, legal and institutional factors must be addressed if marine aquaculture is to realize its potential. Applying the principles of sustainable development to aquaculture projects presents many challenges to the practitioner. Conflicts with other, more traditional uses of coastal waters, social–cultural considerations, environmental concerns, and existing legal and regulatory structures present obstacles to development of the industry. Diversity within the industry and the resources it requires, the locations in which it is practised, and its need for private and public support add to the complexity of the situation.

One of the more important issues facing the marine aquaculture industry is the increasingly negative image that it has today. While great strides have been made in designing environmentally friendly technologies, improving culture practices, developing codes of conduct and best management practices, and enhancing education and outreach programmes, recent negative publicity has set the aquaculture community back on its heels. While we must continue to deal with the realities of improving aquaculture practices, the central issue now is one of perception, as well as reality. It involves people, as well as fish. And it will require a concerted government commitment.

The purpose of this chapter is: (i) to review the issues confronting marine aquaculture development in the United States; (ii) to describe briefly the current state of domestic aquaculture policy and its accompanying institutional framework; (iii) to identify current sustainability challenges; and (iv) to offer a series of suggestions that could enhance the future expansion of a sustainable domestic marine aquaculture industry.

Issues Confronting Sustainable Marine Aquaculture

While the potential for marine aquaculture remains high, the complexity of the industry, the locations in which it is practised, the resources it requires, and its need for private and public support have presented major challenges to its future growth. The United Nations Food and Agricultural Organization

(FAO, 1997) reported that the major challenges facing growth and development of the industry include: (i) increasing competition for limited resources; (ii) environmental degradation of resources used or needed; (iii) lack of recognition of aquaculture as a legitimate resource use; (iv) lack of legal and institutional support; (v) over-regulation; and (vi) harmful publicity due to environmental degradation and social conflicts in isolated situations. The diverse nature of the industry; social–cultural considerations; conflicts with other, more traditional uses of coastal waters; environmental concerns; and the existing legal and regulatory climate all contribute to this situation (DeVoe, 1997, 2000) and must be addressed in a comprehensive and concerted fashion.

Complexity of the marine aquaculture industry

In general, aquaculture practices range from extensive, with few inputs and modest output, to intensive, with high inputs and output. Based on annual yields per hectare of water, increased intensification requires greater resource use, ranging from simple pond culture to intensive tank and closed system aquaculture (Muir, 1985). Dodge and Jolly (1978) identified, in terms of labour costs, the least to most intensive culture systems as: (i) transplantation from poor to better habitats without intervention of hatcheries; (ii) hatchery cultivation where juvenile stages of organisms are reared to a sufficiently large or mature stage of development prior to release; (iii) trapping, as in the case of coastal impoundments; (iv) artificial enclosures to enhance control over stocking and stocks; (v) ocean ranching; and (vi) raft culture. The more recent development of closed recirculating systems and offshore structures has added to the diversity of aquaculture technologies now being employed by the industry. But with them comes a variety of environmental, socio-cultural, economic and policy implications that may require a range of scientific, technological and management responses.

Although aquaculture practices vary widely, the need for water, food, land, labour and capital is common to all forms of aquaculture (Muir, 1985). Aquaculturists, for the most part, also require good water quality, convenient access to the aquaculture site, assertion of exclusive fishing and culturing rights, financial investment and government commitment (DeVoe and Mount, 1989).

The availability and maintenance of a good water quality environment is a critical need for aquaculture. The industry must be assured that current and future uses of the surrounding aquatic and upland environment will not result in a reduction of the quality of waters where species are being cultured.

Marine aquaculture typically requires both an aquatic environment and an adjacent on-land base of operation. Institutional and legal structures in the United States provide the public with the right to use common property resources for navigation, recreation and fishing. However, the variety of aquaculture methods in use now and in the future may require exclusive use of coastal or ocean waters. Such exclusive use could consequently deny, to varying degrees, access to those areas by traditional users.

Establishment of aquaculture operations may require a significant financial commitment; however, aquaculture as an industry is viewed by many prospective investors as a high-risk activity for many reasons. The availability of funding through venture capital, public and private sector loans, or other sources of capital will depend to a large extent on the anticipated stability of the operation and the level of property rights to be vested with the proposed operation.

Government support is most critical. Government at all levels must demonstrate support for the industry by clearly defining its commitment to aquaculture, providing a supporting policy framework, offering financial and non-financial incentives to underscore its commitment, and revising and streamlining its regulatory and legal requirements. Government priorities for aquaculture development should be indicated in national policies, and government should have a nationwide planning and management programme to accommodate its growth (Chua, 1993).

The complex requirements of aquaculture present serious management challenges to industry, government leaders, resource managers, scientists and extension specialists, communities and the public. The issues that result continue to constrain the development of marine aquaculture in the United States and around the world.

Marine aquaculture use conflicts and siting

Finding locations to site aquaculture operations in the marine and coastal environment is one of the most critical challenges facing the industry today. The coastal zones of many countries are experiencing increasing population pressures, industrial and residential development, and additional demands for the use of natural resources. Increased costs of access to coastal lands and marine waters threaten to price many traditional users out of the market.

FAO (1997) provides five criteria for use by government and planning officials for locating aquaculture operations in suitable locations. The code of conduct suggests that aquaculture be established in areas suitable for sustainable production and income generation; where economically and socially appropriate; to prevent or at least minimize conflicts with resource users; where it does not create undue externalities; and with respect to nature reserves, protected areas, and critical and sensitive habitats.

The reality is that the existence of even one such site is questionable, and on this basis it could be argued that there is no room for aquaculture in the coastal environment. However, where no effort is made to intelligently plan aquaculture site selection, the decision of where to locate and the assumption of associated risks are left to the private operator. This autonomy has the potential to create a highly charged situation pitting the individual culturist, and aquaculture as a whole, against all other coastal users. Without planning, the public image of 'rogue aquaculture' can become a self-fulfilling prophecy. To avoid such an occurrence, there are several measures government can adopt to ensure compliance with the FAO recommendations and facilitate local acceptance of marine aquaculture.

Management practices, such as inventories and quality assessments of available coastal areas, are a fundamental step in determining optimal use. With these in place, policymakers can begin the process of designating sites for exclusive use, taking into consideration such factors as the need to preserve navigation rights, the potential for user conflict and minimization of environmental impacts. Local participation and education throughout this development process are important elements of ensuring future compliance. People are significantly more likely to respect limitations originating from a grass roots perspective than those imposed on them from above.

Once appropriate aquaculture sites are selected, culturists must be assured of some property rights over the site to protect their investment. Reiser (1997) cited the elements transferability, duration and renewability, and revocability only for failure to meet specified conditions as ways in which governments can provide security of tenure for aquaculture operators. Government must likewise commit to enforcing the laws that protect the rights of site users. If a site is designated suitable for aquaculture there is an implied provision of certain qualities, such as accessibility and reasonable freedom from pollution. If these conditions are not upheld, government should have some established means of providing restitution for site users.

Difficulties accompanying coastal aquaculture siting are not insurmountable, but their existence has served to promote research into alternatives to coastal use. The National Research Council of the National Academy of Science in the United States predicted that the future of marine aquaculture development will involve the growth of both closed (recirculating) systems and offshore systems (National Research Council, 1992).

Socio-cultural considerations

Sustainable communities seek a balance among social values, market values and ecological values. Ideally, aquaculture development should be planned to improve upon the status quo in these areas. However, studies of the sociological impacts of aquaculture operations are scarce, and those that do exist are primarily focused on less developed countries, which have different social dynamics to those encountered in coastal areas of the United States.

Richard Pollnac of the University of Rhode Island examined the effects of social and cultural characteristics on the success of aquaculture projects in coastal communities around the world, identifying five 'attributes of project components' that must be addressed to ensure social sustainability (Pollnac,

1991). The first attribute addresses the concept of advantage. This is an informal cost/benefit analysis, and answers the community's concerns over how an aquaculture operation will improve or detract from the standard of living. Coastal management plans incorporating aquaculture provide one forum to invite local participation and address this question. Both unintentional and long-range social impacts should receive considerable attention in this analysis (Fiske and Plé, 1992). In the parlance of sustainability, positive advantage should extend for several generations to justify implementation of a project.

Secondly, the target community should not be offered aquaculture systems that are entirely beyond their scope. Developing and operating aquaculture systems effectively is complex. To achieve sustainability in aquaculture systems, participants must be educated in and committed to both their own future and that of the culture operation. Understanding the limits of their collective experience as well as those of the environment will promote the development of projects in scale with local conditions. Education should be a continuous process as the community and the principles of sustainability in aquaculture evolve.

Pollnac's third point is that only aquaculture technologies that meet the traditional work patterns and behaviours of the target population should be used if they are to be adopted with minimal stress (Pollnac, 1991). This requires a thorough understanding of social stratification based on gender, age, ethnicity and other factors, the traditional roles of these groups in the economy, and the potential for aquaculture operations to disrupt those established conditions. Attention should also be given to actual and potential influences outside of aquaculture that may change the existing social dynamic. Technologies should be adaptable to meet with evolving conditions. Aquaculture products must also be compatible with and in demand by the marketplace. The potential for aquaculture to become an enduring part of the local economy is enhanced through a careful match between people and process.

The fourth attribute is that of 'trialability'. Project development is heavily predicated on funding for infrastructure; however, established systems of financial support, such as banks, view investment in aquaculture as a risky proposition. While aquaculture that is not profitable cannot sustain itself, achieving sustainability may demand some sacrifice of initial profits to achieve long-term goals. The establishment of cooperatives and producer associations to address investment issues has had varying levels of success (Pollnac, 1991).

Lastly, communities must be presented with tangible evidence that they can successfully operate aquaculture projects. Unless local residents are involved directly in trial operations and demonstration projects, there is a hesitancy to believe that they will be able to achieve the same results as extension agents, scientists and other 'experts' (Pollnac, 1991). Investment in communities through involvement provides people with the experience needed to become successful and confident culturists.

Environmental and ecological concerns

Ecological concerns related to aquaculture were first raised by a number of authors in the 1970s (Odum, 1974; Ackefors and Rosen, 1979). Since that time much has been published on the environmental impacts of marine aquaculture (e.g. Ackefors and Sodergren, 1985; Weston, 1986; Rosenthal *et al.*, 1988; DeVoe, 1992; Goldburg and Triplett, 1997; Naylor *et al.*, 1998; also see *Estuaries* 18, 1A, 1995). One of the major challenges to the credibility and survivability of the marine aquaculture industry will be how it responds to these environmental sustainability issues (Chamberlain and Rosenthal, 1995).

Aquaculture practices can generate environmental impacts as a function of the technique applied, site location, amount of production and the capacity of the receiving body of water (Ackefors and Sodergren, 1985). The latter can include impacts on water quality, the benthic layer and the native gene pool, as well as impacts from non-native species, disease and chemicals (DeVoe, 1994).

Water quality is of elemental interest to any aquaculture operator. Water exchange has traditionally been required in all forms of marine aquaculture to prevent self-pollution from organic wastes and resulting oxygen depletion. The high concentrations of nutrients, nitrogenous wastes and biochemical oxygen demand (BOD) that can be produced (especially from high-density pond or tank culture systems) and discharged in effluent waters (National Research Council, 1992) have generated concern over siting of marine aquaculture operations along the coast.

The impact of certain aquaculture operations on the benthic layer is a pivotal environmental issue. Some forms of aquaculture can produce organic-rich particulate wastes. Oysters grown in rafts can produce tonnes of faecal and pseudofaecal material (Rosenthal *et al.*, 1988). The impacts of uneaten food and faeces falling on benthic communities beneath salmon cage operations are a worldwide issue (National Research Council, 1992). Finfish operations can generate faecal waste, and feed and medications not ingested add to the particulate load in locations where they are used.

Potential genetic impacts of aquacultured animals on wild populations are a matter of concern as well. Contamination of wild stocks through the escape or release of mariculture organisms might occur. Impacts are believed to be severe (Lester, 1992; National Research Council, 1992); however, very little documentation exists to substantiate this impression. Impacts can be grouped into two categories: (i) overwhelming the 'wild' gene pool with the more restricted gene pool of a hatchery stock through repeated and massive intentional stock enhancement efforts (e.g. salmon and striped bass; National Research Council, 1992); and (ii) weakening of the 'wild' gene pool as a result of interbreeding among native wild stocks and accidentally released non-native culture species (Lester, 1992).

In addition to genetic threats, impacts from non-native species have the potential to impact native populations. Rosenthal (1985) identified four routes by which non-native species can be introduced: (i) translocation beyond their natural range by water traffic; (ii) deliberate transplantation of organisms into new areas; (iii) accidental introductions in connection with the transfer of other species; and (iv) escape of organisms transferred for purposes other than deliberate introduction. In aquaculture, exotic species may exhibit more highly desirable characteristics than native populations in particular locations. However, the benefits derived from such introductions must be balanced with consideration of the political, environmental and cultural implications, as well as production values (National Research Council, 1992). A discussion of these issues in greater detail can be found in Mann (1978) and DeVoe (1992).

Disease remains a major concern within an aquaculture operation and in the surrounding environment. Outbreaks of disease in aquaculture systems can occur with little or no warning and spread rapidly throughout the often highly dense culture population (DeVoe, 1994). Water-borne diseases can be transferred into and out of the production unit via the normal water exchange protocol used in many aquaculture systems. Internal pathogens can be transferred with accidental or intentional, as is the case with stock enhancement, release of organisms into the natural environment. In addition, routine shipments of live oysters, clams and crabs intended for direct sale to consumers are seldom ever checked for diseases, parasites and competitors, nor are most shipments of bait organisms (W. Anderson, personal communication).

Many chemicals including biocides, antibiotics, therapeutics and hormones are used in aquaculture for pest control, treatment of disease, disinfection, anaesthesia and other functions. This chemical 'soup' represents a potential threat to the health of the cultured organisms, the indigenous biota and the human consumer (Rosenthal *et al.*, 1988).

Additional impacts from marine aquaculture have been experienced at a number of locations worldwide. Water use in shrimp culture may affect the surrounding environment through extraction of ground water and discharge of pond wastewater. Heavy water use may draw from freshwater aquifers and reduce the supplies of domestic and agricultural water, aside from causing seawater intrusion, especially of depleted aquifers (Primavera, 1997). Mangrove and wetland destruction has been an unfortunate consequence of aquaculture siting elsewhere in the world (Currie, 1994; Primavera, 1997). Erosion and increased susceptibility to natural hazards are both tangible results of this destruction.

The state of knowledge regarding the environmental and ecological impacts and benefits of aquaculture is rapidly improving. Over the last two decades, significant private and public sector research and monitoring programmes have been initiated to document these effects and to develop the scientific foundation upon which to develop sustainable culture practices and best management techniques. Major studies have examined and are investigating the effluent characteristics of marine shrimp pond culture and salmon netpen culture and the issues regarding species introductions, the use of chemicals in aquaculture, and effluent discharges.

Institutional and legal environment

Both industry and government recognize the current regulatory environment for marine aquaculture in the United States as the major constraint to its development (e.g. National Research Council, 1978, 1992; Joint Subcommittee on Aquaculture, 1993; DeVoe, 1997, 2000). No formal federal framework exists to govern the leasing and development of private commercial aquaculture activities in public waters (National Research Council, 1992).

In a 1981 study commissioned by the US Joint Subcommittee on Aquaculture, the only one of its kind to date, the Aspen Corporation examined the federal and state regulatory framework for aquaculture (Aspen Corp., 1981). The study found that as many as 11 federal agencies are directly involved in regulating aquaculture, with another ten having indirect involvement. Seven federal agencies were identified as the principal ones having regulatory programmes that involved direct interaction with the marine aquaculture industry: US Army Corps of Engineers, US Environmental Protection Agency, US Fish and Wildlife Service, US Food and Drug Administration, US Department of Agriculture, NOAA National Marine Fisheries Service, and US Coast Guard. Further, some 50 federal statutes (with accompanying regulations) were found to have a direct impact on the aquaculture industry, although the actual numbers of statutes that affect an individual operation vary depending on its size, location, the species being cultured and other factors. In total, over 120 statutory programmes of the federal government were found to significantly affect aquaculture development, with approximately one-half requiring direct compliance from the fish farmer.

Federal oversight of the marine aquaculture industry is fragmented as there is no overall federal regulatory framework to address aquaculture development in the coastal zone or offshore waters. Regulations adapted to fill the void in aquaculture policy derive from other sectors such as forestry, fisheries, and agriculture and do little to address the unique requirements of this evolving industry. Further, while recent evaluations of marine aquaculture suggest that offshore locations may represent a viable alternative (National Research Council, 1992), no formal policies have been established to manage aquaculture development in the US Exclusive Economic Zone. Thus, existing federal policies can vary from one agency to another (and among divisions within the same agency in some cases) and the regulatory process can be time-consuming, complex and costly (DeVoe, 2000).

The majority of laws and regulations that specifically authorize, permit, or control aquaculture are usually found at the state level (Aspen Corp., 1981). Thirty-two state regulatory programmes were examined in the Aspen Corp.

study, which found that over 1200 state laws have some significant bearing on aquaculture operations. Policies and regulations were found to affect aquaculture in eight major areas: (i) aquaculture species use; (ii) water quality; (iii) water use; (iv) land use; (v) facility and hatchery management; (vi) processing; (vii) financial assistance; and (viii) occupational safety and health.

The lack of uniformity of laws among the states, the sheer number of permits, licences and certifications that must be obtained, and the difficulty in obtaining them continue to pose significant problems at the state level for the marine aquaculture industry (National Research Council, 1978, 1992). Each state has its own unique legal, political and economic climate for aquaculture, and culturists must navigate the regulatory environment differently in each, and only a few states have the ability to present an applicant with a comprehensive list of all the legal requirements that must be met (DeVoe, 2000). And, since state laws and regulations can be, and sometimes are, more restrictive than those of the federal government, the standards that states apply to aquaculture in each case can vary widely (McCoy, 1989), and some still apply laws designed for other applications, e.g. fisheries management (National Research Council, 1978, 1992).

Policy Framework for Sustainable Marine Aquaculture

Worldwide, the aquaculture industry is calling for transparent and enforceable policies to improve or replace the over-burdensome and largely ineffective policies and legislation that now exists. Many feel that the industry would be able to overcome many of its current institutional limitations with clear policies, empowered lead agencies, and comprehensive and enforceable laws and procedures that encourage sustainable aquaculture and promote trade in aquaculture products (Aquaculture Development Beyond 2000).

Integral components of a policy for sustainable aquaculture

While the differences in aquaculture systems are readily apparent, there are several requirements common to all culture operations (described in Reiser, 1997). These needs can form the basis for a national aquaculture policy. Government authorization in the form of leases, permits or other regulatory permission has to be made accessible to the potential culturist. Within the aquaculture industry, horror stories abound concerning difficulties encountered while attempting to obtain necessary permits. While some governments have taken measures to rectify the situation, many still leave it to the culturist to navigate the bureaucratic maze.

Revised aquaculture policies should identify and provide sources of financial assistance through grants, loans or guarantees (Reiser, 1997). The present stance of the US government with respect to the aquaculture industry is that the industry can benefit from governmental counsel, but does not require governmental dollars. If the United States is sincere in its statements of support for the industry, it should provide financial support to aid in achieving stated objectives.

Policies should also assist in marketing and promotion of aquaculture products (Reiser, 1997). This assistance may range widely, from implementing trade tariffs to funding and publicizing evaluative research on the health benefits of seafood consumption.

Aquaculture polices should address the need for improvements in availability of seed and feed stocks. While research is ongoing in both these areas, objectivity is being compromised by media attention. Current public opinion concerning the environmental impacts of aquaculture has been negatively influenced by reports concerning the feeding of fish meal to farmed fish, or 'farming down the food web' as it has been termed (Pauly *et al.*, 1998; Naylor *et al.*, 2000). Resolving the validity of these concerns will be an essential step in proving aquaculture's capacity to operate sustainably. Policies should provide a conduit for public input into decisions concerning the aquaculture industry. Together with this provision the government and industry should ensure that there is ready accessibility to accurate and unbiased information about all types of aquaculture. As part of this effort, government policies should encourage research and development in concert with industry objectives.

Policies should provide a framework of technical assistance. This can occur through the implementation of educational programmes or provision of access to competent authorities such as extension agents. Lastly, policies should address the subject of planned and unintentional species introductions, and the potential for disease and parasites in culture operations.

The elements of an ideal marine aquaculture policy framework are identified in an industry study commissioned by the provincial government of Nova Scotia, Canada (Wildsmith, 1982). This construct would include designated zones for aquaculture, exclusive property rights, regulatory coordination and streamlining, a risk safety net for aquaculturists, and mechanisms to address seed and feed stocks, species selection and disease.

Respective of these requirements, the framework currently in place in the United States is far from this ideal. In its report on marine aquaculture, the National Research Council (1992) characterized the federal role as a series of general policy statements of support, repeated studies on the obstacles facing aquaculture, and the formation of interagency mechanisms lacking in authority and power. Policymakers are fully aware of the problems facing the industry, but do not seem to be able to move beyond that awareness to provide solutions. Until the limitations inherent in the current US aquaculture policy structure are resolved, many of the problems facing aquaculture are likely to persist.

Aquaculture and US policy

The National Aquaculture Act has been the driving force behind aquaculture policy in the United States since its enactment on 26 September 1980. The intent of the Act was to 'promote aquaculture in the United States' and mandated a declaration of a national policy, development and implementation of a national aquaculture development plan, establishment of a coordinating group of federal agency representatives, establishment of a National Aquaculture Information Center, and encouragement of aquaculture activities and programmes in both the public and private sectors (DeVoe, 2000). The 1980 Act was substantively amended in 1985 and 1990, and was most recently reauthorized in 1998.

The Act declares that it is 'in the national interest, and it is the national policy, to encourage the development of aquaculture in the United States'. The National Aquaculture Act of 1980 gave principal responsibility for the development of aquaculture to the private sector but jointly assigned the Departments of Agriculture, Commerce and Interior responsibilities for carrying out the provisions of the legislation.

A federal interagency coordinating group, the Joint Subcommittee on Aquaculture (JSA), was created to increase the overall effectiveness and productivity of federal aquaculture research, technology transfer and assistance programmes. The JSA, composed of the leadership of more than a dozen federal agencies, is generally regarded as a model coordinating mechanism in the US government. Up until most recently, the JSA received no direct funding. However, it exists now as a statutory committee of the National Science and Technology Council (NSTC) within the Office of Science and Technology Policy in the Office of the Science Advisor to the President. The JSA reports to the NSTC's Committee on Health, Safety and Food Research and Development, which is one of nine research and development committees established by NSTC to prepare coordinated research and development strategies and budget recommendations for accomplishing national goals. Permanent chairmanship of the JSA has been designated to the Secretary of Agriculture as per the 1985 amendments to the National Aquaculture Act.

The JSA completed the first version of a national aquaculture development plan in 1983. No assessment regarding progress on the original plan's recommendations was ever made. It was not until 1996 that revision of the 1983 plan was considered. An updated national aquaculture development plan has now been prepared by the JSA and is under review by the JSA member agencies at the time of writing.

The National Aquaculture Act of 1980 and its amendments provide the only federal policy framework for, and endorsement of, US aquaculture. Nevertheless, although the 1980 Act was reauthorized in 1998 as part of the Farm Bill, the recent failure of legislation explicitly extending and funding the 1980 Act can be directly attributed to the lack of consensus in Congress and the Administration on what the federal government's commitment to aquaculture should be.

The Sustainability Challenge

The history of marine aquaculture development in the United States, while not extensive, reveals a number of issues that continue to plague the industry and constrain its growth. Detractors of the industry have used these factors to target public attention on aquaculture's real and perceived ecological, environmental and social impacts. Addressing the real issues and moving beyond public perceptions will require a heightened awareness of the efforts being made to render marine aquaculture compatible with other coastal activities and the environments within which it is practised.

It is becoming increasingly clear that aquaculture, especially in coastal and marine areas, must be planned, developed, operated, managed and monitored with long-term sustainability in mind. Marine aquaculture should be looked at as an important element of expanding protein supplies, providing economic development opportunities and jobs, and offering communities the ability to diversify. But aquaculture expansion must also accommodate the need to maintain and improve the environment, social norms, and overall quality of life. Effective management of aquaculture in a sustainable fashion requires the involvement and cooperation of government, academia, the private sector, investors, communities and the public at large. The industry is extremely complex and diverse, and the establishment of a sustainable and responsible marine aquaculture sector will require the generation of research information in the areas of culture species, resources, culture systems, economics, ecosystems, and sociology and anthropology (Chua and Tech, 1990). Concerted and coordinated efforts to generate science-based information to support sustainable marine aquaculture planning and development are needed. To achieve these ends, government policies must provide a reasoned and balanced foundation from which to proceed.

Katz (1995, unpublished), for instance, offered a summary of the characteristics of successful and sustainable aquaculture industry development worldwide. To begin with, the objectives of sustainable aquaculture must answer to both the industry and the community within which it will be realized. If either of these components is excluded from consideration, the chances of achieving sustainability are reduced. Sustainable aquaculture promotes interaction and communication between the public and private sectors at all stages of development as key factors in ensuring that the needs of each are met without undue conflict or compromise.

Sustainable aquaculture will occur where aquaculture is a national priority. This commitment goes beyond simply stating that this is so. The government must provide tangible guidance and support for the industry, substantiating its claim. Policies that provide mechanisms to identify sites and establish tenure over those areas must be put into place. In anticipation of providing legal rights of exclusion over culture areas, a means of resolving user conflicts should be developed and implemented. Relationships between national and local authorities should foster cooperation. The process of obtaining governmental permissions to conduct aquaculture should be streamlined, and specific legislation addressing the subject of foreign or multinational corporate ownership of aquaculture operations must be approved.

Sustainable aquaculture benefits from an established industry-driven research agenda (Katz, 1995). Such research can serve in part to disprove the notion that sustainable aquaculture is an unobtainable objective. Planning includes an evaluative means of assessing research products and incorporating technological advances into the industry. In addition to promoting directed research, the industry is also engaged in developing new markets and niches to ensure ongoing industry expansion.

Katz (1995) argued that sustainable aquaculture occurs where 'site-specific policy; legal, social, and environmental infrastructure; and market, technical support, and conditions needed to favor development' are all in place. Achieving sustainable aquaculture is not a spontaneous occurrence, but a directed effort with multiple components.

Sustainable marine aquaculture in the United States – present challenges

There are three principal challenges that the aquaculture industry must overcome to achieve sustainability. Without the support of the public and politicians, marine aquaculture will continue to encounter the same limitations that have hampered its growth for the last 30 years. People need to be educated to believe in the promise of sustainable aquaculture and give their support to policies that will promote this objective.

- Perception as reality. Resolving the negative image that marine aquaculture has today is one of the most important challenges. Many leaders in aquaculture have been calling for action. Pillay (1996) argued that the most serious challenge facing the industry is to change the notion that aquaculture cannot be conducted in a sustainable fashion, a perception he suggested was created by the mistakes of the past and misinformation communicated through the media. The need for continual education of policymakers, planners and the public was emphasized by New (1996), who argued that these constituencies must be made aware of aquaculture's significant contributions to food security, human nutrition, the social and political stability of rural and coastal communities, and local and national economies.

  The creation of an enabling environment has been raised by a number of authors (Barg and Phillips, 1997; FAO, 1999; New, 1999). This concept is based on the need for the industry and its partners to promote aquaculture based on its positive contributions and on information generated by objective, well-designed scientific enquiry (Pillay, 1996). Those involved with aquaculture understand this. Nevertheless, those who oppose aquaculture for its real and/or perceived misuse of natural resources and its impacts on environmental resources and the social fabric of communities continue to object (New, 1999).

- Resource management as people management. An ongoing challenge in aquaculture development has been the extension of resource management to include people who use the resource as well as the resource itself. Several decades ago, aquaculture was touted as the blue revolution in many parts of the world. Its early development was encouraged by its promotion as a clean industry providing many social and economic benefits. The few, but highly publicized, failures that occurred during early stages shook public confidence and demonstrated that with aquaculture's benefits came unanticipated impacts. People became involved in the issue because they were affected or were concerned. Complaints voiced to the government demanded answers to questions that had never before been addressed. Over time the gap between practice and policy has closed considerably. Close interaction between the government and aquaculture communities will ensure comprehensive responses to current and future challenges.
- Duality of government roles. The third significant challenge will be to find a resolution to the inherent conflict of interest in the government's position on aquaculture. The government wears many hats – as a supporter and promoter versus regulator and enforcer – in its aquaculture policies. This lack of consensus permeates both the executive and legislative arms of the government. Even with the National Aquaculture Act, the Joint Subcommittee on Aquaculture mechanism, and other established elements of the federal policy for aquaculture in the United States, difficulties persist in seeking a consensus on a government policy for aquaculture (Becker and Buck, 1997).

Recent executive and legislative actions indicate that the United States is aware of these issues. The revised National Aquaculture Development Plan, which is presently under executive review, includes important references to sustainability issues. Amendments to the Coastal Zone Management Act in 1996 and 1999 provide for the 'adoption of procedures and policies to evaluate and facilitate the siting of public and private aquaculture facilities in the coastal zone . . .' as well as 'development of a coordinated process among state agencies to regulate and issue permits for aquaculture facilities in the coastal zone' (Reauthorization/Amendment of CZMA, 1999). The United States is a signatory to the Food and Agriculture Organization's Code of Conduct for Responsible Fisheries. Section 9 of the code addresses principles and standards for aquaculture. The National Oceanic and Atmospheric Administration and the National Marine Fisheries Service have taken a leadership role in preparing guidelines in the form of a Code of Conduct for Responsible Aquaculture in the US exclusive economic zone, initiated with a series of six regional workshops in September and December 2000 to engage stakeholders in the development process.

Over the last 10 years, there have also been a number of attempts, albeit unsuccessful, to strengthen the National Aquaculture Act and to establish a legislative basis for marine aquaculture. The current draft National Marine Aquaculture Act proposes to establish a federal framework for offshore aquaculture, including the development of a set of environmental guidelines. These positive measures should be used as a springboard by the aquaculture industry to push for further government action to advance sustainable and responsible marine aquaculture.

Sustainable marine aquaculture in the United States – the future

The United States is currently in the position of crafting the future of its marine aquaculture industry. Decisions taken now will determine if the concept of sustainable marine aquaculture becomes a reality or remains an ambiguous ideal. The cycle of the past 30 years of identifying limitations to and of the industry and making suggestions as to how they can be addressed has done little to advance the cause of domestic aquaculture development to date. The biggest challenge in sustainable aquaculture for the future may well be finding a way to move beyond analysis paralysis to begin implementing recommendations and allow the industry to realize its full, sustainable, potential.

There are no quick and easy solutions to the issues related to the sustainable development of marine aquaculture. Progress towards the future will depend in part on having learned from past mistakes and profiting from positive experiences. Marine aquaculture may still prove to provide enduring societal benefits, but only if we are very careful to ask and address the right questions. How we address the following challenges – through government policy-making, integrated planning and management, community and public involvement, and education and outreach – will determine the future for sustainable aquaculture in the United States.

Government policy-making

The National Aquaculture Act of 1980 contains a clear and unambiguous statement in support of aquaculture development in the United States. The United States, through Congress and the Administration, and with the support of industry and the involvement of all constituencies, must take a hard look at the current situation and decide if it wishes to aggressively pursue the policy. Many scholars, academics and industry leaders have offered a wide range of possible solutions to address the constraints limiting marine aquaculture development. But without strong commitment and leadership by government to work towards this goal, which would include the development of comprehensive federal legislation, the current situation will be difficult to change.

In developing legislation pertaining to aquaculture, the US government should consider how and whether the industry could be best served by the existing institutional infrastructure. The extension of rules and regulations developed for fisheries management to aquaculture is often inadequate to address the many variations within the industry. Government must develop a clear vision for aquaculture management, and use this ideal to integrate aquaculture within existing legislation or to formulate a new legal framework.

While it may be extremely difficult to do, the United States should continue to pursue the development of a comprehensive revision of the 1980 National Aquaculture Act *et seq.* that re-establishes the respective roles of the federal agencies involved with the industry, eliminates duplicative and non-germane laws and regulations, and provides the legislative framework for the development of comprehensive and coordinated marine aquaculture policies, programmes and practices.

Mechanisms must be put into place to refine existing and establish new implementation measures to guide growth of the marine aquaculture industry. The fundamental framework to meet this challenge already exists with the Joint Subcommittee on Aquaculture (JSA). There are three areas where the role of the JSA could be strengthened. First, the role of the JSA in the administration could be expanded to include policy development and implementation. Second, the permanence of the JSA could be strengthened through the provision of a stable source of funding and staff assistance to improve coordination and consistency. And third, JSA membership could be expanded to include representatives from key groups such as the marine aquaculture industry, environmental community and other constituencies in its deliberations and decision-making (DeVoe, 2000).

In a 1981 study commissioned by the US Joint Subcommittee on Aquaculture, the only one of its kind to date, the Aspen Corporation examined the federal and state regulatory framework for aquaculture. The JSA could be charged with conducting a follow-up analysis of the current regulatory situation for US marine aquaculture and, based on the analysis, propose a coordinated and streamlined planning, management and regulatory framework for marine aquaculture activities in the US coastal zone and the exclusive economic zone (EEZ). This review, analysis and proposal must be conducted in consultation with the marine aquaculture industry, the states, the academic and policy community, and pertinent constituencies, and the results should emphasize joint local, state and federal coordination.

Planning for marine aquaculture

The draft National Aquaculture Development Plan should be developed to promote, support, regulate and report on the aquaculture industry, to ensure that aquaculture is ecologically sustainable and the use of resources is allocated fairly. The plan should be designed to take into account the needs of the industry, the markets it serves, the communities within which it is based, and the environments where it is conducted.

Policymakers, governments, local communities and others must consider the future profile of coastal and marine areas when developing overall management plans. This analysis should take into account the potential effects of global population increase and global climate change. Without realistic attention to these variables, any management plan will prove to be inadequate for long-term use. Clearly defined property rights and effective legal structures to support them will provide stability to coastal resource-based industries, and increase accountability for environmental degradation and social disruption. A concerted effort to develop operational and sustainable coastal management plans is essential. Existing management plans should be reviewed on a regular basis to ensure that they continue to meet the needs of affected populations and the coastal environment.

An important and necessary institutional and legal component of integrated coastal zone management is the provision of a mechanism for conflict resolution. This is becoming more critical as the availability of coastal resources becomes scarcer, and the prevalence of multiple-use conflicts with commerce and recreation increases. Formal discussions among all parties potentially affected by a planned aquaculture development early in the process may minimize or alleviate conflicts that might arise henceforth.

Planning is not the final objective but the first step in establishing viable aquaculture enterprises. There must be a corresponding commitment by government, communities, industry and the public to see that plans are put into effect. Strategies, with identified action steps, responsible parties and available resources, should be designed and implemented.

Community involvement

Coastal communities share several unifying characteristics, but there is no one formula for aquaculture development that will fit all situations. Flexibility is essential in all planning processes. Consideration should be given to a community's unique cultural, economic, social, gender and other features in aquaculture development. Focus should be placed on the long-term benefits and ramifications of development, and adjustments made in the planning stages to ensure minimal disruption of community attributes. There are multiple ways to pursue aquaculture development, and the plan in hand may not always be the best choice for a given community.

To coexist amicably with other resource users, aquaculture operators must engage in some form of benefit sharing within the local community. The nature of this exchange should be determined through negotiations in all the planning stages of an aquaculture operation. Involvement of all community members in a project during the development stage will identify potential conflicts and determine if necessary compensations will be cost effective for the proposed project. Where benefits do not outweigh costs, the siting or feasibility of the project may need to be re-evaluated.

The problem of urban migration could be abrogated through investment in rural communities. Education, government–community partnerships, improvements in communications and transportation networks, and the availability of private sector and other financing are essential provisions to render rural communities economically viable and socially stable. Aquaculture projects in rural areas that receive this type of support can serve as sources of community and economic revitalization.

The promotion of environmentally friendly aquaculture products, cultivation of high-end market goods, and emphasis on fresh local products are all-important marketing tools that can make small-scale aquaculture more profitable. Where the potential for such markets does not exist, products with regional appeal may cater to local markets. As the concept of the global marketplace becomes more widely accepted, new marketing strategies adapted for rural communities may provide additional product end-points.

Information and education

The importance of education in aquaculture development should not be undervalued. If the objectives of sustainable development and effective coastal management are to be realized, coastal populations must be informed about environmental limits and the consequences of resource exhaustion. Incorporation of sustainability in research and development will enable further definition of these limits, while increased understanding of issues will foster community investment in achieving the goal of sustainable aquaculture development.

There are a number of federal agencies in the United States that support aquaculture development through the provision of research, extension, outreach, and financial assistance, including the US Department of Agriculture, US Department of Commerce, and the US Department of the Interior. These agencies also operate and support a number of research and technology transfer facilities and laboratories. The JSA should conduct an assessment of all current federal funding programmes to assess the nature, scope, and accomplishments of past and current research, development, assistance and technology transfer activities, and whether they are meeting the needs of the industry and the public. The JSA should, based on that assessment, assemble a coordinated, cross-cutting funding plan to ensure that future priority needs and issues related to sustainable marine aquaculture are identified and that future research and outreach efforts are responsive (DeVoe, 2000).

Aquaculture is a unique and complex industry with similarities to many other disciplines. To better serve aquaculture, stronger linkages should be pursued with agriculture, forestry, engineering, chemistry, economics, sociology and other related fields of study. Multidisciplinary connections could result in new techniques and technologies, and provide valuable new sources of funding, insurance, and scientific and social expertise.

Technology transfer is a critical function made available through government and university programmes. Increasing attention should be paid to the suitability of technologies for the target population. It should be stressed that education is not a mandate from above, but an exchange of ideas. Extension agents and others should endeavour to incorporate local knowledge and skills specific to the target community into their work.

Aquaculture has proved itself to be a profitable industry. However, historically some of these benefits have been obtained at the expense of environmental and community integrity. This has led to public vilification of the industry, and very little acknowledgement for the many benefits aquaculture can provide. Communities involved in properly managed aquaculture projects can provide tangible evidence of these values, and dispel the popular notion that all environmental degradation and social disruption near aquaculture operations derive directly from culture practices.

Concluding Remarks

Effective management of aquaculture in a sustainable fashion will require the involvement and cooperation of government, academia, the private sector, investors, communities and the public at large. Marine aquaculture should be looked at as an important component of expanding protein supplies, providing economic development opportunities and jobs, and offering communities the ability to diversify. But aquaculture expansion must also accommodate the need to maintain and improve the environment, social norms and overall quality of life. It is becoming increasingly clear that marine aquaculture must be planned, developed, managed and monitored with long-term sustainability in mind.

Acknowledgements

The authors wish to thank the convenors of the Aquaculture 2001 Special Session on Responsible Marine Aquaculture, Drs Robert Stickney and James McVey, for the opportunity to participate and contribute. We appreciate the support of the South Carolina Sea Grant Consortium and the State of South Carolina without which the preparation of this chapter would not have been possible.

References

Ackefors, H. and Rosen, C.G. (1979) Farming aquatic animals: the emergence of a world-wide industry with profound ecological consequences. *Ambio* 8(4), 132–143.

Ackefors, H. and Sodergren, A. (1985) *Swedish Experiences of the Impact of Aquaculture on the Environment*. International Council for the Exploration of the Seas, C.M. 7 p. ICES, Copenhagen, Denmark.

Aquaculture Development Beyond 2000: The Bangkok Declaration and Strategy (2000) Conference on Aquaculture Development in the Third Millennium 20–25 February 2000, Bangkok, Thailand.

Aspen Corporation (1981) *Aquaculture in the United States: Regulatory Constraints – Final Report*. Contract No. 14-16-009-79-095 to US Fish and Wildlife Service, Washington, DC, 51pp.

Barg, U. and Phillips, M.J. (1997) *Environment and Sustainability. Review of the State of World Aquaculture.* FAO Report No. 886:55-66. Food and Agriculture Organization of the United Nations, Rome.

Becker, G.S. and Buck, E.H. (1997) *Aquaculture and the Federal Role.* Congressional Research Service – The Library of Congress, Washington, DC, 97-436 ENR. 29pp.

Chamberlain, G. and Rosenthal, H. (1995) Aquaculture in the next century: opportunities for growth – challenges of sustainability. *World Aquaculture* 26(1), 21–25.

Chua, T-E. (1993) Environmental management of coastal aquaculture development. In: Pullin, R.S.V., Rosenthal, H. and Maclean, J.L. (eds) *Environment and Aquaculture in Developing Countries.* ICLARM Conference Proceedings 31. International Center for Living Aquatic Resources Management, Manila, Philippines, pp. 199–212.

Chua, T-E. and Tech, E. (1990) Aquaculture in Asia: quo vadis? In: Mohan, M. (ed.) *Aquaculture in Asia.* Asian Fisheries Society, Indian Branch. Mangalore, India, pp. 13–30.

Currie, D.J. (1994) Sustainable aquaculture in developing countries- who can make it happen? *World Aquaculture* 25(4), 20–25.

DeVoe, M.R. (ed.) (1992) *Introductions and Transfers of Marine Species: Achieving a Balance Between Economic Development and Resource Protection.* Proceedings of a Conference and Workshop. Hilton Head Island, South Carolina, 30 October–2 November 1991. South Carolina Sea Grant Consortium, Charleston, South Carolina, 201 pp.

DeVoe, M.R. (1994) Aquaculture and the marine environment: policy and management issues and opportunities in the United States. *Bulletin of the National Research Institute of Aquaculture* (Suppl. 1), 111–123.

DeVoe, M.R. (1997) Marine aquaculture regulation in the United States: environmental policy and management issues. In: *Interactions Between Cultured Species and Naturally Occurring Species in the Environment*, Proceedings of the 24th US–Japan Aquaculture Panel Symposium, 8–10 October 1995. Texas A&M University, Sea Grant College Program, Bryan, Texas, pp. 1–16.

DeVoe, M.R. (2000) Marine aquaculture in the United States: a review of current and future policy and management challenges. *Marine Technology Society Journal* 34, 5–17.

DeVoe, M.R. and Mount, A.S. (1989) An analysis of ten state aquaculture leasing systems: issues and strategies. *Journal of Shellfish Research* 8, 233–239.

Dodge, C.W. and Jolly, W.C. (1978) Aquaculture: status of technology and future prospects. Issue Brief No. IB77099, The Library of Congress, Congressional Research Service, Washington, DC, 12 pp.

Fiske, S.J. and Plé, J. (1992) Sociocultural aspects of domestic marine aquaculture. In: *Marine Aquaculture: Opportunities for Growth.* Report of the Committee on Assessment of Technology and Opportunities for Marine Aquaculture in the United States, Marine Board, Commission on Engineering and Technical Systems,

National Research Council. Washington, DC. National Academy Press, Washington, DC, pp. 253–268.

Food and Agricultural Organization (1997) *FAO Technical Guidelines for Responsible Fisheries*. Fisheries Department. Aquaculture Development, No. 5. Rome, 40pp.

Food and Agricultural Organization (1999) Creating an enabling environment for sustainable aquaculture. In: *The State of World Fisheries and Aquaculture*. FAO, Rome.

Goldburg, R. and Triplett, T. (1997) A salmon in every pot: aquaculture for economic development. In: *Murky Waters: Environmental Effects of Aquaculture in the United States*. Environmental Defense Fund, Washington, DC (http://www.edf.org/pubs/Reports/aquaculture), pp. 87–102.

Joint Subcommittee on Aquaculture (1983) *National Aquaculture Development Plan*, Volume I. JSA, Washington, DC, 67pp.

Joint Subcommittee on Aquaculture (1993) *Aquaculture in the United States: Status, Opportunities and Recommendations*. Report to the Federal Coordinating Council on Science, Engineering and Technology. JSA, Washington, DC, 21pp.

Katz, A. (1995) Aquaculture – international examples of success and failure and lessons for the United States. Associates in Rural Development, Burlington, Vermont (unpublished manuscript). [Taken from Corbin, J.S. and L.G.L. Young (1997) Planning, regulation and administration of sustainable aquaculture. In: Bardach, J.D. (ed.) *Sustainable Aquaculture*. John Wiley & Sons, New York.]

Lester, L.J. (1992) *Marine Species Introductions and Native Species Vitality: Genetic Consequences of Marine Introductions*. In: DeVoe, M.R. (ed.) Proceedings of a Conference and Workshop, Hilton Head Island, South Carolina, 30 October–2 November 1991. South Carolina Sea Grant Consortium, Charleston, South Carolina, pp. 79–89.

Mann, R. (ed.) (1978) *Exotic Species in Mariculture*. Proceedings from a symposium, case histories of the japanese oyster, *Crassostrea gigas* (Thunberg). Woods Hole Oceanographic Institution, Woods Hole, Massachusetts, 18–20 September. MIT Press, Cambridge, Massachusetts, 363pp.

McCoy, H.D. II (1989) Commercial aquaculture zones: a legislative proposal. *Aquaculture* 6, 39–46.

Muir, J.F. (1985) Aquaculture – towards the future. *Endeavour* 9, 52–55.

National Research Council (1978) *Aquaculture in the United States: Constraints and Opportunities*. National Academy Press, Washington, DC, 123pp.

National Research Council (1992) *Marine Aquaculture: Opportunities for Growth*. Report of the Committee on Assessment of Technology and Opportunities for Marine Aquaculture in the United States, Marine Board, Commission on Engineering and Technical Systems, National Research Council. Washington, DC. National Academy Press, Washington, DC, 290pp.

Naylor, R.L., Goldburg, R.J., Mooney, H., Beveridge, M., Clay, J., Folke, C., Kautsky, N., Lubchenco, J., Primavera, J. and Williams, M. (1998) Nature's subsidies to shrimp and salmon farming. *Science* 282, 883–884.

Naylor, R.L., Goldberg, J., Primavera, J.H., Kautsky, N., Beveridge, M.C.M., Clay, J., Folke, C., Lubchenco, J., Mooney, H. and Troell, M. (2000) Effect of aquaculture on world fish supplies. *Nature* 405, 1017–1024.

New, M.B. (1996) Sustainable global aquaculture. *World Aquaculture* 27(2), 4–6.

New, M.B. (1999) Global aquaculture: current trends and challenges for the 21st century. *World Aquaculture* 30(1), 8–13ff.

Odum, W.E. (1974) Potential effects of aquaculture on inshore coastal waters. *Environmental Conservation* 1, 225–230.

Pauly, D., Christensen, V., Dalsgaard, J., Froese, R. and Torres, F.C. Jr (1998) Fishing down marine food webs. *Science* 279, 860–863.

Pillay, T.V.R. (1996) The challenges of sustainable aquaculture. *World Aquaculture* 27(2), 7–9.

Pollnac, R.B. (1991) The role of sociocultural factors in aquaculture development projects. In: Hargreaves, J. and Alston, D. (eds) *Advances in World Aquaculture, Vol. 5, Status and Potential of Aquaculture in the Caribbean.* World Aquaculture Society, Baton Rouge, Louisiana, pp. 165–191.

Primavera, J.H. (1997) Socio-economic impacts of shrimp culture. *Aquaculture Research* 28, 815–827.

Reauthorization/Amendment of CZMA (1999) Aquaculture Related Amendment. (http://ag.ansc.purdue.edu/aquanic/news/reuthmdczma.htm).

Reiser, A. (1997) Defining the federal role in offshore aquaculture: should it feature delegation to the States? *Ocean and Coastal Law Journal* 2(2), 209–234.

Rosenthal, H. (1985) Constraints and perspectives in aquaculture development. *GeoJournal* 10, 305–324.

Rosenthal, H., Weston, D., Gower, R. and Black, E. (1988) *Environmental Impact of Aquaculture.* Report of the ad hoc study group, International Council for the Exploration of the Seas, No. 154. ICES, Copenhagen, Denmark, 83pp.

Weston, D.P. (1986) *The Environmental Effects of Floating Mariculture in Puget Sound.* Report 87-16 to Washington Department of Fisheries. Washington Department of Fisheries, Olympia, Washington, 148pp.

Wildsmith, B.H. (1982) *Aquaculture: the Legal Framework.* Emond-Montgomery Ltd, Toronto, Canada, 313pp.

Marine Mammals and Aquaculture: Conflicts and Potential Resolutions

3

Bernd Würsig and Glenn A. Gailey

Department of Marine Biology, Texas A&M University at Galveston, 4700 Avenue U, Building 303, Galveston, TX 77551, USA

Abstract

Two main types of marine-based aquaculture come into potential conflict with marine mammals (and, in some areas, marine turtles and seabirds): (i) extensive raising of shellfish, such as oysters, mussels and shrimp; and (ii) intensive raising of finfish, such as salmon, sea bass and sea bream. The first takes up space in near-shore waters but does not generally require nets or cages that can entangle or otherwise hurt air-breathing vertebrates. It also does not require supplementary feeding, and therefore is not generally a major attractant for marine mammals and others. However, shellfish aquaculture puts extra nitrogen into the ecosystem, and can change local ecology where tidal and other flushing is minimal. It takes up extensive space in inlets, fjords and the like, and may compete for limited habitat access with foraging, resting, socializing and nurturant mammals. The intensive but generally more localized farming of finfish often requires supplementary feeding, and both the stock in holding pens and the feed serve as powerful attractants especially to pinnipeds (but toothed cetaceans, river and sea otters, marine turtles, and seabirds are often involved as well). As such, major problems are caused to the industry by destruction of gear and the target aquaculture species; and to the marine animals by shooting and other techniques, such as large-scale use of Acoustic Deterrent Devices (ADDs) and Acoustic Harassment Devices (AHDs). No technique has proved highly successful, and the widespread use of ADDs and AHDs is particularly problematic and largely untested. We recommend that owing to potential for entanglement, chemical and sound pollution, habitat loss or gross alteration, traffic, and changes in species interactions, all proposed development of marine aquaculture in nature should be subjected to initial evaluations and – as needed – scientific research relative to interactions between the food being raised by humans and the predators that attempt to take advantage of this. The loss of habitat to marine mammals by both shellfish and fish aquaculture facilities needs to be investigated on a case-by-case basis.

Introduction

Aquaculture is an important industry that produces nearly one-third of the fish and shellfish products commercially available worldwide (FAO, 1999). Owing to the growth of the industry, the amount of habitat utilized along near-shore waters is increasing rapidly, and the number of conflicts between aquaculture and marine mammals is increasing as well. Aquaculturists estimate a loss of 2–10% of their gross production owing to marine mammal predation (Nash *et al.*, 2000). Most of this concerns marine-based aquaculture, except where riverine dolphins and lake seals co-occur with land-based aquaculture in such areas as the Yangtze River, the Amazon basin and several large lakes of Russia. A host of marine and freshwater birds and turtles, as well as river otters and other mammals, can take advantage of human-grown food so readily made available to these predators. In this chapter, we summarize marine-based problems related to marine mammals, with the realization that those of inland water systems can be similar; and recommend potential solutions to minimize conflicts between the aquaculture industry and marine mammals.

Shellfish Aquaculture

One form of aquaculture is that of extensive marine-based growing of shellfish (crustaceans and molluscs), such as mussels, oysters and shrimp. Such aquaculture takes up at times large spaces in inland bays and other waterways, and does not require supplementary feeding. Shellfish farms usually are developed in areas with sufficient water flow owing to tidal and other currents; and therefore potential pollution effects tend to be restricted in space, or only moderate over a large area (but see Páez-Osuna *et al.*, 1999; Kautsky *et al.*, 2000). Sea otters (*Enhydra lutris*), marine otters (*Lontra felina*) and walrus (*Odobenus rosmarus*) are the only marine mammals to habitually feed on shellfish. Therefore, there is usually no problem of marine mammal predation on the aquaculture resource, and little or no direct competition for the harvest between humans and the marine mammals in many parts of the world. However, in the Pacific Northwest of the United States and Canada, shellfish industries experience significant losses especially from river and sea otters. Also, faecal coliform counts from harbour seals (*Phoca vitulina*) have been known to contaminate the aquaculture beds (Nash *et al.*, 2000).

Extensive shellfish aquaculture competes for space in those areas where it occupies substantial portions of inland waterways. For example, in the Marlborough Sounds of the northern South Island of New Zealand, mussel and other shellfish farms are utilizing breeding, resting and foraging space formerly occupied by dusky dolphins (*Lagenorhynchus obscurus*), common bottlenose dolphins (*Tursiops truncatus*), Hector's dolphins (*Cephalorhynchus hectori*) and New Zealand fur seals (*Arctocephalus forsteri*). Dusky dolphins and other mammals are known to avoid the aquaculturally farmed areas, probably in large part owing to the numerous lines-to-buoys that may inhibit movement of schooling fish on which the dolphins feed, and make it difficult or impossible for the dolphins to efficiently aggregate their prey. A terrestrial analogy might be open-savanna lion (*Panthera leo*) prides being forced to chase and corral their prey in a forest. The planned expansion of mussel farming in the Marlborough Sounds indicates that near-shore habitat for marine mammals will become even more scarce, with presently unknown impacts on the health and survivability of the marine mammal species (Fig. 3.1). A recent illegal oyster farm (for harvesting of pearls) that was set up in the Shark Bay Marine Park of Western Australia dramatically affected known movements of Indian Ocean bottlenose dolphins (*Tursiops aduncus*), by excluding mothers and their calves from the farming area (Mann, 1999). Once the oyster lines were confiscated by the authorities, dolphins returned to their former habitat (Janet Mann, personal communication, February 2001).

It is not merely loss of space that causes habitat degradation. Marine mammals rely in large part on acoustic communication, echolocation (by toothed whales) and passive listening for prey in a world that is relatively opaque to sight (Richardson *et al.*, 1995). Extensive farming requires almost constant traffic owing to lighted-buoy maintenance, float and line maintenance, checking for the health and growth of the target crop, and harvesting. The vessels are a source of noise disturbance in formerly relatively pristine habitat, but amounts of habitat degradation and reduction of communication capabilities of marine mammals near shellfish farms are simply unstudied.

Where near-shore or on-shore shrimp farms have resulted in large-scale habitat disruption, coastal dolphins, porpoises, pinnipeds, manatees (*Trichechus* sp.) and dugongs (*Dugong dugon*) may be displaced or otherwise affected. The disruption is perhaps greatest for manatees and dugongs in the subtropical and tropical waters where coastal shrimp farming has led to loss of mangrove forest and other habitat (Dewalt *et al.*, 1996; Primavera, 2000; Páez-Osuna, 2001), but the above-mentioned effects on dolphins are also cause for concern.

Finfish Aquaculture

Intensive farming of marine or anadromous finfish is also of great concern. Such farming often requires supplementary feeding and the providing of medicines, with attendant problems of ecosystem change (Tovar *et al.*, 2000). For example, invertebrate benthic communities were affected at up to about 0.3 km distance radius from fish farm facilities in the North Baltic Sea even after several years of pollution abatement (Kraufvelin *et al.*, 2001). Similar

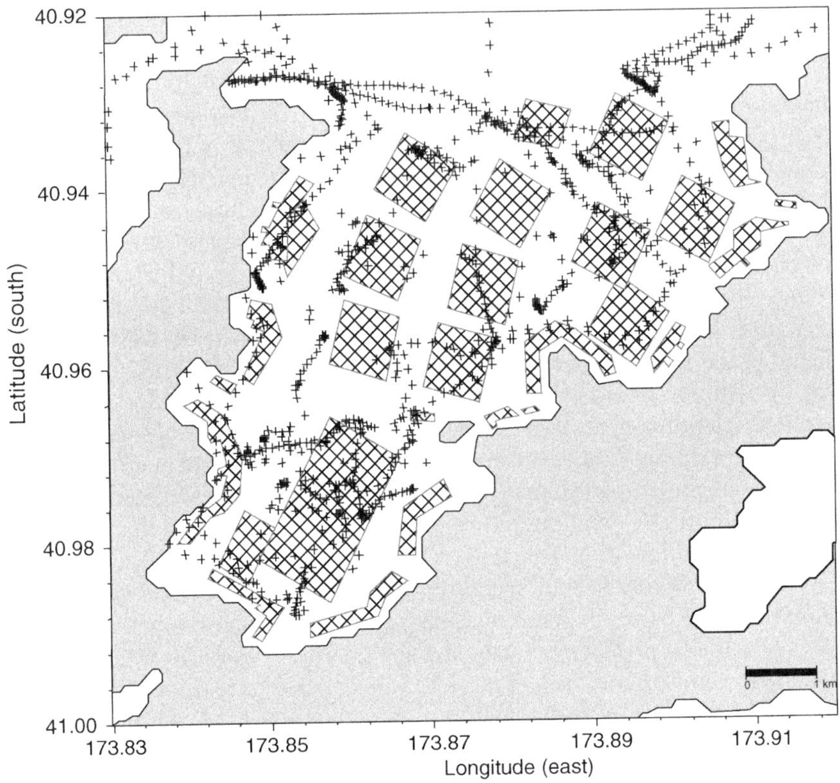

Fig. 3.1. Admiralty Bay in the Marlborough Sounds, South Island, New Zealand. Cross-hatched squares represent proposed shellfish farms, and near-shore cross-hatched longish patterns represent both proposed (in the west) and existing (in the east) farms. The small '+' marks represent dusky dolphin tracks ascertained during 6 total days of study in the winters of 1998–2000. Note that the existing near-shore farms in the east have no dolphin use overlap. It is clear that if even some of the proposed farms are built, dusky dolphin habitat use in the bay will be affected. (Data courtesy of A.D. Harlin and T. Markowitz, personal communication, February 2001.)

results were found for benthic communities down-current of salmon farms in British Columbia. The benthic infaunal ecosystem changes can displace feeding sea otters, near-benthic feeding dolphins and porpoises, and other organisms that rely on such benthic communities.

The providing of food and the very presence of fish being cultivated serve as powerful attractants to opportunistic dolphins and pinnipeds that normally feed on similar or the same fish stocks in nature. This sets the stage for large-scale competition between humans and marine mammals. In a well-known case study at the Ballard Locks, Seattle, Washington, only a handful of California sea lions (*Zalophus californianus*) may have been largely responsible for reducing an annual migration of 2400 steelhead trout (*Oncorhynchus mykiss*) – most of which were themselves raised by aquaculture techniques – to fewer than 200. The main culprit is said to have been a large male sea lion nick named 'Herschel', although other sea lions and human-caused problems of pollution and water quality degradation were likely factors as well (Gearin *et al.*, 1986; Fraker and Mate, 1999). In fact, only 3% of 248 marked sea lions were found to enter the Ballard Locks and feed on steelhead salmon in 1995 (NMFS, 1996a).

It is difficult to blame human aquaculturists for attempting to dissuade the marine mammals (or other predatory vertebrates) from taking fish that are being grown at great expense, or that are themselves endangered and in need of human management (Fraker and Mate, 1999; Nash *et al.*, 2000). Methods of dissuasion consist mainly of anti-predatory nets or other enclosures around finfish aquaculture facilities, noise-making devices intended to chase away predators, and shooting of the interlopers. The latter is almost always illegal, and has apparently been under-reported in the literature and to government agencies. By a recent web-based account of a salmon farm managed in British Columbia, Canada, 431 harbour seals, 38 sea otters, 29 sea lions, one harbour porpoise (*Phocoena phocoena*), 16 herons (family Ardeidae) and one osprey (*Pandion haliaetus*) were killed in a 4-year period. Because of the perceived threat of pinnipeds in the area, any seals encountered around the entire island harbouring the farm were also regularly shot. Their bodies were punctured to hasten sinking in an attempt to hide the evidence (Georgia Strait Alliance, 2000). While presumably a huge problem, especially for endangered populations of sea otters, actual numbers taken in most areas are unknown. Lethal force (generally by shooting) was legal in US waters until 1994, to control predators from damaging gear and finfish, but a more recent amendment of the Marine Mammal Protection Act now prohibits routine killing of marine mammals that prey on aquaculture facilities. However, individually identified pinnipeds that cause significant negative impact to the aquaculture site can still be lethally removed (Fraker and Mate, 1999).

Sporadic entanglement of marine mammals occurs in nets and enclosures designed to house finfish. While published records are few, bottlenose and common dolphins (*Delphinus delphis*) have been entangled in sea pens holding blue-fin tuna (*Thunnus thynnus*) off South Australia (Gibbs and Kemper, 2000); grey whales (*Eschrichtius robustus*) are at times entangled in netpens holding herring (Robin Baird, personal communication, December 2000); and an adult Bryde's whale (*Balaenoptera edeni*) became entangled at a mussel farm in New Zealand (Chris Roberts, personal communication, December 2000). We assume that the latter entanglements of larger whales in aquaculture facilities are relatively rare events, and not as important as other aquaculture-related problems.

Finfish and Predatory Marine Mammals: Major Problems

Interactions between aquaculture and marine mammals are detrimental to all concerned. Marine mammals that take finfish in pens or after their release cause scarring of the fish, with presently unknown impacts on fish survivability and reproduction (Scordino, 1993; Harmon *et al.*, 1994; Fryer, 1998); decimate the target fish (Fraker and Mate, 1999); increase fish susceptibility to disease or decrease growth owing to stress (Morris, 1996); and destroy gear, at times causing massive fish escape through torn pens (Pemberton and Shaughnessy, 1993). Losses to the aquaculture industry can be high, and have been estimated at many millions of dollars for the United States and Canada (Nash *et al.*, 2000). For example, the US National Marine Fisheries Service estimated a loss of over US$50 million in 1 year in the Gulf of Maine alone (NMFS, 1996b), but we are not aware of accurate worldwide estimates. The marine-mammal-induced escape of finfish from aquaculture sites is also of ecological concern. Aquaculture stocks sometimes have commercially bred traits that could have unknown impacts on feral populations. Furthermore, escaped fish can transmit unknown diseases to other natural stock, which would increase conflict between aquaculture and fisheries (Morris, 1996).

Whenever a large problem exists, extensive efforts are instigated to reduce it. In the present case, predators that specialize on food from aquaculture facilities have been removed by trapping and shooting, deterred by firecrackers and electronic acoustic harassment or deterrence devices (AHDs and ADDs, respectively), deterred by such means as playing killer whale (*Orcinus orca*) vocalizations, chased by high-speed vessels, and given distasteful or emetic foods to eat (summarized in Buck, 2000). Large (and expensive) physical barriers also have been set up. The detrimental interaction is usually not alleviated completely, and much of the time the marine mammals are themselves harmed/killed during the interaction. This can present a problem to the predator in cases where populations are limited or endangered, such as for sea otters or Steller sea lions (*Eumetopias jubatus*), and a problem of at least perception in countries such as the United States, Canada, New Zealand and Australia, as major examples, where the public tends to view marine mammals as being special animals that warrant special protection. In the United States, this problem is particularly acute, as all marine mammals are protected from harassment by the Marine Mammal Protection Act of 1972, and its numerous amendments (Baur *et al.*, 1999).

Mitigation Measures for Marine Mammal Conflicts

Aquaculturists have developed a suite of techniques to attempt to reduce marine mammal depredations where they occur. These can be classified under six major topics, from individual dissuasion to overall population-wide effects: (i) harassment; (ii) aversive conditioning; (iii) exclusion; (iv) non-lethal removal; (v) lethal removal; and (vi) population control. We shall briefly discuss each of these broad categories, and then make recommendations of our own from this list.

Harassment

Harassment of marine mammals in order to try to encourage them to stay away faces the general problem that one attempts to dissuade animals from obtaining a relatively easy and nutritious meal. The most primitive techniques of harassment consist of chasing the predators with fast boats, throwing fire-crackers or other incendiary devices (seal bombs) at them, and shooting them with rubber bullets and blunt-tipped arrows. More recent and sophisticated devices consist of emitting loud and noxious underwater sounds. These tend to consist of killer whale vocalizations in areas where killer whales occur and might be predators of the problem-causing marine mammals, and of acoustic pingers or buzzers that make noxious sounds so loud as to cause discomfort in an acoustically sensitive pinniped or dolphin. The sounds are of two major types: acoustic harassment devices, which tend to have shrill-sounding (scream) frequencies of 12–17 kHz, and the similar acoustic deterrent devices, which are usually set around 10 kHz and have a piercingly loud (high decibel level) sound that is also designed to cause discomfort or pain at close distance. The use of these acoustic devices usually depends on the predator species. ADDs are considered to be of sufficient sound disturbance to deter most toothed cetaceans, who are believed to be more sensitive to sound than pinnipeds, while the increased frequency of AHDs mainly targets pinnipeds (Richardson *et al.*, 1995; Olesiuk *et al.*, 1996; Reeves *et al.*, 1996; Kraus *et al.*, 1997). Unfortunately, all such techniques tend to represent short-term solutions (Morris, 1996; Fraker *et al.*, 1998; Fraker and Mate, 1999; Nash *et al.*, 2000). Animals that are chased will generally return, and the loud sounds can even condition animals to perceive the acoustic signal as a dinner bell (Mate *et al.*, 1987; Olesiuk *et al.*, 1996): 'There is food to be had in the region of sound discomfort'. Even the projected vocalizations of killer whales are ignored after some time, for acoustically adept marine mammals quickly learn to distinguish the real alarm from the false one, and at any rate habituate to sounds that are not followed by real danger (Fraker and Mate, 1999).

Nevertheless, AHDs and ADDs show reasonable success in certain areas, and are enjoying widespread use. They have intensity levels up to approximately 194 dB re 1 μPa @ 1 m, and are estimated to be heard in some environments at up to 50 km from source (Haller and Lemon, 1994). We deplore their use, because we believe that they put unacceptably intense noise into the marine environment without a clear understanding of their long-term effects on both target and non-target species. Such noise can reduce the communication capabilities of the very fish that are being grown by the aquaculturists (Tolimieri *et al.*, 2000), and the overall level of noise pollution can adversely affect animals, such as threatened harbour porpoises, that frequent an area, and may or may not be direct targets of the noise-making devices (Olesiuk *et al.*, 1996; Johnston and Woodley, 1998). Recent evidence (Morton and Symonds, 2002) indicates that high-amplitude AHD pulses designed to deter harbour seals from preying on salmon farms in British Columbia, Canada, displaced killer whales from their regular movements, and made large areas of formerly inhabited range unavailable to them (Table 3.1). The AHD pulses were apparently perceived as noxious by the killer whales (which are not harmful to the salmon pens), and the whales avoided the ensonified channels and bays in both the short and long term.

The intense sound production of ADDs also provides potential concern of injury to some marine mammals, and authorities are no longer advising the use of ADDs in Canada (Nash *et al.*, 2000). Furthermore, the so-called seal bombs are believed to deafen some seals, therefore making acoustic deterrent devices ineffective (Morris, 1996). Animals that undergo hearing loss due to exposure to acoustic devices may experience a decrease in their ability to capture prey in nature, thereby making aquaculture sites an even more important resource to fulfil their energetic requirements.

Table 3.1. The number of days that killer whales were detected in two separate areas of Vancouver Island, British Columbia. The area of the Broughton Archipelago was largely ensonified with AHD pulses during all of 1994–1998, and killer whale occurrence patterns were reduced in those same years. There were no fish farms or AHD/ADD sounds in Johnston Strait. (Data presentation after Morton and Symonds, 2002.)

Year	Johnson Strait	Broughton area Archipelago
1985	146	15
1986	166	30
1987	134	61
1988	132	29
1989	184	26
1990	160	30
1991	225	36
1992	175	33
1993	183	38
1994	207	15
1995	195	8
1996	183	4
1997	183	13
1998	186	9
1999	152	35

Aversive conditioning

The association of a food with a later illness was suggested some time ago as a means to induce taste aversion in predators (Garcia *et al.*, 1955), and is used with some success by, for example, sheep farmers who feed sheep carcasses laced with lithium chloride (LiCl) to coyotes. The LiCl induces stomach cramps and vomiting in coyotes, and a single exposure tends to keep an exposed coyote from preying on lambs (Gustavson *et al.*, 1974). The technique has been used with less success in keeping rough-toothed dolphins (*Steno bredanensis*) away from oceanic fishing lines, but with greater success in deterring California sea lions (Kuljis, 1984; Costa, 1986) and Australian fur seals (*Arctocephalus pusillus*; Pemberton and Shaughnessy, 1993). Nevertheless, success is not guaranteed: when there are many predators, a large number has to be trained for taste aversion, and LiCl may not always be the most effective or safest taste aversion agent (Cowan *et al.*, 2000).

Exclusion

Exclusion of predators from fish pens is an obvious but not always attainable goal. Physical barriers have to be high and strong enough to keep large sea lions from causing damage to gear and fish. Nevertheless, construction of barriers has been instrumental in keeping smaller seals and fur seals from large-scale depredations, especially in Australia (e.g. Pemberton and Shaughnessy, 1993), and harbour seals from contaminating shellfish beds in Washington (Nash *et al.*, 2000). Attempts at excluding predators with physical models (scarecrows) of other predators (i.e. life-sized models of killer whales) and with alarms at the pens or at nearby pinniped haul-out sites have been generally ineffective. The use of bubble curtains, a barrier of air bubbles produced from submerged perforated hoses, has been shown to acoustically mask surrounding environments (Würsig *et al.*, 2000) and may deter predators from further enquiry into the enclosed aquaculture site. Bubble curtains have been largely untested, and their masking effects may be reduced in some high-energy areas (Tillapaugh *et al.*, 1994; NMFS, 1996b).

Non-lethal removal, lethal removal, and population control

Removal of offending predators has been widely practised, and runs the gamut from capture and relocation, to capture and permanent holding in captivity, to lethal removal by shooting. The costly method of relocating animals only seems to delay the amount of predation since animals usually return to the same problem area. As with the problem of taste aversion, such removal becomes less effective when there are many predators, as is usually the case with pinnipeds and salmon farms in the Pacific Northwest. On the other hand, non-lethal removal can be highly effective where only several predators of a larger population in an area have learned to take advantage of (or have acquired a taste for) artificially reared food. Large-scale culling of offending predators, often termed population control or population management, can be effective but may be economically or legally difficult to implement. There is growing sentiment in the Pacific Northwest that California sea lions have reached numbers that exacerbate a host of human-use problems, with only one of these being aquaculture related. Consequently, various public sector concerns are arguing for culling of populations as a viable solution to pinniped-related problems (Buck, 2000). The problem is not easy to solve, however, as pinnipeds feed on many fishes and squid other than those commercially important or being raised by aquaculture. Some of these other prey themselves feed on the aquaculturally important food. When reducing the population of a predator (here, the pinniped), one invites unknown eco-system changes, including a potential increase in predatory fishes that had been kept in check by the pinnipeds in the first place (Fraker and Mate, 1999).

Finfish and Predatory Marine Mammals: Suggestions for Solutions

While it is undeniable that especially pinnipeds cause major harm to some types of marine aquaculture facilities, we believe that there is much misperception of the extent and the reasons for the problem. As we view the larger picture, we realize that much of the problem stems from human overutilization and therefore a decrease in the natural food resources of the sea. Aquaculture is a partial attempt to remedy our unwise (or greedy) use of nature, and it is quite natural that the concentration of food resources should attract predators that have evolved to feed on marine prey. This somewhat philosophical argument does not directly help the interaction problem. However, a view of sharing the sea with the predators (or respect for them) might well be used to argue for including potential predation in the design and location of aquaculture facilities. For example, a clear relationship has been found with the extent of pinniped-related problems and distance of facilities from preferred haul-out or mating/pupping sites of the predators (Pemberton and Shaughnessy, 1993). The further away these sites, the fewer the problems (Table 3.2). In terms of foraging, the pinniped predators must invest more energy into travelling to and from the aquaculture site as the distance increases from their haul-out site. Therefore, the potential energetic gains from an aquaculture site decrease as the distance increases, and the site becomes less attractive than natural fish stocks. Distance from haul-out sites has generally not been taken into account in past aquaculture facility placement. Also, erection of high and strong-enough barriers against predation has not been universally factored into the cost of a facility, and quick fixes later on have proven generally insufficient.

Table 3.2. The number of Australian fur seal attacks on fish farm facilities in southern Australia, as a function of distance of the facilities from the pinniped haul-out sites.

Distance to haul-out site (km)	Number of seal attacks
19	108
22	63
45	18
38	11
35	10
60	3
40	2
48	1
40	1
33	0

A part of gauging the extent of predator problems is an accurate assessment of how many predators are involved, and what appear to be the costs of the predation. In Maine, for example, it has been found that rogue seals are often responsible for stealing fish and destroying pens. The removal of such a seal may mean no or very little pinniped problem to the facility for several months, until the next seal acquires a taste for this way of feeding (Morris, 1996). Unfortunately, such removal has generally been by lethally removing the animal, whereas it is possible (albeit generally more labour intensive) to capture and relocate the offending predator. The problem of a few animals creating at times intensive damage is very different (and is solvable) from that of many animals. Therefore, we recommend that: (i) the location of pinniped and other potential prey haul-out and foraging sites be investigated before placement of aquaculture facilities; and (ii) counting, individual identification, and perhaps even marking and radio tracking of select animals be conducted in areas where there is a problem. Only when we know how many animals are involved, can the most effective routes for action be investigated.

There are many different aquaculture facilities worldwide, and many different problems of predation. As far as marine mammals are concerned, we argue that success is most likely by: (i) physical exclusion; (ii) non-lethal removal; (iii) aversive conditioning; and (iv) acoustic or other harassment, as a last resort. The first two of these are known to be effective in many cases, and do not introduce other problems when properly carried out. The improvement of physical barriers between the environment and aquaculture farms seems to hold the greatest promise for a long-term solution (see Tillapaugh *et al.*, 1994; Reeves *et al.*, 1996). However, soft-material nets around finfish facilities are to be avoided, as they can entangle and kill marine mammals (Gibbs and Kemper,

2000), but thick plastic and steel fencing seems to create few entanglement difficulties. Non-lethal removal can be expensive, and is useful only when predator numbers are low. Aversive conditioning holds promise, but has only been tested sporadically and incompletely on marine mammals. Types and dosages of noxious chemicals to be given need to be worked out for enhanced safety and efficiency (Cowan *et al.*, 2000). As for removal, it is unlikely that aversive conditioning is very useful when predator numbers are high. Acoustic harassment is probably the most widely (and variably successful) deterrent presently used. It has only recently become appreciated that the underwater world is of great acoustic importance to all sorts of fauna, and wholescale ensonification of this world is – in our strong opinion – unwise. We have recent evidence that such ensonification can have undesired effects on porpoises (Johnston and Woodley, 1998; Kastelein *et al.*, 2000), dolphins and whales (Morton and Symonds, 2002) and fishes (Tolimieri *et al.*, 2000). An automatic mechanism that turns on the acoustic alarm only upon detection of nearby predators could reduce the amount of noise emitted by ADDs and AHDs into the surrounding environment, and may be less habituating to predators than continuous transmission (Morris, 1996).

Conclusion

It is incumbent on marine aquaculturists and appropriate government agencies to requisition marine mammal studies before allowing new facilities to be placed. Extensive shellfish facilities can clog waterways and displace marine mammals, and intensive finfish facilities are likely to encounter unacceptably high levels of predator problems if located too close to haul-out sites, or other areas of traditionally high marine mammal use.

Acknowledgements

We thank Robert Stickney of the Texas Sea Grant for inviting us to write this paper, Jim McVey of National Sea Grant for support, and John Cooksey of the World Aquaculture Conference Administration for advice and support. The Department of Biology and the Research Development Office of Texas A&M University at Galveston helped with travel support to the World Aquaculture Society conference. Danny and Lyn Boulton, Rochelle Constantine, April Dawn Harlin, Timothy Markowitz, Alexandra Morton, Cath Kemper, Janet Mann and Peter Shaughnessy kindly provided unpublished data, and Ben E. Nana gave encouraging support.

References

Baur, D.C., Bean, M.J. and Gosliner, M.L. (1999) The laws governing marine mammal conservation in the United States. In: Twiss, J.R. Jr and Reeves, R.R. (eds) *Conservation and Management of Marine Mammals*. Smithsonian Institution Press, Washington, DC, pp. 48–86.

Buck, E.H. (2000) *Fishery, Aquaculture, and Marine Mammal Legislation in the 106th Congress*. Congressional Research Service Report, The Committee for the National Institute for the Environment, 1725 K Street, Suite 212, Washington, DC.

Costa, D.P. (1986) *Physiological Effects of Prey Aversion Protocol Using Lithium Chloride on California Sea Lions*. Report to the US Marine Mammal Commission, Washington, DC.

Cowan, D.P., Reynolds, J.C. and Gill, E.L. (2000) Reducing predation through conditioned taste aversion. In: Gosling, L.M. and Sutherland, W.J. (eds) *Behaviour and Conservation*. Cambridge University Press, Cambridge, UK, pp. 281–299.

Dewalt, B.R., Vergne, P. and Hardin, M. (1996) Shrimp aquaculture development and the environment: people, mangroves and fisheries on the Gulf of Fonseca, Honduras. *World Development* 24, 1193–1208.

FAO (1999) *The State of World Fisheries and Aquaculture, 1998*. Fisheries Department, Food and Agriculture Organization, Rome, 128pp.

Fraker, M.A. and Mate, B.R. (1999) Seals, sea lions, and salmon in the Pacific Northwest. In: Twiss, J.R. Jr and Reeves, R.R. (eds) *Conservation and Management of Marine Mammals*. Smithsonian Institution Press, Washington, DC, pp. 156–178.

Fraker, M.A., Duval, W. and Kerr, J. (1998) *Physical Countermeasures against Predation by Seals and Sea Lions at Salmon Farms*. Report of a workshop held 17 September, Campbell River, Canada, for British Columbia Salmon Farmers Association, TerraMar Environmental Research Ltd, Sidney, British Columbia, Canada, 40pp.

Fryer, J.K. (1998) Frequency of pinniped-caused scars and wounds on adult spring–summer chinook and sockeye salmon returning to the Columbia River. *North American Journal of Fisheries Management* 18, 46–51.

Garcia, J., Kimeldorf, D.J. and Koelling, R.A. (1955) A conditioned aversion to saccharin resulting from exposure to gamma radiation. *Science* 122, 157–159.

Gearin, P., Pfeifer, B. and Jeffries, S. (1986) *Control of California Sea Lion Predation of Winter-run Steelhead at the Hiram M. Chittenden Locks, Seattle, December 1985–April 1986*. Fishery Management Report 86-20, Washington Department of Wildlife, Mill Creek, Washington, 108pp.

Georgia Strait Alliance (2000) Impacts on aquatic mammals and other species. http://www.georgiastrait.org/reviewmammals.htm.

Gibbs, S.E. and Kemper, C.M. (2000) Tuna feedlots at Port Lincoln, South Australia: dolphin mortalities and recommendations for minimizing entanglements. International Whaling Commission Working Document SC/52/5M1. International Whaling Commission, Cambridge, UK.

Gustavson, C.R., Garcia, J., Hankins, W. and Rusiniak, K.W. (1974) Coyote predation: control by aversive conditioning. *Science* 184, 581–583.

Haller, D.R. and Lemon, D.D. (1994) *A Study of the Far-field Generated by Seal-scarers*. Canadian Department of Fisheries and Oceans report No. XSA FP941-3-8025/00/A. Department of Fisheries and Oceans, Ottawa, Canada, 52pp.

Harmon, J.R., Thomas, K.L., McIntyre, K.W. and Paasch, N.N. (1994) Prevalence of marine-mammal tooth and claw abrasions on adult anadromous salmonids returning to the Snake River. *North American Journal of Fisheries Management* 14, 661–663.

Johnston, D.W. and Woodley, T.H. (1998) A survey of acoustic harassment device (AHD) use in the Bay of Fundy, NB, Canada. *Aquatic Mammals* 24, 51–61.

Kastelein, R.A., Rippe, H.T., Vaughan, N., Schooneman, N.M., Verboom, W.C. and de Haan, D. (2000) The effects of acoustic alarms on the behavior of harbor porpoises (*Phocoena phocoena*) in a floating pen. *Marine Mammal Science* 16, 46–64.

Kautsky, N., Rönnbäck, P., Tedengren, M. and Troell, M. (2000) Ecosystem perspectives on management of disease in shrimp pond farming. *Aquaculture* 191, 145–161.

Kraufvelin, P., Sinisalo, B., Leppäkoski, E., Mattila, J. and Bonsdorff, E. (2001) Changes in zoobenthic community structure after pollution abatement from fish farms in the Archipelago Sea (N. Baltic Sea). *Marine Environmental Research* 51, 229–245.

Kraus, S., Read, A., Anderson, E., Baldwin, K., Solow, A., Spradlin, T. and Williamson, J. (1997) A field test of the use of acoustic alarms to reduce incidental mortality of harbor porpoises in gill nets. *Nature* 388, 341.

Kuljis, B.A. (1984) *Report on Food Aversion Conditioning in Sea Lions (*Zalophus californianus*)*. NMFS Contract Report. 84-ABC-00167, National Marine Fisheries Service, NOAA, Washington, DC.

Mann, J. (1999) *Recent Changes in Female Dolphin Ranging in Red Cliff Bay, off Monkey Mia, Shark Bay*. Report submitted to West Australian Department of Fisheries and West Australian Department of Conservation and Land Management, Perth, Australia.

Mate, B.R., Brown, R.F., Greenlaw, C.F., Harvey, J.T. and Tempte, J. (1987) An acoustic harassment technique to reduce seal predation on salmon. In: Mate, B.R. and Harvey, J.T. (eds) *Acoustical Deterrents in Marine Mammals Conflicts with Fisheries*. Oregon Sea Grant Publication ORESU-W-86-001. Oregon State University, Corvallis, Oregon, pp. 23–36.

Morris, D.S. (1996) Seal predation at salmon farms in Maine, an overview of the problem and potential solutions. *Marine Technology Society Journal* 30, 39–43.

Morton, A.B. and Symonds, H.K. (2001) Displacement of *Orcinus orca* by high amplitude sound in British Columbia, Canada. *ICES Journal of Marine Science* 1–10.

Nash, C.E., Iwamoto, R.N. and Mahnken, C.V.W. (2000) Aquaculture risk management and marine mammal interactions in the Pacific Northwest. *Aquaculture* 183, 307–323.

NMFS (1996a) *Environmental Assessment on Conditions for Lethal Removal of California Sea Lions at the Ballard Locks to Protect Winter Steelhead*. National Marine Fisheries Service Environmental Assessment Report. Available from Northwest Regional Office, NMFS, NOAA, 7600 Sand Point Way NE, Seattle, WA 98115, USA, 81pp.

NMFS (1996b) *Report of Gulf of Maine Aquaculture–Pinniped Interaction Task Force*. Office of Protected Resources, National Marine Fisheries Service, Silver Spring, Maryland, 70pp.

Olesiuk, P.F., Nichol, I.M., Sowden, P.J. and Ford, J.K.B. (1996) *Effects of Sounds Generated by an Acoustic Deterrent Device on the Abundance and Distribution of Harbour Porpoise (*Phocoena phocoena*) in Retreat Passage, British Columbia*. Draft report for the Department of Fisheries and Oceans, Pacific Biological Station, Nanaimo, British Columbia, Canada, 47pp.

Páez-Osuna, F. (2001) The environmental impact of shrimp aquaculture: a global perspective. *Environmental Pollution* 112, 229–231.

Páez-Osuna, F., Guerrero-Galván, S.R. and Ruiz-Fernández, A.C. (1999) Discharge of nutrients from shrimp farming to coastal waters of the Gulf of California. *Marine Pollution Bulletin* 38, 585–592.

Pemberton, D. and Shaughnessy, P.D. (1993) Interaction between seals and marine fish-farms in Tasmania, and management of the problem. *Aquatic Conservation: Marine and Freshwater Ecosystems* 3, 149–158.

Primavera, J.H. (2000) Development and conservation of Philippine mangroves: institutional issues. *Ecological Economics* 35, 91–106.

Reeves, R.R., Hofman, R.J., Silber, G.K. and Wilkinson, D. (1996) *Acoustic Deterrence of Harmful Marine Mammal–Fishery Interactions.* Proceedings of a workshop held in Seattle, Washington, USA, 20–22 March 1996. NOAA Technical Memorandum NMFS-OPR-10, December 1996, Washington, DC, 70pp.

Richardson, W.J., Greene, C.R. Jr, Malme, C.I. and Thomson, D.H. (1995) *Marine Mammals and Noise.* Academic Press, San Diego, California.

Scordino, J. (1993) *Review and Evaluation of Pinniped Predation on Salmonids in the Columbia River Basin.* NMFS report prepared for the Snake River Salmon Recovery Team. Available from Northwest Regional Office, National Marine Fisheries Service, NOAA, 7600 Sand Point Way NE, Seattle, Washington, 98115, 39pp.

Tillapaugh, D., Brenton, C. and Harrower, B. (1994) *Predation on Salmon Farms in British Columbia – the Impacts of Harbour Seals (*Phoca vitulina*).* The results of a 1991 survey. Commissioned by the British Columbia Ministry of Agriculture, Fisheries, and Food, Vancouver, British Columbia, Canada, 45pp.

Tolimieri, N., Jeffs, A. and Montgomery, J.C. (2000) Ambient sound as a cue for navigation by the pelagic larvae of reef fishes. *Marine Ecology Progress Series* 207, 219–224.

Tovar, A., Moreno, C., Mánuel-Vez, M.P. and García-Vargas, M. (2000) Environmental impacts of intensive aquaculture in marine waters. *Water Research* 34, 334–342.

Würsig, B., Greene, C.R. Jr and Jefferson, T.A. (2000) Development of an air bubble curtain to reduce underwater noise of percussive piling. *Marine Environmental Research* 49, 79–93.

Recreational Fishing and Aquaculture: Throwing a Line into the Pond

4

William D. Harvey and Larry D. McKinney

Resource Protection Division, Texas Parks and Wildlife Department, 4200 Smith School Road, Austin, TX 78744, USA

Abstract

Mariculture is increasing as traditional commercial fisheries production declines (Goldberg and Triplett, 1997). World aquaculture production (freshwater and marine) has doubled since 1984 and reached a record 20,900,000 tonnes in 1995. Total worth exceeded US$36 billion worldwide. In the United States production annually exceeds 400,000 tonnes and a value of $729 million. Recreational fishing in the United States is big business as well, with 35.3 million participants and an economic impact of $37.7 billion (American Sportfishing Association, 1996). Both are industries that make valuable contributions to the US economy. They are especially important to regional economics where they are often concentrated because of favourable conditions. Another factor in common to both industries is their respective growth potential in marine waters. The fastest-growing segment of the Texas recreational fishery, the second largest in the United States, is saltwater fishing. This segment of angling in Texas has seen at least 15% growth over the last 15 years during a period in which total angling participation (as measured by licence sales) has declined over 20%. An expansion of shrimp farms to 1625 acres (660 ha) (Office of Science and Technology Policy, 1997) along the coastal margins of Texas through the 1980s and early 1990s and growing interest in caged aquaculture of estuarine and marine fishes are examples of increased marine focus in that industry.

Marine aquaculture enterprises such as shrimp farms, mollusc culture and netpen systems for raising salmon have generated environmental impacts that are of increasing concern to recreational fishing interests. What are the perceived impacts of mariculture on recreational fishing? Why do these perceptions lead to conflicts? What avenues exist to mitigate or resolve such conflicts before they become confrontations? These are the questions explored in this chapter.

The Common Enemy

> Hug your friends tight, but your enemies tighter – hug 'em so tight they cannot wiggle.
>
> Old Texas maxim

In a very real sense, recreational anglers view mariculture and commercial harvest as a common enemy. Their view, simply stated, is that both of these activities affect the quality of the angling experience – generally in terms of numbers of available fish – either directly (through harvest) or indirectly (through environmental impacts). When commercial and mariculture activities are perceived as a threat to recreational angling, the stage is set for a classic resource conflict between and among these competing uses. It should come as no surprise that recreational interests view with deep concern any commercial activity which: (i) may affect the environment and (ii) is deemed as low in the hierarchy of societal necessity. As a result, mariculture has come under close scrutiny by recreational interests, much closer perhaps than other commercial activities that often represent a more formidable threat to recreational angling.

This scrutiny is predicated on the recognition of ecological externalities common to commercial fisheries and aquaculture that appear as threats to recreational fishing interests. Recreational anglers are acutely aware of the direct relationship between quality environment and quality angling. They are inundated with that relationship every day. Perusal of popular outdoor literature or Saturday afternoon television outdoors-oriented programming will invariably produce a direct reference to the necessity of environmental protection. Recreational interests view mariculture and commercial fishing as part and parcel of the same problem: decreases in available fish for recreational harvest. These include:

1. Environmental impacts (primarily habitat disruption and adverse effects on water quality).
2. Competition for access to – and harvest of – the same resource (mostly finfish species that are susceptible to sport-fishing techniques).
3. Ecological disruptions (loss of baitfish or other ecologically important species through bycatch of the commercial fisheries and to produce fish meal for highly carnivorous species in aquaculture).

Shrimp mariculture and shrimp trawling in Texas represent a perfect example of the common enemy. Both have externalities viewed as harmful to recreational interests and capable of long-term ecological degradation.

Habitat Destruction

Native penaeid shrimp are the principal wild stock species harvested along the coast of Texas. The inshore waters of Texas, for example, have supported a multimillion dollar industry for the better part of a century. But as the trawling fleet expanded and fishing effort increased, the effects of trawling on bay bottoms and non-target species (bycatch) became a ringing alarm for recreational interests.

Anglers clearly see shrimp trawling as a direct threat to stocks of finfish, either as a result of bycatch or destruction of habitat. Recognizing the need to reduce shrimping efforts, Texas saltwater anglers supported a 42% increase in the saltwater stamp fee to provide additional funds for the shrimp licence buyback programme in Texas.

Water Quality

The ecological movement of the late 1960s and early 1970s happened to coincide with a dramatic rise in recreational fishing in the United States. As northern lakes died from pollution and eastern rivers were engulfed in flames, water quality became the environmental poster child for the following decades. But water quality can be a more elusive target than habitat destruction because it is often a much more subtle environmental reality and in some situations poor water quality is viewed as a necessary evil.

To the recreational angler, water quality is largely framed by that which can be seen, for example, industrial effluent disgorged from the end of a pipe into a waterway is clearly water pollution. The same view applies to effluent from a mariculture facility that may periodically discharge water with elevated total suspended solids and low oxygen levels. These discharges may be episodic, but the observer is likely to view them as both unnecessary and highly detrimental to the environment. At the same time, sewage effluent, which may constitute a much more dangerous threat, tends be ignored (except in extreme situations like fish kills). Poor water quality resulting from treated municipal sewage effluent or agriculture runoff may be viewed as a necessary evil, while pond effluent from a marine shrimp farm (much less of an issue in terms of overall nutrient loading) may be viewed as totally unacceptable.

Resource Competition

Although resource competition is a very clear common enemy as it relates to commercial harvest of shellfish and finfish, mariculture is more of a guerilla warrior. In terms of finfish mariculture, red drum (*Sciaenops ocellatus*) is both a marine culture species and the prime target of coastal saltwater anglers. For years, anglers fought legislative battles for removal of nets in Texas bays and to have red drum declared a game fish, a designation that brings considerable regulatory protection.

As wild populations of red drum were used as sources for mariculture facilities and concerns began to arise as to the actual source of red drum sold in the marketplace, attention was once again turned to mariculture of red drum. Clearly, any mariculture species the culture of which requires sourcing of wild, recreationally attractive broodfish will raise the eyebrows of the recreational angler. If sourcing is then coupled with the sale of a mariculture product that can directly or indirectly result in illegal take and sale of wild-caught fish under the claim of mariculture production, the stakes are much higher and the outcome more draconian.

Ecological Disruption

Recreational anglers are more educated and ecologically aware than at any time in the past. They clearly understand the interwoven intricacy of food webs, ecosystem stability and the need for ensuring that the integrity of these systems remains intact. In the eyes of these anglers, commercial activities and mariculture are again the common enemy.

In many aquaculture systems, more protein in the form of fish meal is used to feed the cultured animals than is obtained from the cultured fish. In short, mariculture of carnivorous species can result in a net loss of fish protein, not a net gain (Goldburg and Triplett, 1997). This fact is not lost on recreational anglers, who understand that massive harvest of fish to feed fish can result in a decrease of forage species available to sportfish species, as well as direct loss of sportfish in meal production.

The Uncommon Enemy

While there are numerous shared concerns between commercial fisheries and aquaculture there are a number specific to the latter. Conflicts between aquaculture and recreational fisheries centre on concerns about facilities as sources of disease and parasites as well as so-called genetic pollution.

Disease and parasites

The basic nature of successful commercial aquaculture is dependent upon the ability of the culturist to maintain healthy animals in high density. In short, the more crowded the culture species, the more financially lucrative the enterprise – to a point. Crowding often triggers a series of stress-related physiological responses that can result in disease outbreak. The effects of disease on cultured species are well documented and the causative factors are clearly understood by most culturists, but often less understood by other interests.

The nature of conflict between aquaculture and recreational anglers is focused; diseased animals are viewed as vectors and as such constitute threats to wild stocks. This concern is not altogether unfounded as many western trout fisheries have spiralled into oblivion after outbreaks of whirling disease spread by infected hatchery stocks. Although whirling disease is not related to marine aquaculture, the logic of the recreational angler remains Socratic. Diseased animals transmit disease; aquaculture animals are diseased; therefore, cultured animals transmit disease.

Parasites represent a more subtle issue, one perhaps less well understood or focused, but an issue none the less. Again, concerns of recreational anglers are not without foundation. The African sabellid worm (*Terebrasabella heterouncinata*) was introduced into California abalone farms by the mid- to late-1980s with imported South African abalone. The worms infest and weaken the shells of California abalone, reducing growth rates and production, and causing deformities. Experiments at University of California, Santa Barbara, have demonstrated that the worms can infest many types of native marine snails, not just abalone.

The infestation not only had the potential to affect molluscs, but also had a substantial impact on the mariculture of abalone as well. Infested abalone were freely transferred between facilities, spreading the African worm to virtually all California abalone farms by 1993, with the resulting infestations bankrupting some growers. California officials were slow to respond and regulations regarding spread of the parasite were not instituted until late 1996 (Andrew Cohen, San Francisco Bay Institute, personal communication). Control efforts appear to be succeeding.

Genetics

Without question, recreational interests view disease and parasites as a cause for concern. But potential effects of genetic introgression of cultured stocks into wild stocks represents a whole different level of anguish. The reasons are straightforward: genetic effects are viewed (correctly) as irreversible. Further, much of the scientific community stands shoulder to shoulder with anglers in opposition to genetic pollution of wild stocks.

Genetic effects are only realized when a genetically distinct culture stock is commingled, either as a result of stocking or escape, with a wild stock that is also genetically vulnerable. In a very real sense, the burden of proof related to the real or actualized threat of genetic introgression lies with the culturist, and the prevailing sentiment among those who manage or consume wild stocks is one of genetic conservation. Genetic differences are easily displayed through straightforward biochemical techniques, but the meaning of those differences is difficult, if not impossible, to interpret.

In short, genetic differences can be demonstrated, but determining how these differences translate into loss of fitness in the affected stock is still a mystery. The consensus among ecologists is that discrete genetic make-up of a stock is a function of selective pressures that result in populations suited for specific environments. Introgression from other stocks then results in reduction of fitness in the impacted population. Though loss of fitness is hard to ascertain, the argument is convincing and difficult to refute.

The Enemy of My Enemy . . .

> The art of war is simple enough. Find out where your enemy is. Get at him as soon as you can. Strike him as hard as you can, and keep moving.
>
> Ulysses S. Grant

Environmental impacts of shrimp aquaculture were probably the first significant shared concern (relative to Texas mariculture) between both commercial and recreational fisheries. In what has generally been an adversarial relationship, it may be the one issue upon which both recreational and commercial fishing interests have agreed and actively joined in opposition to at least one aspect of aquaculture: shrimp farming.

Habitat impacts from the siting of aquaculture facilities in coastal wetlands have been a concern because of the size of the Texas facilities. Texas shrimp farms covered approximately 6500 ha in 1997. The low-lying and flat nature of Texas coastal plains is ideally suited to extensive aquaculture development. However, most of these areas are also productive wetlands. Because Texas has lost approximately half of its coastal wetlands (Moulton *et al.*, 1997) additional losses of any kind are of concern.

Early Texas shrimp farms were prodigious water users; in some cases passing more untreated water through their systems than many cities in south Texas, where the majority of the farms were located. There was little or no treatment of discharge water from these high-density, intensive operations. Discharge into coastal waters with limited assimilative capacity (generally because of low flushing rates) raised concerns about over-nitrification and subsequent oxygen depletion. Additional concerns about nutrient stimulation of harmful algal blooms appeared partially in response to the development of what became known as brown tide (Whitledge and Pulich, 1991). This phenomenon persisted for several years and speculation (not substantiated) continued in the recreational fishing community about the contribution of shrimp farms to its persistence. The 1996 *Pfiestra piscicida* (not of aquaculture origin) occurrence in the Chesapeake Bay (Terlizzi, 1997) and increasing red-tide events along the Texas coast contributed to the maintenance of heightened concern about shrimp farm discharges.

Another water quality issue common between commercial fishing and aquaculture is turbidity. Highly turbid discharges from aquaculture facilities were of great concern to adjacent landowners along the Arroyo Colorado and were the visible evidence of the adverse impacts of shrimp farms on the environment. In one case, Texas Parks and Wildlife Department (TPWD) biologists documented the filling of a small adjacent embayment from such discharges. Increased turbidity and subsequent impact on benthic communities and submerged aquatic vegetation (seagrass) have been attributed to commercial trawling and dredging (Onuf, 1994).

The Enemy Attacks

In the late 1980s, shrimp aquaculture began to blossom in the Texas Rio Grande Valley. Access to relatively inexpensive land, saltwater and labour brought Taiwanese interests in search of new areas for expansion of the successful – and profitable – intensive culture systems of Taiwan. But development was made much more attractive by the lenient aquaculture regulatory environment in Texas.

Aquaculture in Texas prior to the investments made into shrimp aquaculture was almost exclusively a 'mom and pop' operation. Small fish farms, with most of the hectareage dedicated to production of catfish and game fish for stocking private impoundments, dotted areas located near plentiful water supplies. The industry was loosely regulated by the TPWD and Texas Water Commission (TWC), since renamed the Texas Natural Resources Conservation Commission (TNRCC). TPWD had limited regulatory authority (licensure) over production and possession of non-indigenous fish species and game fish species. TWC had statewide authority to regulate water quality discharges from the facilities. However, marine aquaculture facilities were exempt from these water quality standards.

The initial concerns of Texas anglers regarding aquaculture actually arose in freshwater as tilapia species were imported, particularly blue tilapia (*Oreochromis aureus*). As blue tilapia began to emerge in freshwater systems (almost exclusively thermally enriched systems associated with electricity generating stations), anglers, power generators, and regulators became concerned. The use of tilapia as both an aquaculture species and as bait led to its wide occurrence in Texas. TPWD eventually banned the use of tilapia, but the agency lacked the statutory authority to regulate other shellfish and plant species.

The ban on tilapia resulted in legislation that transferred regulatory authority of aquaculture to the Texas Department of Agriculture (Acts of the 71st Texas Legislature 'The Fish Farming Act of 1989'). However, TPWD was delegated specific regulatory authority over exotic fish, shellfish and aquatic plants. In the spring of 1990, the Parks and Wildlife Commission adopted rules for regulation of these species. But the adopted rules lacked clear mariculture facility standards and this proved to be a serious misstep.

The fragile balance of the fragmented regulatory authority shared by TDA and TPWD toppled when a major release of exotic Pacific white shrimp (*Litopenaeus vannamei*) occurred in November 1991. In anticipation of an approaching cold front, mariculturists at a large shrimp growout facility near the lower Laguna Madre began to drain ponds and harvest shrimp with little regard for escape from the facility into effluent canals. Within hours, cultured exotic shrimp were appearing in trawl samples in the Arroyo

Colorado. TPWD was notified by interests other than the culture facility and began enforcement and environmental clean-up activities. The facilities came under sharp scrutiny by commercial and recreational interests concerned that future escapes could cause harm to valuable native shrimp stocks. As one would predict, new rounds of regulatory activities began immediately. The result for the mariculture industry was predictable. Costs related to production increased, as new regulatory strategies required re-engineering of existing structures, notification procedures and a series of other operational cost increases.

Although there appears to have been no long-term environmental effects of the escape, the political and social effects were enormous. The Coastal Conservation Association or CCA (the leading fisheries conservation organization in Texas) and the Texas Shrimp Association (its commercial counterpart) joined forces to oppose both legislative and administrative (application for various permits) initiatives of shrimp farm interests. They also joined with a number of environmental organizations such as the Environmental Defense Fund and Sierra Club to oppose them.

In the spring of 1995, came the release of a second threatening enemy, which could not be viewed with the field glass, rather it appeared under the microscope. Several shrimp farms began reporting (as was required under state law) the presence of diseased shrimp. The aetiology of the disease was quickly identified as the Taura syndrome virus (TSV), a particularly virulent shrimp disease. The virus affects moulting physiology and is spread among infected shrimp. Although there is no hard evidence suggesting that the disease is water-borne, it does affect native white shrimp (*Litopenaeus setiferus*) in the postlarval stage. Both recreational anglers and commercial shrimping interests were gravely concerned that effluent from mariculture facilities or escaped diseased shrimp could cause devastation in native shrimp populations.

Not unlike the shrimp releases of 1991, a new round of regulatory activities began, again with regulations that were stricter and more costly than previous iterations. The relationship between shrimp culturists and the regulatory community frustrated both the regulated and regulators.

Towards Conflict Resolution

Always do right. This will gratify some people and astonish the rest.

Mark Twain

If the Texas experience with shrimp mariculture is an example of what can happen when mariculture stares across the battlefield at recreational interests (we believe it is a prime example) and if the perception of mariculture as a threat to recreational fishing is to be vanquished, government, industry, and the scientific community must adopt proactive management and regulatory strategies. Clearly, regulatory strategies that are enacted to cure existing problems are only short-term fixes because these strategies lack the vision necessary to provide a predictable regulatory environment for mariculture and are ineffective at addressing evolution of the industry. Good public policy can at least be defined in part as policy that provides an environment in which economic interests can fully assess the cost–benefit of a business strategy. If regulatory policy – both in intent and in fact – is prone to change, the ability of the mariculturist to assess operating margins is hampered, as is the ability to secure and manage capital necessary to ensure adequate operating cash flows.

Further, such an after-the-fact regulatory strategy does little to alleviate the concerns of those who may not feel the financial burden of the regulations yet have a stake in the ultimate outcome of the regulatory process. This discomfort, in our view, is a symptom of a failure of the regulatory agencies to reflect their authentic role – that is, to allow individuals the comfort and confidence that the decision-making process actually results in an outcome which genuinely reflects the public interest. But the nature of government is such that issues become issues as a result of market or regulatory failure. It is a fundamental fact that government is prone to inaction except in crisis or when segments of society demand proactive measures. In our view, if mariculture and recreational angling interests are to sign meaningful treaties, it becomes the *de facto* responsibility of the industry to initiate the dialogue necessary to form that alliance. It may not seem fair, but it is reality.

Mariculture and recreational fishing interests do not have to be enemies. They can share an environment in which business profits and recreational anglers are assured that mariculture poses little or no risk to game fish species. The shrimp mariculture experience in Texas is an excellent example of both what to do and what not to do. In the final assessment, mariculture was no longer viewed as the enemy. But prior to the signatures being placed on the peace treaty, hard and painful battles were fought between these competing interests.

In almost all situations, mariculture development will take place in proximity to recreational interests and will require access to some biological or ecological component of the recreational fishery. Mariculture may need water, broodstock or receiving waters for effluent, but it will require some asset that is already claimed by sport interests. With that knowledge in hand, we believe that the first step in preventing conflict is up to the mariculture industry. The industry must initiate the dialogue with other stakeholders, particularly the regulators. We suppose that mariculture may not have to take this approach, but we believe that in the long term it is both economical for the industry and more socially acceptable to recreational interests.

There are several advantages to this process. First, and perhaps foremost, educated regulators have much more in-depth knowledge of the proposed mariculture operations and will at least have the opportunity to formulate regulations that are user-friendly to the industry while protecting recreational (and perhaps commercial) fisheries. It is a fundamental fact that regulators, largely as a result of their biological background, will tend to take the most conservative approach to regulations in the absence of information. As a result of this dialogue, mariculture has a reasonable assurance that regulations are not likely to evolve in a fashion that increases costs of operations. This is not to say that this will absolutely not occur, but it is substantially less likely.

The second clear advantage of proactive dialogue is that of assuaging the concerns of the recreational anglers. These stakeholders understand their relationship to the regulators very well; in fact, they often argue that recreational activities pay the salaries of the regulators (a claim that is often accurate). They know that regulators usually will err on the side of conservative management. The second step, then, is that of engaging the recreational interests directly as stakeholders. Again, it may be more time consuming in the short run than just drawing up the plans and starting the earthmoving equipment, but it is likely to save a great deal of time and effort after the ponds are filled with product.

As a result of our experiences in both the mariculture arena and other areas of resource conflict, we are convinced that bringing stakeholders into the decision-making process from the very beginning is much more likely to facilitate a peace treaty before a shot is fired. Our battle-by-battle approach to shrimp mariculture regulation in Texas did result in a workable solution, but in a real sense, there were far more dollars lost as casualties than were necessary. It is an unerring characteristic of Americans, we believe, that they are willing to live with the outcome, if they believe the process is fair and that they have had a chance to be included in that process. We think they are right.

References

American Sportfishing Association (1996) *The Economic Importance of Sportfishing.* 10pp. http://www.asafishing.org/statistics/reports/economicimpact.htm

Goldburg, R. and Triplett, T. (1997) *Murky Waters: Environmental Effects of Aquaculture in the United States.* Environmental Defense Report. No. 00297S. New York, USA, 195pp. http://www.edf.org/pubs/Reports/Aquaculture/

Moulton, D.W., Dahl, T.E. and Dall, D.M. (1997) *Texas Coastal Wetlands; Status and Trends, Mid-1950s to Early 1990s.* US Department of the Interior, Fish and Wildlife Service, Albuquerque, New Mexico, 32pp.

Office of Science and Technology Policy (1997) *An Evaluation of Potential Shrimp Virus Impacts on Cultured Shrimp and Wild Shrimp Populations in the Gulf of Mexico and Southeastern US Atlantic Coastal Waters.* A Report to the Joint Subcommittee on Aquaculture by the JSA Shrimp Virus Work Group, Washington, DC, 65pp.

Onuf, C.P. (1994) Seagrasses, dredging and light in Laguna Madre, Texas, USA. *Estuarine, Coastal and Shelf Science* 39, 75–91.

Terlizzi, D.E. (1997) Pfiesteria piscicida *Associated with Massive Fish Mortality in a Maryland Hybrid Striped Bass Farm.* National Aquaculture Extension Conference, Annapolis, Maryland, USA, 3–18 April 1997. [Conference Summary.] http://www.asafishing.org/statistics/reports/economicimpact.htm

Whitledge, T.E. and Pulich, W.M. (1991) *Report : Brown Tide Symposium and Workshop,* 15–16 July 1991. Marine Science Institute, The University of Texas at Port Aransas, 44pp.

Aquaculture: Opportunity or Threat to Traditional Capture Fishermen?

5

Rollie Barnaby[1] and Steve Adams[2]

[1]Extension Educator, University of New Hampshire, Cooperative Extension/Sea Grant, 113 North Road, Brentwood, NY 03833, USA; [2]Communications Coordinator, New Hampshire Sea Grant, Kingman Farm, Durham, NH 03824, USA

Abstract

What are the opportunities, issues and threats for traditional commercial fishermen with the development of marine aquaculture? In 1997, Robert Robertson, University of New Hampshire Department of Resource Economics and Development, conducted a survey entitled, 'Open ocean aquaculture and commercial fishing in the northwest Atlantic: a study of northern New England's commercial fishermen, aimed at determining the respondents views toward aquaculture'. One of the areas examined was their interest in and knowledge of open ocean aquaculture. Fishermen holding permits for vessels between 6 and 15 m (20 and 50 feet) were targeted. There were 570 permit holders contacted. Of these, 311 interviews were completed. The survey instrument was a 93-question telephone questionnaire.

Of those interviewed 89% said they wished to remain traditional capture fishermen, but when asked about their willingness to work in the aquaculture industry 53% said they would be willing and 83% expressed a desire to learn more about aquaculture. When asked about characteristics of aquaculture 73% of the fishermen said small privately owned and financed operations were desirable and 57% indicated that large corporate owned and financed operations were undesirable. The results of a similar study describing the characteristics of those fishermen who would be more apt to enter aquaculture were released in July 2000. Fishermen with fewer years of experience and those that fish in several different fisheries expressed more interest in entering the aquaculture industry.

An informal phone survey of extension educators working in marine aquaculture revealed that many have actively targeted commercial fishermen with education programmes on aquaculture. While results are mixed the most successful project may be the integration of capture fishermen into hard-clam aquaculture in Florida. In Florida, inshore fishermen forced out of business through legislative action have adapted and are successful farmers. On the other hand, fishermen in Alaska remain very opposed to any finfish aquaculture other than that which is done to enhance wild stocks.

In addition to the more traditional aquaculture endeavours, there is hope that open ocean or offshore aquaculture will provide unique opportunities for commercial fishermen either as a new occupation or as a business that could complement their present fishing practices. Fishermen will already own a vessel plus have the maritime skills and knowledge of local oceanic and weather conditions. Efforts are currently underway in New Hampshire and elsewhere to take advantage of the skilled commercial fishermen workforce and introduce them to a new career that is consistent with their experiences, life skills and expectations.

Introduction

Most aquaculturists support the notion that aquaculture has the potential to help segments of the commercial fishing industry. Not all commercial fishermen agree. There are definitely clear geographic differences in how fishermen feel about aquaculture. The contrast between Alaskan commercial fishermen and New England fishermen on this issue is a case in point. Robert Robertson, a professor in the Department of Resource Economics and Development at the University of New Hampshire, interviewed 311 inshore fishermen from Massachusetts, New Hampshire and Maine (Robertson *et al.*, 1999). He found that 53% were willing to be employed in the aquaculture industry; this would not be the case in Alaska, where commercial fishermen are adamantly apposed to any netpen culture of fish. In fact, the state of Alaska does not allow any netpen culture of finfish.

Different segments of the commercial fishing industry also differ greatly: shellfish harvesters support aquaculture because most of them are using aquaculture techniques to enhance productivity, while New England lobster fishermen are very concerned about losing valuable fishing grounds. In the Cedar Key area of Florida, there is a US$10 million clam aquaculture business with the majority of the 300 growers being displaced net fishermen.

In the Cobscook Bay region of Maine, commercial fishermen have been successful at contracting their services – delivering feed, transferring fish, harvesting fish and carrying out general maintenance, such as mooring work and towing cages to the salmon farms. (Chris Bartlett, University of Maine Sea Grant, personal communication). Commercial fishermen have received financial compensation for working in open ocean aquaculture research projects in Hawaii and New Hampshire. If marine aquaculture successfully moves offshore, commercial fishermen, their vessels and their expertise will be needed.

The Environmental Defense Fund's publication, *Murky Waters: Environmental Effects of Aquaculture in the United States*, acknowledges that 'aquaculture fulfills socioeconomic goals to help sustain communities in some areas of the United States' (Goldberg and Triplett, 1997). It goes on to cite Florida and Massachusetts shellfish aquaculture, where commercial harvesters have made the switch, as positive examples.

Sixteen extension professionals from 13 coastal states completed a survey on commercial fishermen and aquaculture; 15 of the 16 said they believed that commercial fishermen could benefit from marine aquaculture in the future, but 13 of the 16 also said aquaculture could harm commercial fishermen in the future. The main issues affecting fishermen are lower prices for wild harvest products because of added product on the market and privatization of productive fishing grounds for aquaculture operations.

Aquaculture's Impact on Seafood Prices

In Alaska, USA, salmon prices have declined and the influx of farmed salmon on the world market is certainly one reason. In the areas where Alaskan salmon fishermen's landings are enhanced by local hatcheries, the effect has not been as severe as in those areas dependent on just wild returns (Dodd, 2000). This, of course, has created a serious division among salmon fishermen and their attitudes towards aquaculture.

The Alaskan fishing industry is still very opposed to any netpen culture of finfish. The opposition to finfish aquaculture has carried over to shellfish aquaculture. In an article in *Northern Aquaculture*, Ray RaLonde, University of Alaska Marine Advisory Programme, says, 'Not viewing shellfish aquaculture as an environmentally friendly industry with the potential for a multimillion dollar contribution to coastal economic development, commercial fisheries and state agencies regularly fabricate and spread horror stories about the damages that aquaculture can cause' (RaLonde, 2000).

There is some evidence that hard clam prices have gone down in some areas of the country because of the addition of cultured clams into the market. Millions of clams coming out of the Cedar Key area of Florida, USA, are in direct competition with older established producing states like Massachusetts and New Jersey, but in both these states many traditional commercial fishermen are now farming their product and competing with Florida.

Gulf of Mexico shrimp fishermen have also complained that farm-raised shrimp have negatively impacted the prices that they receive for their catches. The Texas Shrimp Association has taken a position opposing shrimp aquaculture even though one university researcher sees stock enhancement as a real opportunity that could help commercial shrimp fishermen. One Texas vessel owner and processor has purchased a couple of shrimp farms to diversify his business (Granvil Treece, Texas Sea Grant College Program, personal communication).

Adding marine cultured species to markets is not all bad for commercial fishermen. The important infrastructure and markets that commercial fishermen are dependent upon have been drastically eroded owing to the decrease in landings of wild stocks. Nowhere is this more evident than in New England, USA. New England groundfish processors and wholesalers have been crying out for farmed cod, haddock and flounder because they have lost and are losing many of their markets for those species owing to drastic reductions in landings. New England processors are afraid that once they lose their Midwest market for cod to an aquaculture product, e.g. catfish, they will never get it back. If that large market is lost, where will the cod go when the landing levels start going up?

This phenomenon was seen during the spring of 2001. When yellowtail flounder landings started to come back, the price declined to levels not seen in more than 10 years. The reason was the lack of a market: many restaurants have not had flounder on their menus for years, institutional buyers had switched to imported frozen fillets, and consumers had switched to other species. The only way to get consumers back was to offer the fish at low prices. If there had been farm-raised flounder available, processors and wholesalers might not have lost the market and commercial fishermen might have received a fair price.

Traditionally dependent on local groundfish, New England seafood retailers have made the transition to farm-raised salmon, tilapia, catfish, mussels, clams, oysters and imports. The Sgiegie Market in Tokyo, Japan, displays and sells farm-raised flounder, wild-harvested flounder, and hatchery flounder released and caught by commercial fishermen. All bring good prices. However, it will be a long time before farm-raised marine finfish species will be able to compete on the basis of price with wild-harvested fish in the United States because of the low prices being paid to fishermen.

Loss of Fishing Grounds

One important issue confronting wild harvesters is the loss of fishing ground to aquaculture leases and uses. This issue has been mitigated in most shellfish fisheries because the wild harvesters have become the farmers. There are many examples of wild harvesters embracing culture techniques to enhance their business. Examples along the east coast of the United States include soft-shell clam harvesters in Maine; hard-clam harvesters in Massachusetts, Rhode Island, New York and New Jersey; and oyster producers in Connecticut. Underemployed fishermen in Canada are now growing blue mussels successfully.

Other than shellfish, it is difficult to find any situation where aquaculture operations have seriously impacted commercial fishermen by taking over traditional fishing grounds. However, it is a serious matter for fishermen and the development of marine aquaculture. Fishermen use the 'foot in the door' argument to oppose marine aquaculture. Even though they have not been really impacted yet, they certainly are afraid they will be. When lobster fishermen complain to Sonny Sprague, a former lobster fisherman and now manager of a salmon farm in Swan's Island, Maine, that his three netpens are on good lobster bottom, he asks them if they would rather have him and his three workers each put their 800 lobster traps back in the water. This stops the complaining: you might be able to squeeze 100 traps into the area where the pens are located.

In Alaska, commercial fishermen and other opponents to aquaculture describe shellfish aquaculture as 'privatization of the resource away from common use' (RaLonde, 2000). This is in a state that has limited entry fisheries, individual fisheries quotas and community fisheries quotas.

Stock Enhancement

Though a controversial issue at this time, stock enhancement has been supported by commercial fishermen and has the potential to help commercial stocks in the future. Alaskan salmon fishermen have benefited from salmon stock enhancement and have supported the private hatcheries that release salmon to be harvested when they return to spawn. In 1999, 21% of the common property salmon harvested (42,030,000 fish) in Alaska were hatchery-reared fish and 30% of the commercial harvest was hatchery reared (Alaska Department of Fish and Game, 1999). The private, non-profit enhancement associations are supported by landing levies, or taxes, of up to 3% that local commercial fishermen pay.

Stock enhancement has been accepted and used successfully in most shellfish fisheries. Other than the salmon fishery in Alaska, there are no successful US finfish stock enhancement projects that have had a significant impact on commercial fisheries, but commercial fishermen have supported stock enhancement research for obvious reasons. (Enhancement programmes, such as the one in the state of Texas with red drum (*Sciaenops ocellatus*) have been established to support recreational fisheries.) In Maine, commercial fishermen supported a licence fee increase dedicated to cod stock enhancement research. Japanese fishermen are harvesting released hatchery flounder and a number of other species. The potential for finfish stock enhancement other than salmon enhancement to have a positive effect on commercial fisheries is certainly real, but more research is needed for most species in the United States (see Chapter 6 for additional information on this topic).

Offshore Aquaculture

Significant growth in nearshore marine aquaculture has slowed because of a lack of good sites. Traditionally, netpens have been placed in protected sites in bays or behind islands, usually in rural areas. Norway and Chile have large protected fjords in fairly remote areas that are home to large corporate salmon farms. There are very few areas in the United States that can support this kind of aquaculture operation. Cobscook Bay in Maine may be the only area comparable. It is also blessed with strong tidal currents that provide important flushing action. New marine finfish aquaculture operations will probably have to take place in warehouses where recirculating systems are employed or offshore in high-energy environments.

If marine aquaculture moves offshore, there will be business opportunities for fishermen. Their vessels, knowledge of ocean and weather conditions, rigging skills, boating skills and work ethic will be needed. Whether fishermen become fish farmers or contract with fish farms will depend on factors such as capital needed, species, and size of farms. Many people believe inshore fishermen have the right work habits and mentality to switch to aquaculture. They start early in the morning, spend long days tending their gear, are at the mercy of the elements, and are used to great fluctuations in their income. The New England lobster fishery is viewed as a large offshore lobster farm; fishermen feed the lobsters, provide shelter, protect juveniles, and mark and protect a brood stock.

A survey of northern New England commercial fishermen was conducted in 1997 by Robertson *et al.* (1999). One goal of the project was to determine the views of traditional capture fishermen towards open ocean aquaculture. Fishermen holding permits for vessels of between 6 and 15 m (20 and 50 feet) were targeted. Of 570 permit holders contacted, 311 interviews were completed. The survey instrument was a 93-question telephone questionnaire.

Of those interviewed, 89% said they wished to remain traditional commercial fishermen. When asked about their willingness to work in the aquaculture industry, 53% said they would be willing and 83% expressed a desire to learn more about aquaculture. When asked about characteristics of aquaculture, 73% of the fishermen said small privately owned and financed operations were desirable and 57% indicated that large corporate owned and financed operations were undesirable.

In 1999, Rob Robertson and Torene Tango-Lowy (unpublished) conducted a second study entitled: 'Preadoption of Open Ocean Aquaculture by Northern New England's Inshore Commercial Fishermen'. This study investigated whether and under what conditions northern New England inshore fishermen were willing to adopt open ocean aquaculture. Quantitative data were collected through a self-administered, 105-question mail survey. Of the 767 fishermen contacted, 186 completed the survey yielding a 24.3% response rate. The survey found that fishermen with fewer years of experience and those that fish in several fisheries were more willing to enter aquaculture. More than 70% of the fishermen agreed that they could easily tend to the operational aspects of aquaculture, but more than 50% said it would involve too much political risk among peers. More than 60% felt aquaculture would limit the times and places where they could fish, while 58% said aquaculture should be used to supplement fishing to meet demand.

Conclusion

If capture fishermen are going to become major stakeholders in marine aquaculture, there will have to be a conscious effort to prevent aquaculture from developing into large corporate businesses. Shellfish aquaculture has done a good job of promoting small individually owned farms. The economy of scale may prevent that in finfish aquaculture.

Fishermen's cooperatives could provide the capital and infrastructure needed for larger operations and still be owned by fishermen. A 50-member fishermen's cooperative in Marina Di Camerota, Italy, set up a seabream fish farm to offset decreasing income from declining catches. The cooperative bought the cages and provided the operational workforce and support boats, while outside investors financed operating expenses for such things as fish fry and feed. According to the president of the cooperative Beniamino Esposito, all their fish, both from the cages and the wild, are marketed under the same brand (Anonymous, 1997). The Japanese fishing cooperatives are additional examples of very successful large fishermen-owned and operated aquaculture businesses that combine aquaculture and wild harvest.

The Portsmouth, New Hampshire, fishermen's cooperative is a partner in the University of New Hampshire Open Ocean Aquaculture Demonstration Project. If marine aquaculture moves offshore, fishermen's vessels, knowledge and skills will be needed.

When asked if he thought fishermen would be harmed by aquaculture in the future, one extension professional from Louisiana said, 'No, gear restrictions and political trends are far greater threats' (C. Gregory Lutz, personal communication). The point is that there are so many issues that threaten the commercial fishing industry, it is surprising that aquaculture is even on their radar screen. When asked if aquaculture will harm the fishing industry, one East Coast extension specialist said, 'Only if they fail to act on the opportunity to get involved and help shape the industry' (Donald Webster, personal communication). In Japan, there is no distinction between aquaculturists and fishermen; if you harvest fish you are a fisherman, whether the product is wild or cultured. What a great concept.

References

Alaska Department of Fish & Game (1999) *1999 Alaska Enhanced Salmon Contribution to Common Property Commercial Fisheries*. www.cf.adfg.state.ak.us

Anonymous (1997) Fishermen become farmers. *Fish Farming International*, 24, 18. American Sportsfishing Association, October.

Dodd, Q. (2000) Alaskan enhancement hatcheries the focus of state power struggle. *Northern Aquaculture* 18, 3.

Goldburg, R. and Triplett, T. (1997) *Murky Waters: Environmental Effects of Aquaculture in the United States*. Environmental Defense Report. No. 00297S, New York, 195pp.

RaLonde, R. (2000) Shellfish culture: the fight for legitimacy in Alaska. *Northern Aquaculture* 6(11).

Robertson, R., Nichols, R., Lindsay, B. and Barnaby, R. (1999) Atlantic fishermen contemplate open ocean aquaculture. *Fish Farmer* 24(2), 34.

Advances in Marine Stock Enhancement: Shifting Emphasis to Theory and Accountability

6

Kenneth M. Leber

Center for Fisheries Enhancement, Mote Marine Laboratory, 1600 Ken Thompson Parkway, Sarasota, FL 34236, USA

Abstract

After over a century of use in the United States, marine stock enhancement is still at an early stage of development as a fisheries management tool in coastal environments. The principal constraint to rapid development of enhancement's potential is clear – the rush to use stock enhancement, and use it now, has nearly always led to emphasis on hatchery production at the expense of developing the science of stock enhancement. Until we develop, test and demonstrate a clear predictive capability about the outcome of hatchery release effect, premature conclusions will prevail and continue to hamper the development of stock enhancement as a fishery management tool. Multidisciplinary teams are needed, coupled with an experimental approach and adaptive management to resolve critical uncertainties. Two of the most important uncertainties are 'does stock enhancement increase production' and, if so, 'is this at the expense of wild individuals?' The theoretical background needed to stimulate hypothesis testing has yet to be transferred from related fields, such as aquatic ecology and fisheries economics. Since 1989, development of stock enhancement theory has accelerated, based on several basic hypotheses, which are beginning to appear in scientific papers and symposia discussions. The basic hypotheses are now being tested by several research groups around the world, and the need for a responsible approach to enhancement has been recognized. Within the past decade, several new examples of effective stock enhancement have emerged, along with a cautious optimism about the potential to manage enhancement effectively. However, there remains a compelling need to identify research priorities to help focus global research on the key issues that will resolve critical uncertainties about stock enhancement potential.

(eds R.R. Stickney and J.P. McVey)

Introduction

The State of World Fisheries and Aquaculture (FAO, 2000) confirms that 47% of the 441 major fish stocks in the world are fully exploited and are producing catches at or near their maximum limit. About 28% are overexploited and have no room for expansion, and only 4% are underexploited (FAO, 2000). A similar situation exists in the United States, where many of our traditionally harvested species are at record low levels. Production from aquaculture is clearly needed to maintain per capita consumption of seafood (New, 1997). Declining growth in capture fishery landings has precipitated a new interest in marine stock enhancement worldwide. We must ask two important questions, because of the desire to use new aquaculture technologies that are available for producing marine organisms suitable for stocking: 'What is the potential of marine stock enhancement as a fishery management tool?' and 'Is there a scientific basis for implementing large-scale stock enhancement programmes to supplement or enhance seafood catch?'

As knowledge about the gloomy status of our fishery stocks has increased in the past decade, so too has awareness of the pitfalls of inadequate development of fisheries management strategies, including those related to hatchery-based stock enhancement. This awareness, coupled with lack of development of a science of stock enhancement, has fuelled growing controversy about effects and effectiveness of hatchery releases. The controversy has had a healthy effect in fostering new research to evaluate the issues, along with a divisive effect on fisheries scientists and resource managers.

The Harvest from Production-oriented Stocking: Polarized Fisheries Management

After over a century of use in America, stock enhancement is still at an early stage of development as a fisheries management tool in coastal environments (Blankenship and Leber, 1995; Grimes, 1995; Munro and Bell, 1997; Hilborn, 1999; Leber, 1999). In the United States, hatcheries have been stocking to enhance populations of marine fishes since the 1880s (Grimes, 1995). The emphasis though, for over a century, has been on the magnitude of hatchery production. Salmonids, Spanish mackerel, cod, haddock, pollock and flounder were all stocked and biologists working with marine fishes were still stocking only newly hatched fry when the federal hatcheries were shut down at Woods Hole in 1948 and in Gloucester, Massachusetts, in 1953 (Richards and Edwards, 1986).

The principal constraint to rapid development of enhancement's potential is clear – the rush to use stock enhancement, and use it now, has nearly always led to emphasis on hatchery production at the expense of developing a scientific basis for how or whether to use stock enhancement. In fact, with marine fishes (i.e. those that spawn in seawater), little or no attention was given even to evaluating hatchery-release impact on fisheries or wild stocks during the first 90 years of stocking programmes. No one knew if it worked – the biologists actually did not know if stocking marine fish had any effect on fisheries! There is little wonder why, in the late 1940s and 1950s, the US Bureau of Commercial Fisheries (now the National Marine Fisheries Service, NMFS) rightly took a definitive stance against the use of marine hatcheries as a fisheries management tool and has never looked back. Today, NMFS scientists are among the strongest opponents of marine stock enhancement in the United States, and dismiss the notion that enhancement might be effective with the attitude, 'we tried it and it didn't work'. This is understandable in light of the huge resources that were poured into hatchery boondoggles during the formative days of US fishery management. However, now, in the absence of NMFS stewardship, marine stock enhancement programmes are coming back into use in several states in the United States and some of these are well funded. This is because the public response to stocking has generally been very favourable, which encourages political support for stocking, regardless of whether any assessment of impact is conducted. The cycle of production hatcheries leading to denial is repeating itself and there are few data available to evaluate the claims of either the opponents or the supporters of hatchery-based enhancement. Given the public support for stocking and lack of information to evaluate enhancement impact and cost effectiveness, perpetual demand for stocking should be expected.

The lack of effectiveness of early stocking programmes in the United States curtailed much of the funding for assessment of stock enhancement effect on marine organisms until the advent and relatively recent proliferation of modern tagging technology, which provided a way to track released juveniles, albeit with varying degrees of success (e.g. the binary coded-wire tag, genetic 'tags', otolith marking, chemical marks, and various external tags; see Parker *et al.*, 1990; Bergman *et al.*, 1992). Because of the slow pace of research on marine enhancement following the closure of marine hatcheries in the northeastern USA, advances in marine stock enhancement effect have not kept pace with advances in marine aquaculture technology. Thus, we now have the capability to produce many species of marine fish but we lack basic knowledge about how or even whether to use hatchery-reared fish as a resource management tool in marine environments (Leber, 1999).

The Pros and Cons of Stock Enhancement

There is an allure to stocking that has captured our interest. The proponents of stock enhancement view it as a management tool with enormous potential to rapidly replenish depleted fisheries and supplement weak year classes; as a means for recovering endangered species; as a quick fix that can provide disaster relief from a host of environmental calamities, such as fish kills from red tide, hard winter freezes and various toxins introduced into aquatic environments; and by targeting hatchery fish, as a way to transfer fishing pressure away from wild stocks. As an experimental tool, stocking can clearly increase knowledge about wild stocks.

On the other hand, there is very little development of the scientific background that is needed to make sound decisions about stocking and, thus, there are many critical uncertainties. There are very few good examples of marine enhancement to build upon and the track record is poor. Once started, stocking programmes are usually difficult to stop. Stocking can cause overfishing in mixed-stock fisheries, where abundant hatchery fish can result in elevated fishing rates on the wild stock. Some view stock enhancement as expensive, even arrogant use of technology. Stocking is potentially harmful to wild stocks, which could be adversely affected by transfer of disease, reduction in genetic diversity and fitness, and competitive displacement by hatchery fish. Also, public reaction to expected benefits from stocking can postpone or, even worse, subvert inevitable use of tough restrictions on fishing effort.

Assessing Stocking Impact and its Potential as a Fishery Management Strategy

Finally, after a century of neglect, researchers have begun to quantify survival of hatchery-reared fishes in the wild and gauge the effects of stocking programmes (e.g. Hager and Noble, 1976; Bilton *et al.*, 1982; Olla and Davis, 1988; Tsukamoto *et al.*, 1989; Svasand and Kristiansen, 1990a,b; Svasand *et al.*, 1990; Peterman, 1991; Kitada *et al.*, 1992, 1994; Nordeide *et al.*, 1994; Olla *et al.*, 1994; Ray *et al.*, 1994; Yamashita *et al.*, 1994; Secor *et al.*, 1995; Munro and Bell, 1997; Hilborn, 1998; Okouchi *et al.*, 1998). The authors cited were among the first to highlight key scientific and economic issues about salmonid and marine fisheries enhancement in peer-reviewed scientific journals.

The new scientific discoveries made in the past two decades about enhancement are just the tip of the iceberg of what must be understood to evaluate enhancement's potential as a fisheries management tool (Hilborn, 1999; Leber, 1999). There has been little or no development of the theoretical basis for stock enhancement. Without this – without a paradigm – many of the obvious issues seem equally important. Thus we need to develop and test stock enhancement theory and prioritize the key questions, if we are to cope as scientists or fishery managers with the rash of unanswered questions about how to control enhancement impact. Systematic tests of sequential hypotheses are needed to resolve the critical uncertainties (Table 6.1). Until we develop, test and demonstrate a clear predictive capability about hatchery release effect, premature conclusions will prevail and continue to hamper the development of stock enhancement as a fishery management tool (Leber, 1999).

Clearly, a responsible and scientific approach is needed to evaluate stock enhancement potential (Cowx, 1994; Blankenship and Leber, 1995; Hilborn,

Table 6.1. Critical uncertainties in stock enhancement.

Does it work?
- Effects of release strategies on growth and survival
  - Fish size at release
  - Release habitat
  - Release timing
  - Release magnitude
- Actual impact on production
  - Is there surplus productive capacity (carrying capacity) to support additional organisms?
  - Are there positive effects on fishery yields and reproductive output?

Conservation issues
- Effects on wild stocks
  - Displacement
  - Cannibalism
  - Genetic diversity and fitness
  - Health
- Effects on ecological interactions
  - Community dynamics
  - Predator–prey interactions

Accounting issues
- Cost effectiveness
  - Yield per stocked recruit
  - Optimal size at release
  - Cost–benefit ratio

Sustainability versus dependency on constant stocking
- Are yields achieved beyond yields from alternatives?
  - Regulations
  - Habitat protection and restoration

1999). Marine stock enhancement is now at an intermediate stage of development. Although survival, growth, and entry of stocked fish into fisheries are now well documented for several species (see Munro and Bell, 1997), multidisciplinary teams are needed, coupled with an experimental approach to resolve the key critical uncertainties (Leber, 1999). Two of the most important uncertainties are: 'Does stock enhancement increase production?' and, if so, 'Is this at the expense of wild individuals?' (Hilborn, 1999). Why have so few studies focused on these basic issues? There are several reasons. One is that basic and applied research on stock enhancement is underfunded. Another reason is the lack of theoretical development in the field.

To resolve the paradox of enhancement research (we have not developed much theory), we must: (i) seek to develop a more theoretical approach to stimulate hypothesis testing and a predictive capability in this field; (ii) identify and focus on species that are most suitable for resolving critical uncertainties about stocking; (iii) seek multidisciplinary collaborations, because the range of issues is complex and crosses many fields of study; (iv) focus research on the question(s) of the day; and (v) publish more of our research findings in the peer-reviewed scientific literature.

Since 1989, evaluation of marine stock enhancement has accelerated, based on several basic hypotheses, which are beginning to appear in scientific papers and symposia discussions. These hypotheses are now being tested by several research groups around the world, and the need for a responsible approach to enhancement has been recognized. Within the past decade, some examples of effective stock enhancement have emerged, along with a cautious appeal to fisheries scientists and managers about the need to understand how to manage enhancement wisely and effectively (e.g. Blankenship and Leber, 1995; Grimes, 1995; Munro and Bell, 1997; Hilborn, 1998; Masuda and Tsukamoto, 1998). However, there remains a compelling need to identify research priorities to help focus global research on the key issues that will resolve critical uncertainties about stock enhancement potential. Many of these issues have surfaced in review papers and in the proceedings of stock enhancement symposia published over the past decade by Lockwood (1991), Danielssen *et al.* (1994), Schramm and Piper (1995), Munro and Bell (1997), Coleman *et al.* (1998) and Howell *et al.* (1999).

Accountability in Stock Enhancement Programmes

Cowx (1994) and Blankenship and Leber (1995) summarized several issues that should be considered to conduct stock enhancement initiatives responsibly. Munro and Bell (1997) expanded on the development of some of these issues. Hilborn (1999) highlighted the greatest bottleneck to resolving the issue of whether stock enhancement can be effective: 'The responsible scientific fisheries community needs to step in and realistically evaluate fish stocking versus other management tools'.

The search for information to evaluate stock enhancement has motivated new research to determine stocking effects on coastal fisheries and wild stocks. Within the past decade or so, marine stock enhancement began to move beyond the early fact-finding stage that Kuhn (1970) observed to characterize new fields of science. To resolve the unanswered questions about stock enhancement, fishery scientists must now work closer with the fishery managers who implement stocking. This is a key issue, highlighted below from Leber (1999).

> Now, we must deal with the lack of theoretical development in the field and the clear need to reduce uncertainty about the effects of hatchery releases in coastal environments. Wider use of the scientific method and 'strong inference' (Platt, 1964) would advance knowledge in this branch of fisheries science considerably faster than its current pace. Much of the new research is focused on evaluating

post-release survival. We have entered a passive–adaptive assessment phase of marine stock enhancement (Leber, 1999), which is best described by Walters and Hilborn (1978) and Hilborn and Walters (1992).

However, given the poor track record in developing the scientific basis for stock enhancement, we can ill afford the time it takes to make progress with a passive approach. We need to increase the rigour of scientific studies of stock enhancement potential. Platt (1964) argued that, for exploring the unknown, there is no faster method than 'strong inference' – the systematic application of the age-old scientific method of inductive inference that dates back to Francis Bacon. What makes strong inference so effective is, according to Platt (1964), systematically '. . . recycling the procedure, making subhypotheses or sequential hypotheses to refine the possibilities that remain; and so on'. A key component of strong inference is acknowledging the competing alternative hypotheses (major uncertainties) that could explain an observation, and then rigorously weeding out the false alternatives through experimentation.

Walters and Hilborn (1978) added a caveat for fishery management. 'We learn most rapidly by introducing large disturbances and much monitoring, but we incur high risks and costs by doing so . . . the dual control problem'. Because of uncertainty about genetic, health, and ecological risks to wild stocks, large-scale experimental stocking can pose risks to the very stocks we are trying to replenish. And to advance this field we must experiment, yet funding for marine stock enhancement research lies largely within the management agencies that are implementing hatchery releases. By mandate, the agencies must manage resources (i.e. implement enhancement, not study it). The solution to both issues (risks and costs imposed by large-scale experiments) is active–adaptive management (Walters and Hilborn, 1978; Hilborn and Walters, 1992), where risks of failure are restricted to substocks of the stocks being managed.

With a modest research budget, active–adaptive management in pilot release–recapture experiments should be applied as part of every release in stocking programmes (Leber, 1999). When there is a logical flow of scientific queries about the effectiveness of stock enhancement, which should always be made when resource agencies are engaged in stocking, this form of active–adaptive management is essentially strong inference adapted to fishery science. A quick scan of Platt's (1964) paper reminds us also that it is the systematic application of a logical tree of hypothesis tests and exclusions that produces much more rapid progress than in fields of science that use other approaches. Coupling strong inference and active–adaptive management principles to marine stock enhancement research is the logical next phase in this field (Leber, 1999).

Research in Hawaii to evaluate the potential of marine stock enhancement is one example of the utility of the adaptive management approach. The focus of that research from the outset was mainly on testing critical assumptions about stock enhancement effects in pilot release experiments. The pilot experiments were designed to gain information needed to design release strategies for a test of the marine stock enhancement hypothesis with striped mullet, *Mugil cephalus* (Leber, 1995; Leber *et al.*, 1996, 1997; Leber and Arce, 1996), and with a highly valued, inshore marine fish – Pacific threadfin,

Polydactylus sexfilis (Leber *et al.*, 1998; Ziemann *et al.*, 2002). Leber *et al.* (1996) showed how hatchery contribution to juvenile recruitment was increased 600% over 3 years using an active–adaptive management approach. Results from field experiments indicated that there was a realized increase in production in nursery habitats, rather than replacement of wild recruits (Leber *et al.*, 1995, summarized by Leber and Lee, 1997).

In the meantime, while research is investigating enhancement potential, how do we conduct existing stock enhancement programmes knowing full well that the science needed to guide them is incomplete, or in many cases does not exist? History has shown us that without considerable effort to adopt a scientific approach, stock enhancement programmes will continue to suffer from overemphasis on a production approach where most, if not all, of the success and accountability of enhancement is judged simply on the basis of production efficiency and numbers of animals stocked. Resource agencies need to evaluate their approach to stocking, and if they are not already enhancement oriented in such programmes (Fig. 6.1), they need to take all measures needed to adopt active–adaptive management strategies (Walters and Hilborn, 1978; Hilborn and Walters, 1992) into their operating procedures. For only through posing questions can we gain answers, and active–adaptive management is the single most important measure that can be taken to improve the potential for success in existing stock enhancement programmes. Responsible enhancement programmes need to pose questions about critical uncertainties, by marking a subset of the animals released to evaluate enhancement effect

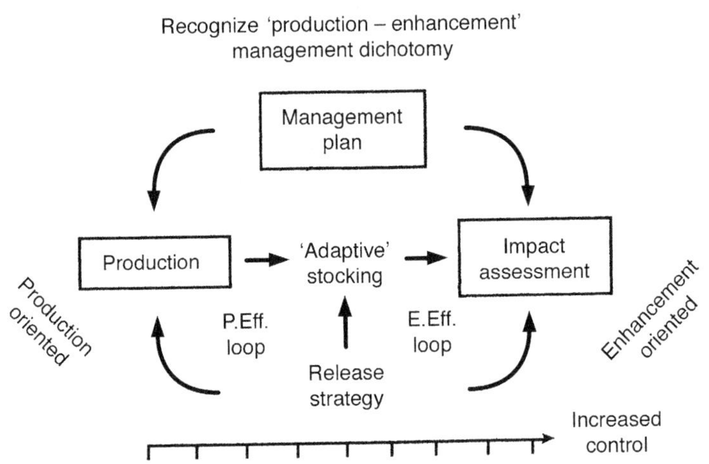

Fig. 6.1. To increase control over stock enhancement effect, stocking programmes need to seriously question whether they are 'enhancement oriented', which involves adaptive management as a key component of every stocking opportunity, or 'production oriented', which yields little if any control over enhancement success. P. Eff, production efficiency; E. Eff, enhancement efficiency.

and develop improved stocking strategies, such as examining effects of stocking density to evaluate carrying capacity; size at release to optimize this variable; effects of other release strategies, and so on (Blankenship and Leber, 1995). Major gains can be made in improving enhancement efficiency and effectiveness by embedding tests of critical assumptions within each release (e.g. Leber *et al.*, 1996; Leber, 1999). Without this approach, stocking programmes can maximize growth and survival in the hatchery, only to suffer great losses of hatchery organisms, and even wild ones, after releases. Why post-release mortality does not get the same attention in stocking programmes that mortality in the hatchery receives is inexplicable.

In summary, stocking's baggage is that lack of consensus on key research issues during 90 years of stocking, and failure to treat stock enhancement as a science, have constrained advances in this branch of the fisheries discipline. The result has been: (i) splintering of fishery scientists into camps for and against stock enhancement; (ii) little, if any, development of stock enhancement theory; and (iii) few data to evaluate either camp's position. Marine stock enhancement research over the past decade shows some potential for stocking to be an effective management tool, but we cannot understand the actual effectiveness of stocking in coastal environments until we develop and test stock enhancement theory. Multidisciplinary research partnerships and an active–adaptive management approach are needed to resolve the critical uncertainties, test key assumptions and evaluate stock enhancement potential as a fishery management tool.

Acknowledgements

Many of the ideas expressed herein are the results of discussions over the past decade with colleagues too numerous to mention without leaving someone out. Particular credit is due, though, to discussions with Lee Blankenship, Devin Bartley, Johann Bell, Ray Buckley, Churchill Grimes, Bill Halstead, Bill Hawkins, Ray Hilborn, Hunt Howell, Don Kent, Shiuichi Katada, Rich Lincoln, Conrad Mahnken, John Miller, Stan Moberly, Bill Neill, Jeff Polovina, Gary Pruder, Erlend Moksness, Dan Simberloff, Ted Smith, Terje Svasand, Masaru Tanaka, Katsumi Tsukamoto, Yoh Yamashita, and David Ziemann. Thanks also to Bob Stickney and Jim McVey for encouraging an increase in the awareness and emphasis on responsible use of aquaculture technology.

References

Bergman, P.K., Haw, F., Blankenship, H.L. and Buckley, R.M. (1992) Perspectives on design, use, and misuse of fish tags. *Fisheries* 17(4), 20–24.

Bilton, H.T., Alderdice, D.F. and Schnute, J.T. (1982) Influence of time and size at release of juvenile coho salmon (*Oncorhynchus kisutch*) on returns at maturity. *Canadian Journal of Fisheries and Aquatic Science* 39, 426–447.

Blankenship, H.L. and Leber, K.M. (1995) A responsible approach to marine stock enhancement. *American Fisheries Society Symposium* 15, 167–175. American Fisheries Society, Bethesda, Maryland.

Coleman, F., Travis, J. and Thistle, A.B. (eds) (1998) Marine stock enhancement: a new perspective. *Bulletin of Marine Science* 62, 303–714.

Cowx, I.G. (1994) Stocking strategies. *Fisheries Management and Ecology* 1, 15–30.

Danielssen, D.S., Howell, B.R. and Moksness, E. (eds) (1994) *Aquaculture and Fisheries Management* 24 (Suppl. 1), 264pp.

FAO (2000) *The State of World Fisheries and Aquaculture, 2000*. Food and Agriculture Organization of the United Nations, Rome. http://www.fao.org/DOCREP/003/X8002E00.htm

Grimes, C.B. (1995) Perspective of the AFS marine fish section on uses and effects of cultured fishes in aquatic ecosystems. In: *American Fisheries Society Symposium 15*. American Fisheries Society, Bethesda, Maryland, pp. 593–594.

Hager, R.C. and Noble, R.E. (1976) Relation of size at release of hatchery-reared coho salmon to age, size, and sex composition of returning adults. *Progressive Fish Culturist* 38, 144–147.

Hilborn, R. (1998) The economic performance of marine stock enhancement projects. *Bulletin of Marine Science* 62, 661–674.

Hilborn, R. (1999) Confessions of a reformed hatchery basher. *Fisheries* 24(5), 30–31.

Hilborn, R. and Walters, C.J. (1992) *Quantitative Fisheries Stock Assessment*. Chapman & Hall, New York, 570pp.

Honma, A. (1993) *Aquaculture in Japan*. Japan FAO Association, Tokyo, Japan, 98pp.

Howell, B.R., Moksness, E. and Svasand, T. (eds) (1999) *Stock Enhancement and Sea Ranching*. Fishing News Books, Blackwell Science, Oxford, UK, 606pp.

Kitada, S., Hiramatsu, K. and Kishino, H. (1992) Effectiveness of a stock enhancement program evaluated by a two-stage sampling survey of commercial landings. *Canadian Journal of Fisheries and Aquatic Science* 49, 1573–1582.

Kitada, S., Hiramatsu, K. and Kishino, H. (1994) Estimating mortality rates from tag recoveries: incorporating over-dispersion, correlation and change points. *Journal of Marine Science* 51, 241–251.

Kuhn, T.S. (1970) *The Structure of Scientific Revolutions*. University of Chicago Press, London.

Leber, K.M. (1995) Significance of fish size-at-release on enhancement of striped mullet fisheries in Hawaii. *Journal of the World Aquaculture Society* 26, 143–153.

Leber, K.M. (1999) Rationale for an experimental approach to stock enhancement. In: *Stock Enhancement and Sea Ranching*. Fishing News Books, Blackwell Science, Oxford, UK, pp. 63–75.

Leber, K.M. and Arce, S.M. (1996) Stock enhancement in a commercial mullet, *Mugil cephalus* L., fishery in Hawaii. *Fisheries Management and Ecology* 3, 261–278.

Leber, K.M. and Lee, C-S. (1997) Marine stock-enhancement potential with striped mullet, *Mugil cephalus*, in Hawaii. *Bulletin of the National Research Institute of Aquaculture* (Suppl. 3), 117–134. Nansei, Mie, Japan.

Leber, K.M., Brennan, N.P. and Arce, S.M. (1995) Marine enhancement with striped mullet: are hatchery releases replenishing or displacing wild stocks? In: *American Fisheries Society Symposium 15*. American Fisheries Society, Bethesda, Maryland, pp. 376–387.

Leber, K.M., Arce, S.M., Sterritt, D.A. and Brennan, N.P. (1996) Test of marine stock-enhancement potential with striped mullet in Hawaii: growth, survival and release impact in nursery habitats. *Fishery Bulletin* 94, 452–471.

Leber, K.M., Blankenship, H.L., Arce, S.M. and Brennan, N.P. (1997) Influence of release season on size-dependent survival of cultured striped mullet, *Mugil cephalus*, in a Hawaiian estuary. *Fishery Bulletin* 95, 267–279.

Leber, K.M., Brennan, N.P. and Arce, S.M. (1998) Recruitment patterns of cultured juvenile Pacific threadfin, *Polydactylus sexfilis* (Polynemidae), released along sandy marine shores in Hawaii. *Bulletin of Marine Science* 62, 389–408

Liu, J.Y. (1990) Resource enhancement of Chinese shrimp, *Penaeus orientalis*. *Bulletin of Marine Science* 47, 124–133.

Lockwood, S.J. (ed.) (1991) *The Ecology and Management Aspects of Extensive Mariculture ICES Marine Science Symposium 192*. International Counsel for the Exploration of the Sea, Copenhagen, 248pp.

Masuda, R. and Tsukamoto, K. (1998) Stock enhancement in Japan: review and perspective. *Bulletin of Marine Science* 62, 337–358.

Munro, J.L. and Bell, J.D. (1997) Enhancement of marine fisheries resources. *Reviews in Fisheries Science* 5, 185–222.

New, M.B. (1997) Aquaculture and the capture fisheries: balancing the scales. *World Aquaculture* 28, 11–31.

Nordeide, J.T., Fossa, J.H., Salvanes, A.G.V. and Smedstad, O.M. (1994) Testing if year-class strength of coastal cod, *Gadus morhua* L., can be determined at the juvenile stage. *Aquaculture and Fisheries Management* 25 (Suppl. 1), 101–116.

Okouchi H., Iwamoto, A., Tsuzaki, T., Fukunaga, T. and Kitada, S. (1998) Economic returns from hatchery released flounder *Paralichthys olivaceus* in Miyako Bay – evaluation by a fish market census. In: Howell, B.R., Moksness, E. and Svasand, T. (eds) *Stock Enhancement and Sea Ranching*. Fishing News Books, Blackwell Science, Oxford, UK, pp. 569–586.

Olla, B.L. and Davis, M. (1988) To eat or not be eaten. Do hatchery-reared salmon need to learn survival skills? *Underwater Naturalist* 17, 16–18.

Olla, B.L., Davis, M.W. and Ryer, C.H. (1994) Behavioral deficits in hatchery-reared fish: potential effects on survival following release. *Aquaculture and Fishery Management* 25 (Suppl. 1), 19–34.

Parker, N.C., Giorgi, A.E., Heidinger, R.C., Jester, D.B., Prince, E.D. and Winans, G.A. (eds) (1990) *Fish-Marking Techniques. American Fisheries Society Symposium 7*. American Fisheries Society, Bethesda, Maryland, 879pp.

Peterman, R.M. (1991) Density-dependent marine processes in North Pacific salmonids: lessons for experimental design of large-scale manipulations of fish stocks. *ICES Marine Science Symposium* 192, 69–77.

Platt, J.R. (1964) Strong inference. *Science* 146, 347–353.

Ray, M., Stoner, A.W. and O'Connel, S.M. (1994) Size-specific predation of juvenile queen conch *Strombus gigas*: implications for stock enhancement. *Aquaculture* 128, 79–88.

Richards, W.J. and Edwards, R.E. (1986) Stocking to restore or enhance marine fisheries. In: Stroud, R.H. (ed.) *Fish Culture in Fisheries Management*. American Fisheries Society, Bethesda, Maryland, pp. 75–80.

Schramm, H.L. Jr and Piper, R.G. (eds) (1995) *Uses and Effects of Cultured Fishes in Aquatic Ecosystems. American Fisheries Society Symposium 15*. American Fisheries Society, Bethesda, Maryland.

Secor, D.H., Houde, E.E. and Monteleone, D.M. (1995) A mark–release experiment on larval striped bass *Morone saxatilis* in a Chesapeake Bay tributary. *ICES Journal of Marine Science* 52, 87–101.

Svasand, T. and Kristiansen, T.S. (1990a) Enhancement studies of coastal cod in western Norway. Part II. Migration of reared coastal cod. *Journal du Conseil* 47, 13–22.

Svasand, T. and Kristiansen, T.S. (1990b) Enhancement studies of coastal cod in western Norway. Part IV. Mortality of reared cod after release. *Journal du Conseil* 47, 30–39.

Svasand T., Jorstad, K.E. and Kristiansen, T.S. (1990) Enhancement studies of coastal cod in western Norway. Part I. Recruitment of wild and reared cod to a local spawning stock. *Journal du Conseil* 47, 1–13.

Tsukamoto, K., Kuwada, H., Hirokawa, J., Oya, M., Sekiya, S., Fugimoto, H. and Imaizumi, K. (1989) Size-dependent mortality of red sea bream, *Pagrus major*, juveniles released with fluorescent otolith-tags in News Bay. *Journal of Fish Biology* 35, 59–69.

Walters, C.J. and Hilborn, R. (1978) Ecological optimization and adaptive management. *Annual Reviews of Ecology and Systematics* 9,157–188.

Yamashita, Y., Nagahora, S., Yamada, H. and Kitagawa, D. (1994) Effects of release size on survival and growth of Japanese flounder *Paralichthys olivaceus* in coastal waters off Iwate Prefecture, northeastern Japan. *Marine Ecology Progress Series* 105, 269–276.

Ziemann, D.A., Friedlander, A. and Cantrell, B. (2002) Carrying capacity for hatchery-reared Pacific threadfin (*Polydactylus sexfilis*) in native nursery habitats in Hawaii. *Transactions of the American Fisheries Society* (in press).

Aquatic Polyculture and Balanced Ecosystem Management: New Paradigms for Seafood Production

7

James P. McVey,[1] Robert R. Stickney,[2] Charles Yarish[3] and Thierry Chopin[4]

[1]National Sea Grant College Program, SSMC-3, Room 11829, 1335 East-West Highway, Silver Spring, MD 20910–3226, USA; [2]Director, Texas Sea Grant College Program, 2700 Earl Rudder Highway South, Suite 1800, College Station, TX 77845, USA; [3]Department of Ecology and Evolutionary Biology, University of Connecticut at Stamford, 1 University Place, Stamford, CT 06901–2315, USA; [4]University of New Brunswick, Centre for Coastal Studies and Aquaculture, PO Box 5050, Saint John, New Brunswick E2L 475, Canada

Abstract

Chinese aquaculture has employed a balanced ecoystem approach for freshwater aquaculture for several thousand years. Utilizing species that feed at different levels of the food web has permitted China to have the largest freshwater aquaculture production in the world. This production has proved to be sustainable in the long run because there is balance in this system. This concept is just starting to be thought of for broader aquaculture, including marine operations at sea or on land, and fishery communities around the world.

The National Oceanic and Atmospheric Administration has developed a Sustainable Fisheries Implementation Plan that recognizes three key elements – fisheries, aquaculture and coastal communities – for obtaining sustained production of seafood in the United States. The concepts of carrying capacity for biological activities in a hydrographic system; ecological balance between primary producers, primary and secondary consumers; and nutrient flows in ecosystems are essential elements for the future development of world aquaculture and fisheries.

This chapter documents the present status of selected polyculture systems being employed by the aquaculture industry, provides examples of balanced ecosystem approaches to aquaculture and fisheries, and examines the question of how to develop models for maximizing the production of seafood through fisheries and aquaculture working in harmony to minimize environmental impacts.

Introduction

Clear signs of overexploitation of important fish stocks, modification of ecosystems, significant economic losses, and international conflicts on management and fish trade threaten the long-term sustainability of fisheries and the contribution of fisheries to the world food supply (FAO, 1997). The collapse of New England fisheries stocks on George's Bank and the economic hardship that this has imposed on coastal communities is a case in point. The National Marine Fisheries Service (2001) estimated that 92 fisheries stocks in the United States are overfished compared with 148 stocks that are not overfished. It is clear that in many fisheries the level of exploitation exceeded the ability of the stocks to replace themselves. Future fishing efforts will have to be at lower levels in New England and throughout the world in order to be sustainable.

Aquaculture is one of the only ways to make up for the reduced fisheries yields that are inevitable in the future. According to FAO (1997), sustainable aquaculture development will need to recognize the diversity of aquaculture practices as well as the social, economic and (we would add) environmental conditions in which they will take place. Aquaculture is taking place in coastal oceans already eutrophied and impacted by human activities. Jewell (1994) stated that a sewage system that serves 10,000 people discharges more than 250 kg of suspended and biodegradable matter daily. Walsh (1988) found that human-related loadings of nitrogen have increased tenfold during the last century. Bouman and van Vuuren (1999) and Smith *et al.* (1999) provided the following statistics: (i) between one-third and one-half of the land's surface has been transformed by human activity; (ii) global production of agricultural fertilizers has increased from less than 10 million tonnes in 1950 to over 80 million tonnes in 1990; (iii) burning of fossil fuels provides the emission of more than 20 million tonnes of nitrogen into the atmosphere on an annual basis; and (iv) phosphorus eroded from the landscape and carried in human wastewater into the world's rivers has increased global fluxes of phosphorus to the oceans almost threefold, from historic levels of about 8 million tonnes to current loadings of about 22 million tonnes per year. Seitzinger and Kroeze (1998) estimated that observed levels of dissolved inorganic nitrogen are due worldwide primarily to agricultural fertilizers (58%), human sewage (24%) and atmospheric deposition (18%). These environmental consequences of human activity were not planned and are seldom taken into account when developing management strategies for coastal ecosystems. The impact of aquaculture relative to the scale of impacts that have already occurred is relatively minor at this time, but it is increasing and needs to be factored into the management strategies that are being proposed for environmental quality and sustainability.

Higher nutrients in estuaries and in coastal locations where deep water upwelling occurs are not inherently bad and, in fact, provide the enrichment necessary for high productivity. Walsh (1988) estimated that 95% of the world's fisheries yield comes from coastal zones because of the nutrient-rich conditions that exist there. The Pacific Ocean averages 1.25 kg ha^{-1} of fishery products annually, but the maximum yield of 280 kg ha^{-1} occurs off Peru where upwelling of nutrient-rich water occurs (Anderson and Gaucher, 1967). This enrichment is the basis for one of the world's most productive fisheries. Only when nutrients outstrip the ability of natural processes to handle them are problems encountered.

Natural aquatic systems have a built-in carrying capacity for handling nutrients, which is dependent on the biological processes that occur in those systems. Christensen *et al.* (2000) observed that denitrification processes in a Norwegian fjord can generally remove 50% of the nitrogen loading from the land. The utilization of nutrients by microbes, phytoplankton, macroalgae, and seagrasses; the consumption of microbes and phytoplankton by filter feeders; and the capture of marine species in fisheries, among other processes, all contribute to the processing or removal of nutrients such as nitrogen and phosphorus from coastal waters.

Considerations for the Balanced Ecosystem Approach

The living resources in coastal waters have a profound influence on the assimilative and recycling capacity for nutrients. Each phylum or species plays a unique role in a productive ecosystem. By ecological function, there are primary producers (phytoplankton, macroalgae, seagrasses), which utilize photosynthesis to capture energy from the sun and fix nitrogen and carbon, secondary consumers (filter-feeding shellfish; filter-feeding fish; herbivorous fish, crustaceans and molluscs; bacterial species), and tertiary consumers (predatory fish, molluscs, birds and mammals, including man). The balance of these species and functions is critical to a well-functioning ecosystem. As mentioned above, man's impact on the functioning and carrying capacity of these systems has been significant and we are now at a point where we must consider human impacts in all of our approaches to resource management.

The United States National Oceanic and Atmospheric Administration (NOAA) is charged with the management of the nation's marine resources. The National Marine Fisheries Service (NMFS) manages marine fisheries and coordinates with several other NOAA agencies including Oceanic and Atmospheric Research and the National Ocean Service to do so. NOAA has a 5-year strategic plan that includes rebuilding marine fisheries as a primary objective. In the NOAA plan, rebuilding fisheries has three major components or considerations, including the management of fisheries, aquaculture, and the coastal communities that will depend on these industries. This view of a more holistic management system incorporates the use of geographic information systems, social sciences, environmental models based on hydrographic and nutrient profiles, and an interconnected understanding of the roles of the biotic factors in coastal ecosystems. The tools for such ecosystem management should be the biological functions available to us and the possibility of using aquaculture to put these tools into the correct temporal and spatial locations. Polyculture of selected organisms in chosen locations in the ecosystem can provide a better balance of ecosystem function than exists presently.

Microbial Role in a Balanced Ecosystem

Microbial communities play a significant role in reducing the negative impacts of nitrogen enrichment in coastal oceans. Certain bacteria are capable of converting ammonia to nitrite and others convert nitrite to the less toxic nitrate. Anaerobic microbial communities are capable of changing nitrate to nitrogen gas, which then escapes back to the atmosphere. The combination of these processes helps develop the balance between nitrogen input and outflow or loss to the ecosystem. Boynton *et al.* (1995) estimated that the natural microbial process of denitrification removes approximately 25% of the nitrogen inputs to Chesapeake Bay under present conditions.

Bacteria can be attached to substrates or associated with free-floating organic particles in an aquatic system. Shieh and Yang(1997) found that denitrifying bacteria were always greater in number in the root complexes of seagrasses than in control mud or sand substrates. Eighty-five strains of denitrifying bacteria were isolated from the plants. Michotey and Bonin (1997) found that there was a constant expression of bacterial nitrate dissimilation (nitrate converted to ammonium) processes associated with particles in the water column. Both denitrification and dissimilatory nitrate ammonification were associated with organic particles from 30 m down to 615 m in coastal waters. In the North Sea, Livingstone *et al.* (2000) found that denitrification decreased with depth in the sediments with the highest values in the 0–5 cm fraction. Also, the highest denitrification potential of 2100 μmol N m^{-2} h^{-1} was found in areas where there was a high anthropogenic input of nutrients. In addition, Gran and Pitkaenen (1999) found that in the nutrient-rich North Sea the highest denitrification rates occurred in the outer estuary and the open Gulf, where bioturbation fauna were present in high numbers. Highest nitrification also occurred in these areas.

The cited observations clearly show that there is a carrying capacity for nitrogen in aquatic systems and that higher nutrient levels lead to higher processing rates for nitrogen. In addition, other flora and fauna, such as plants and benthic burrowers, improve the biological functions of microbes and their ability to process nitrogen. Therefore, species diversity and balance are important to the functioning of microbes in these systems.

Eutrophication can cause hypoxia and anoxia as the result of the breakdown of organic material by the microbial community and the use of oxygen in these breakdown processes. Low oxygen conditions have resulted in significant losses of fish and shellfish resources. Eutrophication is also associated with loss of diversity both in the benthic community and among planktonic organisms, as manifested by the incidence of nuisance algal blooms in many estuaries. The Chesapeake Bay and the upper Gulf of Mexico have anoxic conditions yearly because of the breakdown of organic products by microbes and this results in loss of productivity and species diversity.

Marine Plants in a Balanced Ecosystem

Plants photosynthesize during daylight, producing oxygen and taking up nutrients. Cultivation of seaweeds and animals complement one another. Plants should be integrated with other species to develop a balanced ecosystem approach to responsible aquaculture (Chopin *et al.*, 2002). In China, which produces more than 4.8 million tonnes of brown and red algae annually, seaweeds are considered to be nutrient removers. Production of *Laminaria* sp. alone is estimated at 4 million tonnes, which is equivalent to 2 million tonnes dry weight or 60,000–100,000 tonnes of nitrogen removed each year (Fei, 1998).

The red alga *Porphyra yesoensis*, or nori, is a very valuable commodity. The value of nori was estimated at US$2 billion worldwide in 1992 (Jensen, 1993) and at nearly that amount in Japan alone in 1999 (Kito and Kawamura, 1999). Nori responds to higher levels of phosphorus and nitrogen in the environment by absorbing more into its tissues. In Maine, USA, phosphorus and nitrogen reached 7.9 mg P g^{-1} dry weight and 66.4 mg N g^{-1} dry weight, respectively, in nori plants when water phosphorus and nitrogen reached 0.93 μmol and 11.23 μmol, respectively. Lower levels of phosphorus and nitrogen in the water resulted in lower levels in the plant tissue. Nori culture in China is optimal when total ammonia levels are greater than 100 mg m^{-3}, but the quality and yields go down significantly below 50 mg m^{-3} nitrogen and fertilization becomes necessary (Fei, 1998).

Porphyra sp. can be considered an extremely efficient nutrient pump. The plants reach harvest size on nets in only 40 days and can be harvested thereafter every 9 to 15 days. Nori requires a constant supply of nutrients, particularly during summer when natural levels of nutrients are low. This corresponds to the time when fish culture operations are resulting in the highest levels of nutrient output.

Realizing that algae could provide a sparing of nutrient impact in waters around salmon farms, researchers in Maine have been experimenting with the polyculture of *Porphyra* sp. and Atlantic salmon. Approximately 72% of the nitrogen and 70% of the phosphorus in modern feeds is not retained by the fish. Earlier estimates of phosphorus and nitrogen output from salmon farms of 9.5 kg phosphorus and 78 kg nitrogen per tonne of fish have probably been reduced to 7 kg phosphorus and 49.3 kg nitrogen per tonne of fish because of improved feeds that support better assimilation rates. The new value for the number of nori nets needed to mitigate a tonne of fish per year is 27 standard nets because of the availability of improved fish feeds (Chopin and Yarish, 1999). It is not necessary to completely absorb all nutrients from a salmon farm. It is only necessary to keep below threshold limits for deleterious harmful algal blooms or levels that result in reductions in dissolved oxygen levels. Ongoing research is trying to determine the threshold limits for algal blooms and develop the management models for balancing aquaculture and nutrients in the coastal zone.

There is a gradual reduction in dissolved nutrient levels away from finfish aquaculture facilities. Placement of macroalgal farms close to marine fish farms would provide the optimal location for the plants and could help balance nitrogen levels in the ecosystem. In contrast, phytoplankton, which also depend upon dissolved organic and inorganic nutrients, would probably reach their greatest abundance at some distance from a fish farm, depending upon hydrographic conditions and the reproductive rates of the phytoplankton species.

Wear and Moore (1994) found that epiphytes in enriched seagrass beds were significantly greater than in control beds. Increased epiphytes led to coverage of blades and stems and loss of photosynthetic efficiency resulting in a loss of bottom coverage for the grass beds. Phytoplankton respond more rapidly than macroalgae or seagrasses to excess nutrients in coastal ecosystems because of their reproductive rate. The loss of water clarity in many estuaries, and the corresponding loss of seagrass beds, has been due to significant increases in phytoplankton levels.

Recirculating systems in aquaculture still have to contend with high levels of nutrients, and polyculture systems are being considered for mitigating increases in phosphorus and nitrogen in such systems. One recirculating aquaculture system company in the USA, which produces 8000 kg of summer flounder per year in recirculating systems, estimates that the fish generate 82 g phosphorus and 547 g nitrogen per day. The company has calculated that algal tanks put in line with flounder tanks would need to be 287 m^3 and 217 m^3 in volume to handle all the phosphorus and nitrogen (Chopin and Yarish, 1999). Even though the company can safely and responsibly dispose of nutrient-rich sludge in a municipal sewage plant, it is considering adding nori culture to the system to take advantage of the high value of the product and the natural cleansing of nutrients that would result. The fact that we can calculate the amount of phosphorus and nitrogen produced in aquaculture and the ability of plants like *Porphyra* sp. to take up those nutrients allows us to model and design systems, whether completely controlled by man or in nature, that are in balance.

The Role of Shellfish in a Balanced Ecosystem

Plants address the issue of inorganics produced by fish culture but they do not address that of organic matter. Suspension feeding organisms and filter feeders are more important for that function.

In the Chesapeake Bay, Newell (1988) estimated that the oyster population was once so numerous that it could filter the water in the bay in less than a week. Because of overfishing and disease, the oyster resource has been reduced to the point that it is now estimated to require over a year to filter the water volume of the bay. The ecological role of oysters in Chesapeake Bay is so important that the decline of water quality and fishery resources in the bay is considered to be linked to the collapse of the oyster populations. A multimillion US dollar, decade-long effort is focused on increasing the oyster biomass by tenfold in order to restore the ecological role of the oyster.

Individual bivalves have the capacity to filter from 1 to 4 litres of water per individual per hour. Communities of bivalves have the capacity to filter considerable volumes of water. Bivalves filter particles including silt and clay, phytoplankton and detritus (Jorgensen, 1966). This entrainment of organic material from the water column and deposition on the bottom in the form of faeces and pseudofaeces from bivalves is an integral and essential part of ecosystem function. Rice (2001) calculated that the population of quahog clams (*Mercenaria* sp.) in the Providence River is able to filter 21.3% of the tidal prism volume of 2.58×10^6 m^3 day^{-1} on each tidal cycle. Several studies have shown that filter-feeding bivalves can increase water clarity, thereby increasing light penetration. Cohen *et al.* (1984) studied the effects of *Corbicula*, an introduced freshwater clam from Asia, on light penetration in the Potomac River. Increased clarity in the river led to the re-establishment of submerged aquatic vegetation in the area where the clams were most plentiful. The impact of the zebra mussel on water clarity in Lake Erie is also well documented, with clarity increasing by an order of magnitude. In both cases this has resulted in increased macroalgal production and dramatic shifts in ecosystem function and the appearance of the water.

Mussel farms can remove sufficient phytoplankton so that they compete with zooplankton for the consumption of phytoplankton cells (Rodhouse and Roden, 1987). Mussels convert phytoplankton into rapidly falling organic particles, which can increase sedimentation rates threefold (Kaspar *et al.*, 1985). The hard clam, *Mercenaria mercenaria*, excretes about 9.35 mg NH_3 kg^{-1} of soft tissue per day and oysters produce about 4.76 mg NH_3 kg^{-1} of soft tissue per day. So it is important to understand the processing capabilities of benthic communities for additional organic material and the actual primary productivity of phytoplankton relative to the abundance of filter feeders, including zooplankton. We do not have sufficient understanding of these relationships to perfectly model their functions, but we do know enough to be vigilant with respect to both the positive and negative aspects of bivalve culture and to make rough estimates to be included in management models.

Rice (1999) stated that the dry weight of the soft tissues of most bivalves is typically around 30% protein. For each kilogram of shucked shellfish meats harvested there are 16.8 g of organic nitrogen removed from the estuary. Nutrient removal from estuaries can by maximized through management of shellfisheries for maximum biomass production and harvest, and development of aquaculture projects where growing shellfish are harvested regularly (see Chapter 14). An average person excretes 3.8 kg of nitrogen annually. Rice (1999) also calculated that it would take 5600 oysters (225 kg of meats) harvested from an estuary to counter the nitrogen deposition of a single human. It will take better management of human waste, increased populations of wild oyster stocks through the establishment of sanctuaries, better management of existing public beds, and the use of oyster aquaculture as a filtering mechanism and method of nitrogen removal by harvest to move towards better ecosystem function in Chesapeake Bay and other water bodies around the world.

The Role of Fish in a Balanced Ecosystem

Every species of fish has its own unique role within an aquatic ecosystem. The evolution of thousands of fish species has depended upon the ecological and environmental conditions that exist on this planet. All fish have evolved morphologically and behaviourally to the feeds and habitats that are available. Forage fish species, such as menhaden and mullet, are essential for healthy populations of more commercially important species, but menhaden are dependent on zooplankton, and mullet are more dependent on detrital materials rich in microbes and other small biota. Some species of fish have adapted to extremely cold temperatures by having blood proteins that are resistant to freezing and others have adapted to the warm conditions of tropical oceans and may be totally intolerant of winter water temperatures that are commonly experienced by temperate species.

Some coral reef fish species have become adapted to eat the skeletal material and polyps of hard corals. These fishes can process the protein for growth and eliminate the calcium carbonate that ends up as sand. Several coral reef species have developed elongated mouthparts to be able to access the many crevices and small holes in a coral reef where prey may be found. Many fish have evolved special grinder mechanisms that allow them to consume molluscs with their hard shells or crush plant material to release the nutrients from within the plant cells. Many fishes are equipped with fine gill rakers that are used to filter food from the water. Others have fleshy lobes on their heads that act as lures to attract prey. In the ocean depths, fish may be equipped with bioluminescent organs that can be used to attract prey (or mates). Fish have adapted morphologically to their feeding niche by having mouths that open downward, mid-line or up depending upon their feeding strategy. Many additional examples could be provided, but those related here give some idea of the diversity of feeding mechanisms and habits that exist.

All of this diversity and adaptation provides aquaculturists with an extensive tool box in terms of developing a balanced ecosystem approach. Fish species can be polycultured in the same containment system in such a way that one species can help graze down algae on the cage and another can reduce encrusting worms or barnacles. However, in an aquaculture context most fish will be receiving feeds based upon their ability to process animal or plant proteins and the result will be some form of nitrogen, phosphorus and other nutrient enrichment of the water mass. This enrichment can be either positive or negative depending upon the carrying capacity of the culture system for nutrients and its location within the hydrographic and ecological system. In coastal waters, heavily impacted by man's activities and nutrient inputs, any additional nutrients can lead to harmful algal blooms. On the other hand, in nutrient-poor waters, additional nutrients can lead to higher productivity and increased fishery resources. Inputs of nutrients from aquaculture into nutrient-poor waters can increase food-web interactions and careful placement of aquaculture facilities can contribute to ecosystem balance. Phytoplankton derived from such nutrient inputs can provide food for zooplankton species that are important to larvae and juveniles of commercially important species. Fish that consume zooplankton, such as alwives and menhaden, which are commonly preyed upon by commercially important species such as mackerel and cod, could benefit from properly placed nutrients resulting from aquacultural operations.

There have been a number of polyculture investigations in both marine and freshwater aquaculture systems in which secondary species were employed to utilize or recycle waste products from the primary species. Recent examples are production of the alga *Gracilaria parvispora* in a shrimp farm effluent in Hawaii (Nelson *et al.*, 2001) and production of *G. chilensis* in conjunction with salmon cages in Sweden (Troell *et al.*, 1997). Polyculture of such species as mussels (Stirling and Okumus, 1995) and sea urchins (Kelly *et al.*, 1998) with Atlantic salmon has also been evaluated.

Ahlgren (1998) stocked sea cucumbers in salmon netpens in Alaska to graze on fish faeces, excess feed and fouling organisms. A review of integrated polyculture systems designed to meet the balanced ecosystem approach was prepared by Brzeski and Newkirk (1997).

Enell and Ackefors (1991) estimated that 9.5 kg of phosphorus and 78 kg of nitrogen per tonne of fish per year are released into the water column in a typical salmon farm operation. These nutrient additions have to be considered in the context of overall nutrient processing and removal on an ecosystem scale. However, the nutrient processing abilities of algal and filter-feeding species, which we have already discussed, suggest that balance between nutrient inputs and utilization by other species, both in culture and in the wild, is possible. It is important to remember that the existing populations of microbes, plants, invertebrates and fish represent reservoirs of nutrient processing capability. The biological functions of this diverse biota serve as a buffer to nutrients and we have to understand these complex relationships as we begin to factor in human-induced processes.

The removal of nitrogen from the Chesapeake Bay by fisheries activities has been estimated at nearly 10% of nitrogen removal for the bay (Kemp, 1997). This would include harvest of bivalves, crustaceans (crabs) and fish. If we promote aquaculture of bivalves and enhancement of crabs and fish, the harvest of these resources would also remove nutrients from the bay. Human harvesting activity as well as nutrient generation through domestic waste, agriculture and industrial waste are driving forces in the balanced ecosystem equation.

Human Community Role in a Balanced Ecosystem

The profound impact of human communities on the natural environment has been described earlier in this chapter. Coastal waters have suffered the most from industrial wastes, agricultural runoff of pesticides and herbicides, as well as nutrients and domestic waste. In many cases the nutrients flowing into estuaries have overwhelmed the ability of the natural systems to process them and water quality has deteriorated. Many billions of dollars are now being spent to improve domestic-waste treatment, reduce atmospheric pollution, reduce toxic inputs, improve agricultural practices, treat existing toxic sites and correct water flow problems caused by dredging and filling.

We are at a transition in human use and dependence on the oceans of the world. The world human population has grown to the point where we can no longer expect to obtain additional protein from the sea without moving into the husbandry of the food species that are desired in the human marketplace. The capture fisheries have decimated many species of fish, crustaceans and molluscs leading to disruption of the natural balances in nature. Understanding these balances and managing the wild catch and the technology used to capture these resources is the responsibility of fishery resource managers around the world and new management procedures are being implemented for many of the fishing sectors.

We have also learned that unrestrained aquaculture can have negative environmental impacts and that it is essential to balance aquaculture production with ecosystem function and ability to handle additional nutrients. This we have described as the natural carrying capacity for nutrients in the ecosystem. But it is also becoming apparent that certain types of aquacultural production, e.g. filter-feeding species of oysters, clams, mussels and scallops, or commercial algal species, can be used to balance the nutrient additions that result from most finfish production systems. The approach of using these extractive types of aquaculture, coupled with an understanding of the natural carrying capacity of natural systems, can lead to sustainable aquaculture and the adding of desired fishery products to the human food-distribution system.

Proper placement of aquaculture facilities is very important in maintaining ecosystem function. Extractive aquaculture, such as bivalve and algae production, should occur in high-nutrient areas where aquaculture can serve to reduce nutrient levels. Finfish culture should be placed in areas that have assimilative capacity for nutrients. Appropriate areas should have high current-flow rates, be sited in offshore locations that have low nutrient levels, or be located on land, where nutrients can be processed or recaptured through filters or recirculating technologies.

Human populations have an obligation to use all natural resources in such a way that we do not overload the assimilation capacity of natural systems. There should be no such thing as fishery or aquaculture waste and we need to think more of recycling the waste produced in our agricultural and aquacultural industries. The food yield from processed fish is generally less than 50%, leaving the greatest amount of protein and bone tissue as waste. Proper treatment of this waste can provide the fish, crustacean and molluscan meals that can be utilized in the aquaculture industry (see Chapter 16). This is particularly important as fish meal supplies will eventually limit the growth of the aquaculture industry because of its value as feed for not only fish but also for cows, pigs and chickens. Integrating aquaculture and fishery industries with processors and meal production facilities is one way that human communities can reduce their impact on the natural environment.

In the future, fisheries resources will continue to be allocated among several different users. Recreational fishermen and commercial fishermen are already in conflict over the allocation of the most popular sport fish and the commercial fishermen are losing to the larger number of recreational fishermen (see Chapters 4 and 5). Many states, such as Florida, Texas and Louisiana, have enacted laws that prohibit commercial fishing for certain sport fish and this trend is expanding. The US consumer will be denied many of the most popular fish species unless they are cultured.

Other fish species and populations, such as cobia (*Rachycentron canadum*) in the South Atlantic and Gulf of Mexico, or groupers and snappers in coral communities, are not plentiful enough to withstand high levels of commercial exploitation. These top predators have in some cases been fished to near extinction, thus eliminating their important ecological function of keeping other fish species in balance. Yet, they are extremely popular foodfish for human populations. It would seem far better to culture these valuable species while maintaining the wild populations for their ecological function and contribution to the subsistence fisheries and carefully regulated recreational fisheries.

In the past, the random and piecemeal modifications of the natural environment by human populations has had the cumulative impact of upsetting the natural balance of both terrestrial and aquatic ecosystems. Human activities have been credited with the extinction of many terrestrial species. Agriculture evolved when human populations grew too large to maintain the hunter and gatherer cultures that predominated 10,000 years ago. The difference between the number of humans that existed then and now can primarily be attributed to agriculture technology development.

We have reached the same point in terms of the wild harvest of the seas. The wild fishery harvest has stabilized over the past decade (FAO, 1997). Many species are commercially extinct and the function of coastal and oceanic ecosystems compromised by commercial and recreational fisheries. Our response to these obvious changes in species abundance and balance is to impose quotas and restrictions on capture and harvest that reduce the amount of individuals captured.

New food for human consumption can only occur through aquaculture, just as it did for terrestrial systems through agriculture. In order to do that in an ecologically balanced way we must understand and utilize the natural functions of both cultured and wild species so that energy flow and distribution of nutrients is utilized through biological activity. Modelling these relationships in the context of the hydrographic and environmental conditions found in different regional contexts is our challenge.

References

Ahlgren, M.O. (1998) Consumption and assimilation of salmon net pen fouling debris by the red sea cucumber *Parastichopus californicus*: implications for polyculture. *Journal of the World Aquaculture Society* 29, 133–139.

Anderson, A.G. and Gaucher, T.A. (1967) Engineering for human ecology in the marine environment. In: Konecci, E. (ed.) *Ecological Technology: Space, Earth and Sea. Technology Transference Symposium Series No. 1*. College of Business Research, University of Texas, Austin, Texas, pp. 261–271.

Bouman, L. and van Vuuren, D. (1999) Global assessment of acidification and eutrophication of natural ecosystems. *Environmental Information Assessment Technical Report, 2000*, 64pp.

Boynton, W.R., Garber, J.H., Summers, R. and Kemp, W.M. (1995) Input, transformations and transport of nitrogen and phosphorus in Chesapeake Bay and selected tributaries. *Estuaries* 18, 285–314.

Brzeski, V. and Newkirk, G. (1997) Integrated coastal food production systems – A review of current literature. *Ocean and Coastal Management* 34, 55–71.

Chopin, T. and Yarish, C. (1999) Seaweeds must be a significant component of aquaculture for an integrated ecosystem approach. *Bulletin of the Aquaculture Association of Canada* 99-1, 35–37.

Chopin, T., Yarish, C. and Sharp, G. (2002) Beyond the monospecific approach to animal aquaculture. The light of integrated aquaculture. In: Bert, T. (ed.) *Ecological and Genetic Implications of Aquaculture Activities*. Kluwer Academic Publication Series. Reviews in Fishery Biology and Fisheries.

Christensen, P.B., Rysgaard, S., Sloth, N.P., Dalsgaard, T. and Schwaerter, S. (2000) Sediment mineralization, nutrient fluxes, denitrification and dissimilatory nitrate reduction to ammonium in an estuarine fjord with sea cage trout farms. *Aquatic Microbial Ecology* 21, 73–84.

Cohen, R.R.H., Dresler, P.V., Philips, E.J.P. and Cory, R.L. (1984) The effect of the Asiatic clam *Corbicula fluminea* on phytoplankton of the Potomac River. *Limnology and Oceanography* 29, 170–180.

Enell, M. and Ackefors, H. (1991) Nutrient discharges from aquaculture operations in Nordic countries into adjacent sea areas. ICES Council Meeting Papers, ICES CM 1991/f. International Council for the Exploration of the Sea, Copenhagen, 56pp.

FAO Fisheries Department (1997) *Aquaculture Development. FAO Technical Guidelines for Responsible Fisheries, No. 5*. Food and Agriculture Organization of the United Nations, Rome, Italy. http://www.fao.org/fi/agreem/codecond/gdlines/abs5.asp.

Fei, X. (1998) Seaweed cultivation in large scale – a possible solution to the problem of eutrophication by the removal of nutrients. *World Aquaculture* 29(4), 22–24.

Gran, V. and Pitkaenen, H. (1999) Denitrification in estuarine sediments in the eastern Gulf of Finland, Baltic Sea. *Hydrobiologia* 393, 107–115.

Jensen, A. (1993) Present and future needs for algae and algal products. *Hydrobiologia* 260/261, 15–23.

Jewell, W.J. (1994) Resource recovery wastewater treatment. *American Scientist* 82, 366–375.

Jorgensen, C.B. (1966) *The Biology of Suspension Feeding*. Pergamon Press, Oxford, 357pp.

Kaspar, H.F., Gillespie, P.A., Boyer, I.C. and Mackenzie, A.L. (1985) Effects of mussel aquaculture on the nitrogen cycle and benthic communities in Kenpura Sound, Marlborough Sounds, New Zealand. *Marine Biology* 85, 127–136.

Kelly, M.S., Brodie, C.C. and Mckenzie, J.D. (1998) Somatic and gonadal growth of the sea urchin *Psammechinus miliaris* (Gmelin) maintained in polyculture with the Atlantic salmon. *Journal of Shellfish Research* 17, 1557–2562.

Kemp, W.M. (1997) Eutrophication, habitat dynamics and trophic feedbacks: understanding and managing coastal ecosystems. In: *Stockholm Water Symposium 3*. SIWI Report No 2. Stockholm International Water Institute, Stockholm, pp. 93–99.

Kito, H. and Kawamura, Y. (1999) The cultivation of *Porphyra* (nori) in Japan. *World Aquaculture* 30(2), 35ff.

Livingstone, M.W., Smith, R.V. and Laughlin, R.J. (2000) A spatial study of denitrification potential of sediments in Belfast and Strangford Lough and its significance. *Science of the Total Environment* 251–252, 369–380.

Michotey, V. and Bonin, P. (1997) Evidence for anaerobic bacterial processes in the water column: denitrification and dissimilatory nitrate ammonification in the northwestern Mediterranean Sea. *Marine Ecology Progress Series* 160, 47–56.

National Marine Fisheries Service Report to Congress, January 2001. *Status of Fisheries in the United States* (January 2001). www.nmfs.noaa.gov/SFA/statuws%20%fisheries.htm#overfished

Nelson, S.G., Glenn, E.P., Conn, J., Moore, D., Walsh, T. and Akutagawa, M. (2001) Cultivation of *Gracilaria parvispora* (Rhodophyta) in shrimp-farm effluent ditches and floating cages in Hawaii: a two-phase polyculture system. *Aquaculture* 193, 239–248.

Newell, R.I.E. (1988) Ecological changes in Chesapeake Bay: are they the result of overharvesting the American Oyster, *Crassostrea virginica*? In: *Understanding the Estuary: Advances in Chesapeake Bay Research*. Chesapeake Research Consortium Publication 129, Gloucester Point, Virginia, pp. 536–546.

Rice, M.A. (1999) Control of eutrophication by bivalves: filtration of particulates and removal of nitrogen through harvest of rapidly growing stocks. In: Milford Aquaculture Seminar Abstracts, Milford, Connecticut, p. 275.

Rice, M.A. (2001) Environmental impacts of shellfish aquaculture: filter feeding to control eutrophication. *Aquaculture and the Environment. Proceedings of a Symposium*, 10–11 January 2001, Boston, Massachusetts.

Rodhouse, P.G. and Roden, C.M. (1987) Carbon budget for a coastal inlet in relation to intensive cultivation of suspension-feeding bivalve molluscs. *Marine Ecology Progress Series* 36, 255–263.

Seitzinger, S. and Kroeze, C. (1998) Global distribution of nitrous oxide production and nitrogen inputs in freshwater and coastal marine ecosystems. *Global Biogeochemical Cycles* 12, 93–113.

Shieh, W.Y. and Yang, J.T. (1997) Denitrification in the rhizosphere of the two seagrasses *Thalassia hemprichii* (Ehrenb.) Aschers and *Halodule uninervis* (Forsk.) Aschers. *Journal of Experimental Marine Biology and Ecology* 218, 229–241.

Smith, V.H., Tilman, G.D. and Nekola, J.C. (1999) Eutrophication: impacts of excess nutrient inputs on freshwater, marine and terrestrial ecosystems. *Environmental Pollution* 100, 179–196.

Stirling, H.P. and Okumus, I. (1995) Growth and production of mussels (*Mytilus edulis* L.) suspended at salmon cages and shellfish farms in two Scottish sea lochs. *Aquaculture* 134, 193–210.

Troell, M., Halling, C., Nilsson, A., Buschmann, A.H., Kautsky, N. and Kautsky, L. (1997) Integrated marine cultivation of *Gracilaria chilensis* (Gracilariales, Bangiophyceae) and salmon cages for reduced environmental impact and increased economic output. *Aquaculture* 156, 45–61.

US Department of Commerce (1996) *NOAA Strategic Plan: a Vision for 2005*. US Department of Commerce, Washington, DC, 210pp.

Walsh, J.J. (1988) Use of satellite ocean color observations to refine understanding of global geochemical cycles. In: Rosswall, T., Woodmansee, R. and Risser, P. (eds) *Scales and Global change*. John Wiley & Sons, Chichester, UK, pp. 287–318.

Wear, D.J. and Moore, A.D. (1994) Effects of water column enrichment on the production dynamics of three seagrass species and their epiphytic algae. *Marine Ecology Progress Series* 179, 201–213.

The Role of Marine Aquaculture Facilities as Habitats and Ecosystems

8

Barry A. Costa-Pierce[1]* and Christopher J. Bridger[2]

[1]Mississippi–Alabama Sea Grant Consortium, 703 East Beach Drive, Ocean Springs, MS 39564, USA; [2]College of Marine Sciences, The University of Southern Mississippi, 703 East Beach Drive, Ocean Springs, MS 39564, USA

Abstract

Too often the public is provided with a bleak image of marine aquaculture facilities as industrial waste areas, depleting the natural environment and its biodiversity, and creating a desert from an ocean oasis. However, this image frequently has little rigorous scientific basis. Environmental problems have been found only in aquaculture settings with poor management plans, wasteful feeding strategies and where overproduction exceeds the carrying capacity of the natural environment causing the degradation of clean water – the very basis of a successful aquaculture venture. Cage aquaculture facilities provide habitats and nursery areas for juvenile and adult wild fish, and numerous invertebrate and algal species essential to sustaining healthy marine ecosystems and wild fish stocks. In addition, there is an unbalanced focus on marine animal husbandry causing a concomitant lack of appreciation for the positive environmental attributes of marine agronomy, a vital economic sector of global aquaculture. Indeed, tidal wetland, mangrove forest and seagrass restoration aquaculture – in addition to establishment and maintenance of oyster reefs – are important examples of aquaculture creating, enhancing and maintaining productive marine ecosystems, habitats and water quality in a long-term, sustainable manner. There is an urgent need for additional research to generate *primary* data on the positive and negative roles of marine aquaculture in the biogeochemical cycles, habitats and ecosystems of coastal oceans worldwide. The little research that has been done to date has documented numerous examples of marine aquaculture facilities that revitalize natural habitats, ecosystems and marine fisheries, as opposed to degrading the natural environment and competing with the wild fisheries sector. Without more comprehensive assessments

* See contributors list for current address.

and additional research, plans for the sustainable expansion of marine aquaculture will suffer from a lack of a scientific basis for rational planning and policy, and continue to be replaced by heresy, junk science and advocacy.

Introduction

> A new revolution or philosophy is required to promote sustainable development that will more equitably allocate resources among the world population. The Western philosophy in which humanity is considered as above and not part of nature, with a mandate to exploit it, should be replaced by the Oriental philosophy of man being a part of nature.
>
> Edwards and Demaine (1997)

Aquaculture's environmental issues are related to its impacts on nature's goods and services. Closed, recirculating systems potentially have little or no waste impacts on the environment. By contrast aquaculture systems such as sea cages depend upon pelagic and benthic ecosystems and other habitats for waste treatment. Some forms of intensive aquaculture are causes of severe social and environmental impacts (Pullin *et al.*, 1993; Black, 2000). Poorly sited and managed fish and shrimp farms are of particular concern (Hall and Holby, 1986; Weston, 1986; Brown *et al.*, 1987; Ritz *et al.*, 1989; Gowen *et al.*, 1994; Silvert and Sowles, 1996; Bundell and Maybin, 1996; Hagler, 1997; Naylor *et al.*, 1998, 2000).

Fish habitat is the area 'where they [fish] may feed, rest, breed and find shelter from both predators and inhospitable environmental intrusions' (Ryder and Kerr, 1989). Wild fish populations are distributed within an acceptable zone of tolerance of numerous interrelated water quality parameters (e.g. temperature, salinity, dissolved oxygen, etc.). In an open ocean environment fish must either tolerate changes to these parameters or display avoidance behaviour to perturbations creating an inhospitable environment. In an aquaculture setting, fish are contained in a given spatial environment and have no choice but to tolerate environmental changes. Aquaculture operations must be sited to ensure the local natural environment experiences no decreases in its assimilative capacities for absorption of fish farm wastes – excess feed and metabolic products – with no degradation of water quality. To this end, aquaculture is arguably the only user industry of the oceans that requires a constant, clean supply of water and intact ecosystem services for industry success. Poor site selection and farming practices will result in a stressed ecosystem, stressed fish and decreased production (Table 8.1).

The degraded state of many aquatic ecosystems combined with public concerns about adding any new sources of aquatic pollution to already overburdened ecosystems will require aquaculture to develop ecosystem approaches and sustainable operating procedures; and to articulate a sustainable, ecological pedagogy. In the 21st century, aquaculture developers will need to spend as much time on designing ecological approaches to aquaculture

Table 8.1. Symptoms and trends of a stressed ecosystem (from Odum, 1985).

Energetics
Community respiration increases
Production/respiration becomes unbalanced
Maintenance: biomass structure ratio increases
Importance of auxiliary energy increases
Exported or unused primary production increases
Nutrient cycling
Nutrient turnover increases
Horizontal transport increases and vertical cycling of nutrients decreases
Nutrient loss increases
Community structure
Proportion of *r*-strategists increases
Size of organisms decreases
Life span of organisms decreases
Food chains shorten because of reduced energy flow at high trophic levels
Species diversity decreases and dominance increases
General systems-level trends
Ecosystem becomes more open
Autogenic successional trends reverse
Efficiency of resource use decreases
Parasitism and other negative interactions increase and mutualism and other positive interactions decrease
Functional properties are more robust than species composition and other structural properties

development that clearly exhibit stewardship of the environment as they do on technological advances coming to the field. Clear, unambiguous linkages between aquaculture and the environment must be created and fostered, and the complementary roles of aquaculture in contributing to environmental sustainability, rehabilitation and enhancement must be developed and clearly articulated to a highly concerned, increasingly educated and involved public (Costa-Pierce, 1997a).

An alternative model of aquaculture development – called ecological aquaculture – is needed that not only brings the technical aspects of ecological principles and systems thinking to aquaculture, but incorporates, at the outset, principles of natural and social ecology; planning for community development; and concerns for the wider social, economic and environmental contexts of aquaculture (Kautsky and Folke, 1991; Folke and Kautsky, 1992; Grove and Edwards, 1993; Edwards and Demaine, 1997; Edwards, 1998; Costa-Pierce, 2002). Ecological aquaculture research is oriented to the design, development and monitoring of aquatic farming systems that preserve and enhance the form and functions of the natural and social environments in which they are situated. For example, aquaculture depends upon inputs from various food, processing and transportation industries, and can produce valuable, uncontaminated wastewaters and fish wastes, all of which can be a vital part of an ecological system that can be planned and organized for community-based aquatic foods production – and natural ecosystem rehabilitation, reclamation and enhancement, not degradation (Costa-Pierce, 1997b). Lastly, ecological aquaculture takes a global view, integrating ecological science and sharing technological information in a sophisticated, knowledge-based manner, promoting innovation in the global marketplace. Using ecological analyses, aquaculture production systems are analysed holistically – incorporating social and environmental costs, not externalizing them.

Sustainable Aquaculture Development

The Brundtland Commission (WCED, 1987) defined sustainable development as: '. . . the ability to meet the needs of the present without compromising the ability of the future generations to meet their own needs'. The Food and Agriculture Organization of the United Nations (FAO, 1995) define sustainable development in their *Code of Conduct for Responsible Fisheries*, as:

> . . . the management and conservation of the natural resource base and the orientation of technological and institutional change in such a manner as to ensure the attainment and continued satisfaction of human needs for present and future generations. Such sustainable development (in the agriculture, forestry, and fisheries sectors) conserves land, water, plant, and animal resources, is environmentally non-degrading, technically appropriate, economically viable, and socially acceptable.

Sustainability can be defined simply as the maintenance of capital (Goodland and Daly, 1996). From an economic perspective, maintenance of capital is the ultimate goal of investment; but economic development quite often comes at the expense of social and environmental sustainability. Goodland and Daly (1996) state that sustainable development can only occur: '. . . without growth in throughput of matter and energy beyond regenerative and absorptive capacities'. Carrying capacity may be defined as the maximum level of resource consumption and waste discharge that can be maintained without degrading the functionality and productivity of natural ecosystems (Rees and Wackernagel, 1994). Simply stated, wastes associated with human activities must not exceed the absorptive capacity of the environment. Following the proposed 'input–output rules' of Goodland and Daly (1996), renewable resources cannot be harvested greater than the regeneration rate for that resource, and non-renewable resource depletion must remain below the rate at which new renewable substitutes are developed.

The area impacted by a species – based on resource consumption and waste discharge associated with the activities of that species – is defined as the ecological footprint (Rees and Wacknernagel, 1994). An ecological footprint extends the impact humans have on nature to an area several times greater than the land directly inhabited. For example, salmon farming in Sweden was estimated to have an ecological footprint 40,000–50,000 times greater than the area of the cages, whereas mussel aquaculture had a footprint 20 times greater than the mussel production area (Folke *et al.*, 1998). However, Black *et al.* (1997) pointed out that ecological footprint analyses fail to analyse how salmon farming reduces the benefits obtained for society and challenge the use of marginal values by Folke *et al.* (1998) from the present scale of production to increased levels in the future.

Aquaculture Habitats and Ecosystems

> A successful aquaculture system does not have wastes, only by-products, to be used as positive contributions to the surrounding ecosystems and the economy.
>
> Folke and Kautsky (1992)

Poorly sited and managed marine cage aquaculture operations have caused environmental impacts but assessments of impacts have too often been based upon outdated literature, scientific misinterpretation and advocacy (Boyd and Schmittou, 1999). Environmental concerns centre on aquaculture solid wastes and nutrient discharges, overuse of antibiotics, use of fish meal and destruction of natural habitats that existed prior to siting the aquaculture facility. Despite these concerns, the current body of scientific literature has numerous examples portraying marine aquaculture facilities as preserving and enhancing the form and function of the natural environment, serving as vital habitats for enhanced wild fish populations – both juvenile and adult forms – and providing extensive structure to support productive biofouling communities. In fact, in a survey of 17 aquaculture farms in Europe, capture fisheries were found to be little affected by fish farm effluents, and in one case there was an improvement in capture fisheries after aquaculture was initiated (Alabaster, 1982).

Excessive negative focus on marine cage and shrimp pond aquaculture has deflected the positive environmental attributes to natural marine ecosystems by the expansion of marine agronomy. Marine agronomy is a sophisticated form of higher- and lower-plant aquaculture which not only produces commodities for sale, but also assimilates nutrients and creates and restores disturbed natural habitats worldwide. Tidal wetland, mangrove forest and seagrass restoration, plus establishment and maintenance of oyster reefs are all examples of aquaculture creating, enhancing and maintaining productive ecosystems and habitats in a long-term, sustainable manner.

Marine Cage Habitat

An aquaculture cage (Figs 8.1 and 8.2) is a volume of the water column enclosed with some type of mesh forming a container for aquatic animals (Huguenin, 1997). Huguenin (1997) provided a classification scheme for

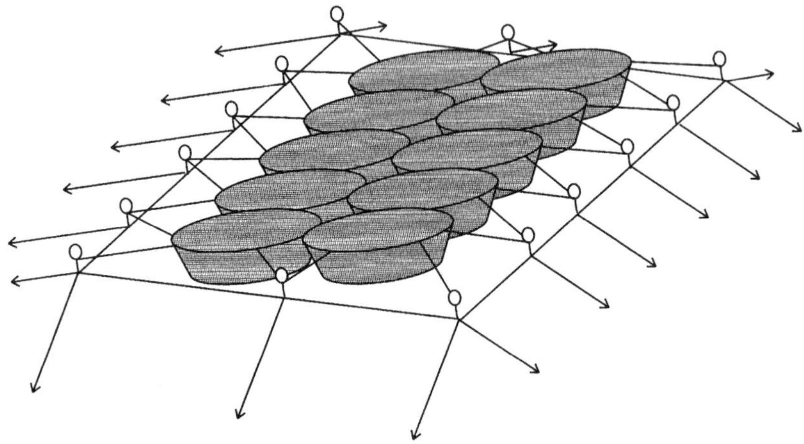

Fig. 8.1. Nested cage arrangement creating a flotilla typically used throughout the world for finfish aquaculture.

cage systems based on location, operation parameters and environment (Table 8.2). Cages have also been classified according to structure, ranging from traditional surface-oriented gravity cages to structurally rigid sea cages, comparable to attempts to utilize barges (Fig. 8.2) for cultivating fish (Loverich and Gace, 1998). The typical gravity cage structure has a surface collar from which a net is attached and placed into the water column to contain the cultured fish stock. The net mesh size is dependent on the fish size, target species, desired water flow through the cage, and operational protocols for net cleaning (or net changing) to remove sessile organisms fouling the mesh. After deployment, an aquaculture cage becomes an integral part of the natural ecosystem in which it is placed.

The marine cage aquaculture industry is changing rapidly. It is in the industry's best interest – from technological, market and regulatory perspectives – to be viewed as an environmentally friendly industry, with highly sophisticated farming practices and feeding technologies to avoid negative impacts on the environment (Myrseth, 2000). Adoption of sustainable practices not only mitigates threats of self-pollution following excessive feed inputs (Lumb, 1989), but also increases the economic potential of the operation. Scientific review of impacts associated with excess feed and faeces on benthic ecosystems shows that wastes are contained within the local farm area in a small 'halo' region under the cage (Table 8.3; see Chapter 10). Even in poorly sited and managed situations, environmental degradation of benthic ecosystems from intensive marine cage aquaculture is reversible after a period of fallowing (Mearns and Word, 1982; Kaspar *et al.*, 1988; McGhie *et al.*, 2000). Previous abuse of antibiotics has greatly diminished following the advent of effective vaccination protocols and conscientious farm operators concerned with fish stress, disease and decreased production (Fig. 8.3).

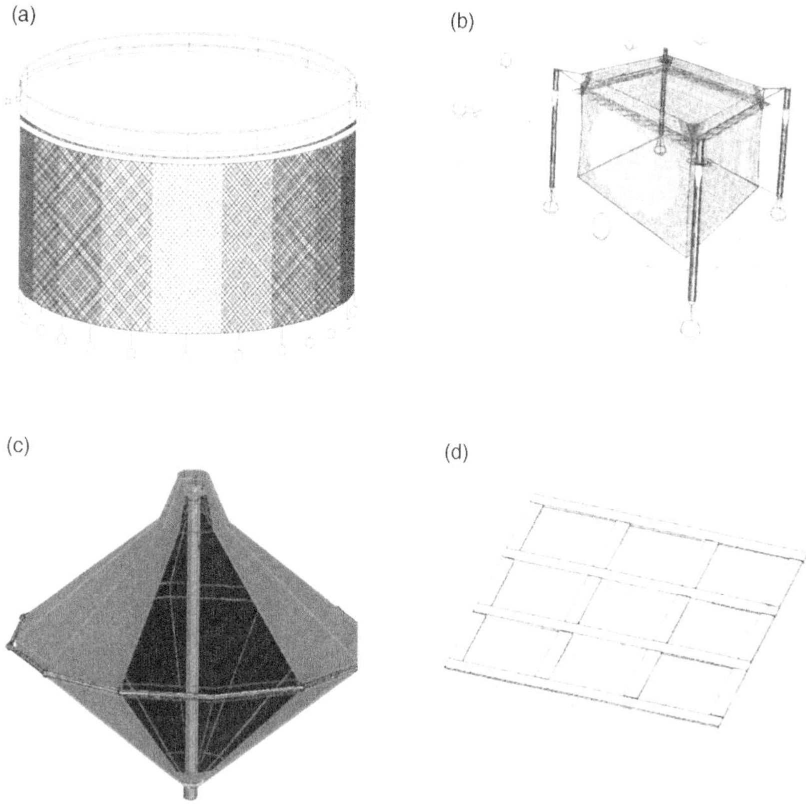

Fig. 8.2. Currently available commercial cage designs representative of surface-oriented gravity cages (a), anchor-tensioned cages (b), self-tensioned, semi-rigid, submersible cages (c), and barge-type cages (d) (Loverich and Forster, 2000).

Table 8.2. Aquaculture cage classification schemes based on location, operation, environment and structure (from Huguenin, 1997).

Where operated	Surface
	Submerged
	Marine, estuaries, freshwater
Means of support	Fixed to bottom
	Floating
Type of structure	Rigid
	Flexible
Access for servicing	Catwalk
	Boat or barge serviced (no catwalk)
Operating parameters	Biomass loading
	Species
	Feeding practices
Environmental severity	Sheltered/exposed/open water

Table 8.3. Degree of benthic 'halo' region impact associated with aquaculture operations.

Reference	Reported benthic impacts
Mattsson and Linden (1983)	Species composition changed up to 20 m away from mussel farm
Brown *et al.* (1987)	Species composition changed up to 15 m away from cage edge
Gowen *et al.* (1988)	Species composition changed up to 30–40 m away from cages
Lumb (1989)	Impacts restricted to within 50 m of cage edges and dependent on seabed type
Ritz *et al.* (1989)	Macrofaunal community under the farm adopted an undisturbed condition 7 weeks postharvest of farm stock
Kupka-Hansen *et al.* (1991)	Species composition changed up to 25 m away from cages
Weston (1990)	Farm effects on sediment chemistry evident up to 45 m from the farm; species composition changed to at least 150 m away from cages
Johannessen *et al.* (1994)	No influence of fish farming could be detected 250 m away from cages
Krost *et al.* (1994)	Affected area extended 3–5 m from the fish farm margin
Wu *et al.* (1994)	Impacted area extended to 1000 m where industry uses trash fish as feed and poor water flushing exists
McGhie *et al.* (2000)	Farm wastes largely restricted to area beneath sea cages; most of the sediment organic input from faeces; a 12-month fallowing period was sufficient to return the site to pre-farm oxygen conditions
Morrisey *et al.* (2000)	Large temporal and spatial variabilities depending on water velocities; recovery times estimated at between 3 and 12 years
Dominguez *et al.* (2001)	No effect on physical and chemical sediment characteristics owing to fish farm operation in high average water current velocity (6 cm s^{-1}) site

Aquaculture cages are exposed to the ambient water conditions and organisms present in the water column. Fouling in the sea is a common occurrence, with the amount of suitable, hard-surface structure being the principal limiting factor for colonizing organisms. Fouling of aquaculture cage netting – and its associated parts – is no exception. With the exception of shellfish spat collection for growout, fouling is an undesirable nuisance in aquaculture. None the less, marine cages provide complex structures to create an ecosystem for sessile fouling organisms.

Regardless of cage classification, the presence of substantial water column structure is certain. Additional structure in the water column will serve as a fish aggregating device (FAD) comparable to that often used to attract fish

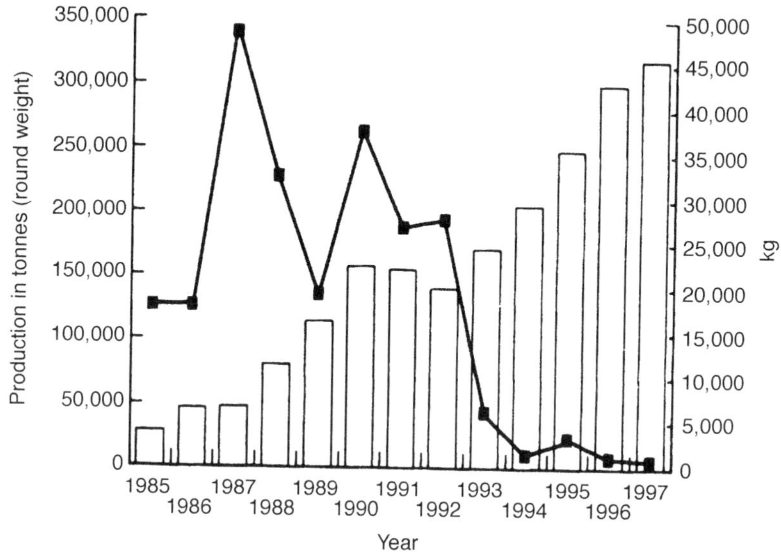

Fig. 8.3. Historical antibiotic use (line plot, right axis), with corresponding annual production (bar graph, left axis), in the Norwegian salmon industry (Willoughby, 1999).

species for commercial and recreational harvests. In fact, most aquaculture sites have a large structural presence in the water column, with numerous individual cages comprising a commercial cage flotilla (Fig. 8.1). Species and quantity of fouling organisms on the cage net will reflect in some way the added nutrients associated with aquaculture operations (Ruokolahti, 1988). Fouling organisms may, in turn, influence the community succession and fish assemblage attracted to the cage (Bray and Ebeling, 1975; Prince and Gotshall, 1976; Love and Ebeling, 1978; Stephens and Zerba, 1981; Laur and Ebeling, 1983; Buckley and Hueckel, 1985).

Biofouling community

Colonization of new structures has been described as a four-phase process that occurs in overlapping time sequence (Wahl, 1989):

- Biochemical conditioning.
- Bacterial colonization.
- Unicellular fouling.
- Multicellular fouling.

Biochemical conditioning occurs immediately following immersion of a new substrate into the water column. Conditioning results in attainment of an equilibrium between the macromolecules present in the water and on the structural surface within a few hours post-immersion (Wahl, 1989). Approximately 1 h post-immersion, a primary film is initiated from bacterial colonization (Wahl, 1989). Bacterial colonization is extremely dynamic with respect to composition (Marshall *et al.*, 1971) and is dictated by species succession, physical disturbance, competition, predation and species availability in the water column (Little, 1984). Unicellular species fouling occurs several days following initial immersion and is comprised mostly of protozoans and diatoms (Cuba and Blake, 1983; Ferreira and Seeliger, 1985). Several days to weeks following immersion, multicellular fouling occurs with invertebrate planktonic larvae and algal spore settlement (Evans, 1981; Hadfield, 1986; Butman, 1987).

The fouling sequence proposed by Wahl (1989) may be based solely on the presence of fouling organisms in the water column at the time of immersion or represent a continuous progression of conditioning required on the structural surface. The latter scenario is most likely. Evidence is provided for bacterial improvement of a hard surface with vitamin production (Lynch *et al.*, 1979) and nitrogen fixation (Goering and Parker, 1972) in preparation for the subsequent phase of colonization. Existence of numerous invertebrate adaptations to allow probing and selection of optimal surfaces with appropriate degrees of biochemical conditioning, bacterial colonization and unicellular fouling (e.g. blue mussel, *Mytilus edulis*, at the pediveliger stage of development (Bayne, 1965)) occur. Species composition of the bacterial film can influence settlement of invertebrate species by creating an avoidance behaviour to certain bacterial species and chemicals (Bonar *et al.*, 1986; Maki *et al.*, 1988, 1990). Physical characteristics of the hard surface can also affect multicellular fouling. Structural colour will attract different types of multicellular forms with invertebrate larvae exhibiting an affinity to darker surfaces (Dahlem *et al.*, 1984). Algal fouling has been observed to be more severe on lighter surfaces (Hodson *et al.*, 2000). Surface roughness, chemical composition and energy have also been observed to influence settlement (Harlin and Lindbergh, 1977; Becker, 1993; Callow and Fletcher, 1994; Roscoe and Walker, 1995).

Numerous shellfish aquaculture industries (e.g. blue mussel longline culture) are based on the successful collection of natural spat that may be considered fouling organisms to other aquaculture sectors (e.g. finfish cage culture). In this case, the aquaculture industry, through the use of collection lines, provides habitat for juvenile invertebrates, which grow to harvestable size (Fig. 8.4). Spat collectors may be polypropylene rope (or polyrope), polyethylene mesh band (or Vexar), Italian sleeving or any other type of mesh tubing (Mallet and Myrand, 1995). Although the species selected for growout is collected from the planktonic community, fouling may still be an issue for shellfish aquaculture. Fouling may be the result of inadvertent collection of predatory species such as starfish and crabs from the water column (O'Neill *et al.*, 1983, Mallet and Myrand, 1995).

Increased fouling of aquaculture cages can be expected owing to the increased levels of nutrients in the farm vicinity due to feeding, excretion

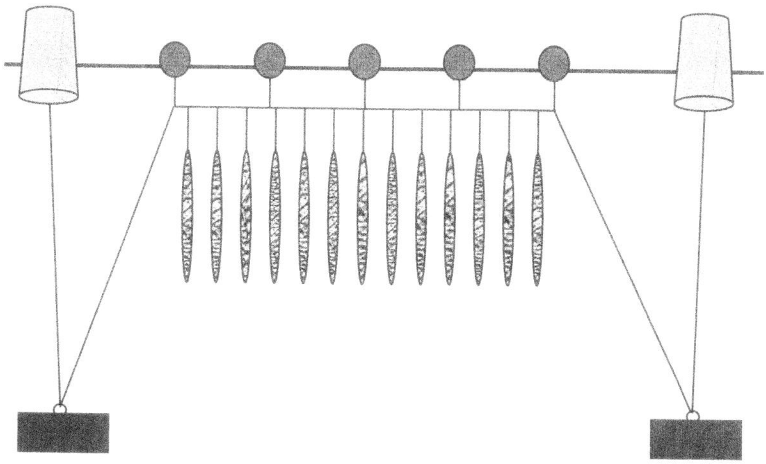

Fig. 8.4. Typical long-line growout system utilized in the Atlantic Canada blue mussel (*Mytilus edulis*) aquaculture industry (Mallet and Myrand, 1995).

and respiration. Fouling of cage netting may be composed of numerous taxa, dependent largely on the amount of available nutrients, season, and composition of regional flora and fauna of settling algal and invertebrate species. In most instances, cages use nylon netting which has a rough micro-surface that enhances colonization by increasing the potential surface area and providing depressions that may protect colonizing species. New net material will also experience the four sequential fouling phases described by Wahl (1989). Following biochemical conditioning, microfouling will occur in the typical bacterial/unicellular sequence. Bacterial populations increase dramatically inside a cage just 10 min post-feeding, owing to increased fish feeding activity. Bacterial populations from the cage netting are dislodged, possibly stressing the fish (Brown *et al.*, 2000). Unicellular fouling may be influenced by light levels, resulting in a high abundance of diatoms on the upper portion of the net owing to more direct lighting, with protists occupying lower mesh surfaces as a result of shading (Hodson and Burke, 1994). Diatom species composition may also change seasonally in a classic community succession dependent on abundance and seasonal suitability for each species (Hudon and Bourget, 1981; Hoagland *et al.*, 1982; Hodson and Burke, 1994). Protists are predatory. Elevated numbers of carnivorous protists on cage netting (Hodson and Burke, 1994) indicate a high level of nutrients near sea cages (Kent, 1980; Hoagland *et al.*, 1982). Multicellular species fouling of the netting, both invertebrate and algal, poses the largest problem to cage aquaculture operations (e.g. blue mussels in Newfoundland and Scotland salmonid aquaculture, and brown and green algae in the Gulf of Mexico, C.J. Bridger, personal observations).

Fouling of the net mesh is undesirable in cage aquaculture. During peak settlement, fouling organisms may rapidly clog net meshes and subsequently limit the flow of high-quality oxygenated water. Net cleaning may be required as often as every 5–8 days for each cage during peak summer fouling (Hodson and Burke, 1994) and can require up to 20–38% of the total aquaculture labour requirement (Huguenin and Ansuini, 1978). Using current nested cage practices (Fig. 8.1), Aarsnes *et al.* (1990) demonstrated that the water flow to interior cages in a flotilla may be 10–20% of that outside the cages owing to increased structural complexity and extensive biofouling. Stresses on cultured animals will increase, growth will decrease, and removal of metabolites will be limited to the point that biofouling may be detrimental to fish health (Brown, 1993). In addition, fouling organisms may harbour fish pathogens (Kent, 1992) which, coupled with increased stress due to poor water quality, could result in disease outbreaks. Increased biofouling could also lead to structural failures and sinking of aquaculture cages (Huguenin, 1997; Bub, 2000), leading to fish escape and subsequent adverse environmental, economic and social implications.

To decrease fouling impacts, gravity cages with a structural presence at the sea surface may have nets changed and cleaned periodically with high-pressure washers on land. *In situ* net cleaning of gravity cages has proved difficult owing to the non-rigid nature of the netting (Hodson *et al.*, 1997). Cleaning cages with rigid (taut) netting has also proved problematic, with removed fouling organisms falling through the mesh and accumulating on the cage bottom (Bub, 2000). Failure to remove fouling debris left on the cage bottom could cause build-up of farm wastes, resulting in decreased oxygen concentrations and stress to the fish stock. Fish in both cage designs will be exposed to fine particulate matter during *in situ* cleaning which may irritate fish gills and disperse potential pathogens (Hodson *et al.*, 1997). Researchers have also reported that some fouling organisms will re-colonize more quickly from remnant rhizoids and reproductive cells following *in situ* cleaning (Nickels *et al.*, 1981), especially from the creation of large amounts of fouling waste and survival of macroalgal remnants in the netting crevices (Hodson *et al.*, 1997).

Cages as fish aggregating devices (FADs)

Natural midwater, mobile structures can be extremely important to fisheries recruitment, and frequently have a very complicated, self-contained pelagic ecosystem associated with them (e.g. Mitchell and Hunter, 1970 (*Macrocystis* sp.); Dooley, 1972 (*Sargassum* sp.)). Artificial structures either in the water column or on the seabed, attract numerous pelagic and benthic fish species (Klima and Wickham, 1971; Wickham *et al.*, 1973). Fisheries managers have utilized artificial structures to increase fisheries production, create new habitat, and improve harvesting efficiencies by concentrating fish assemblages with FADs. Relini *et al.* (1994) observed the effectiveness of a midwater meteorological buoy as a FAD in the Ligurian Sea. Other investigators have observed the ability of midwater FADs to attract pelagic fish species (Workman *et al.*, 1985; Hair *et al.*, 1994). The degree of FAD mobility – stationary versus vertically mobile – has no significant impact on juvenile fish attraction quality, total abundance, or number of species (Hair *et al.*, 1994).

Several studies have documented high concentrations of wild fish in the vicinity of aquaculture cages and their moorings, demonstrating that marine aquaculture facilties are FADs (Collins, 1971; Loyacano and Smith, 1976; Carss, 1990; Anonymous, 1999). Attraction of wild fish to a sea cage structure may be comparable to a shoaling effect found in conjunction with artificial reefs, presumably due to the benefits from the associated aquatic community and increased shelter. 'Acadja-enclos' are used in West African lagoons to attract, capture and grow fingerlings (Hem and Avit, 1994). These are wood and bamboo structures with netting that attract wild fish fingerlings to enter and feed on natural periphyton. Fish eventually grow too large to leave the 'acadja-enclos' (owing to the small mesh size employed) and are captured upon reaching optimal harvest size.

Designing ecologically compatible artificial reef habitats near fish farms may attract specific fish and invertebrate/algal species that can serve as biological and physical filters for farm-associated nutrients and wastes. Cage mooring systems may be ecologically designed to serve not only to hold the cages in place but also to increase structural complexity, provide additional habitat for invertebrates, and act as low profile FADs for benthic fish species. Bougrova and Bugrov (1994) developed a structurally enhanced cage mooring system to increase wild reef-fish populations and reduce environmental impacts. Porter *et al.* (1996) stocked striped mullet (*Mugil cephalus*) in small cages below a commercial fish farm cage to reduce the environmental impact associated with aquaculture operations on the seafloor. Proper design of these types of FADs could provide new habitats and enhance natural ecological services to minimize the impact of aquaculture. Artificial reefs will increase the available surface area for microbial colonization in the oxygenated water column above the seafloor and facilitate aerobic microbial activities (Angel and Spanier, 1999). Members of the newly established reef community will consume associated farm wastes such as detritus, particulate organic matter and bacteria (Manahan *et al.*, 1982; Reiswig, 1985; Brusca and Brusca, 1990; Manahan, 1990).

In addition to enhancing the natural ecosystem and attracting wild fish, compelling evidence has been collected using an advanced telemetry system (Bridger *et al.*, 2001b) that showed the existence of a high degree of aquaculture site fidelity by aquaculture escapees. Bridger *et al.* (2001a) observed escaped triploid steelhead trout (*Oncorhynchus mykiss* Walbaum) displaying a high degree of site fidelity following an on-site release (Fig. 8.5a). Escapees also exhibited directed homing back to the growout site when released off-site (Fig. 8.5b), with subsequent fidelity to the growout site (Fig. 8.5c). Decreased fidelity was observed during the winter season (Fig. 8.5d). Cage fidelity – and the ability to determine fish behaviour from telemetry research – can be

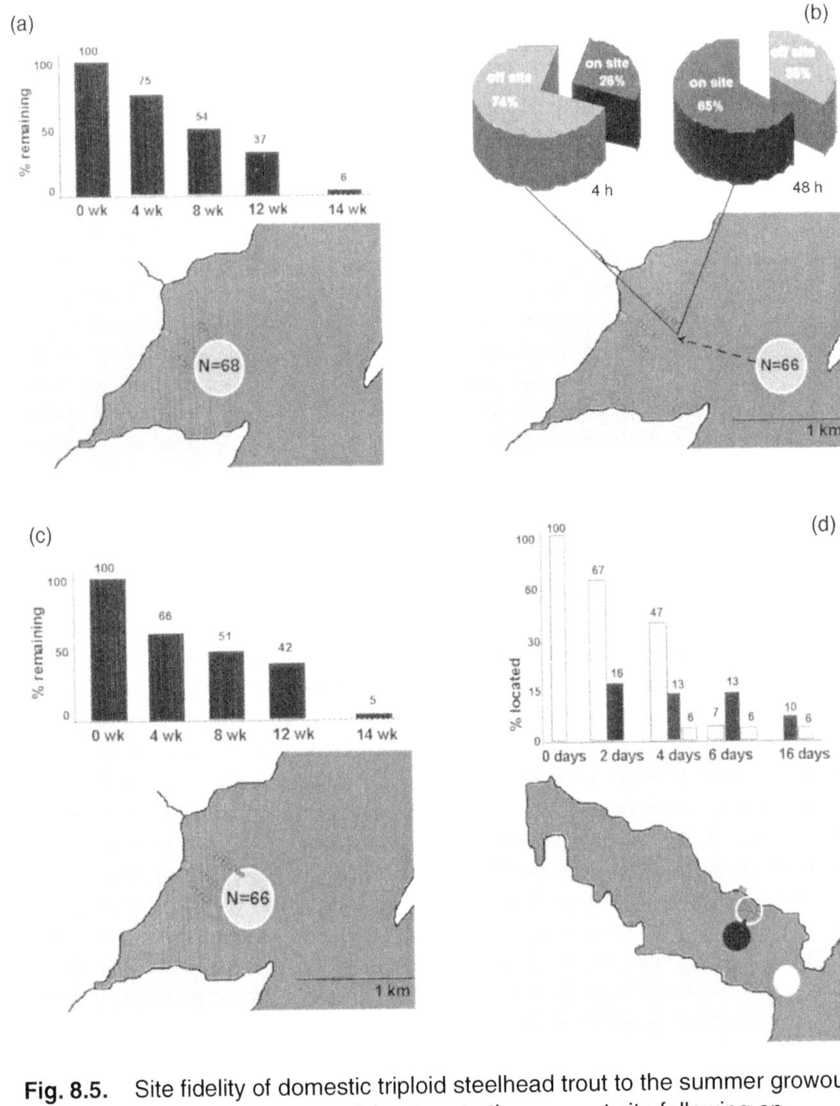

Fig. 8.5. Site fidelity of domestic triploid steelhead trout to the summer growout site following an on-site release (a), return to the growout site following an off-site release (b), subsequent fidelity of returned off-site released steelhead to the growout site (c), and fidelity of winter-released steelhead trout to the overwintering site (d) (adapted from Bridger *et al.*, 2001a).

used to mitigate potential aquaculture escapee impacts on natural ecosystems (Hansen *et al.*, 1991; Hutchinson, 1997) and allow for escapee recapture (Anonymous, 1999). Recovery of escapees will also decrease the economic impact of unfortunate loss of farm stock by returning fish to cages.

Aquaculture facilities supply additional nutrients to the surrounding area (Costa-Pierce, 1997b). Nutrient inputs may increase the quantity of otherwise limiting nutrients in oligotrophic areas, enhancing growth and structure of the biofouling community, and supplementing the diet of attracted fish. Studies have shown that freshwater aquaculture can cause the increased growth rate of wild fish species, in part because of the natural ingestion of uneaten feeds and faeces (Kilambi *et al.*, 1976; IOA *et al.*, 1990). Spanier *et al.* (1985) enriched a FAD with frozen fish meat to enhance recruitment and compared the results with a FAD that was not enriched. Enrichment decreased the time required for the artificial reef to attain its carrying capacity. Similarly, stomach content analyses of fish attracted to aquaculture facilities portrayed a large dependence of attracted individuals on excess farm feed (Bridger and Garber, 2002).

Ecological Engineering for Marine Cage Aquaculture

Placing a salmon cage in the ocean may be considered a throughput system (Folke *et al.*, 1994), requiring a large ecological footprint for production, and releasing its wastes into the environment without treatment, thereby decreasing overall ocean productivity (Naylor *et al.*, 2000). However, by understanding the ecological structure, relationships and functions of the natural ecosystem, and the farmed species, aquaculture may utilize wastes for increased efficiency (Folke *et al.*, 1998) and maximize economic returns using the concept of ecological engineering (Mitsch and Jorgensen, 1989).

'Ecological engineering implies the design of human activity with its natural environment for the benefit of both, and it is, therefore, different from engineering and technology that try to substitute or conquer the natural environment' (Folke and Kautsky, 1992). Polyculture of shellfish (e.g. *Mytilis edulis*) and seaweeds (e.g. brown kelp, *Laminaria digitata*) near salmon cages (Fig. 8.6a) will decrease the impact and quantity of wastes and increase efficiencies and overall productivity (see Chapter 7). Polyculture operations have also been demonstrated in tank (Buschmann *et al.*, 1996; Neori *et al.*, 2000), pond (Shpigel *et al.*, 1997), and cage systems (Stirling and Okumus, 1995). Nearshore intensive cage aquaculture systems can be integrated with sustainable land use and enhanced recreational fisheries (Fig. 8.6b).

Aquaculture in Mangroves*

The pedagogy shift

Only in the last 20 years has the world begun to value wetlands as natural assets to society. Until recently, mangroves and marshlands were considered

* The topic of mangroves and various impacts on mangrove ecosystems is considered in detail in Chapter 9.

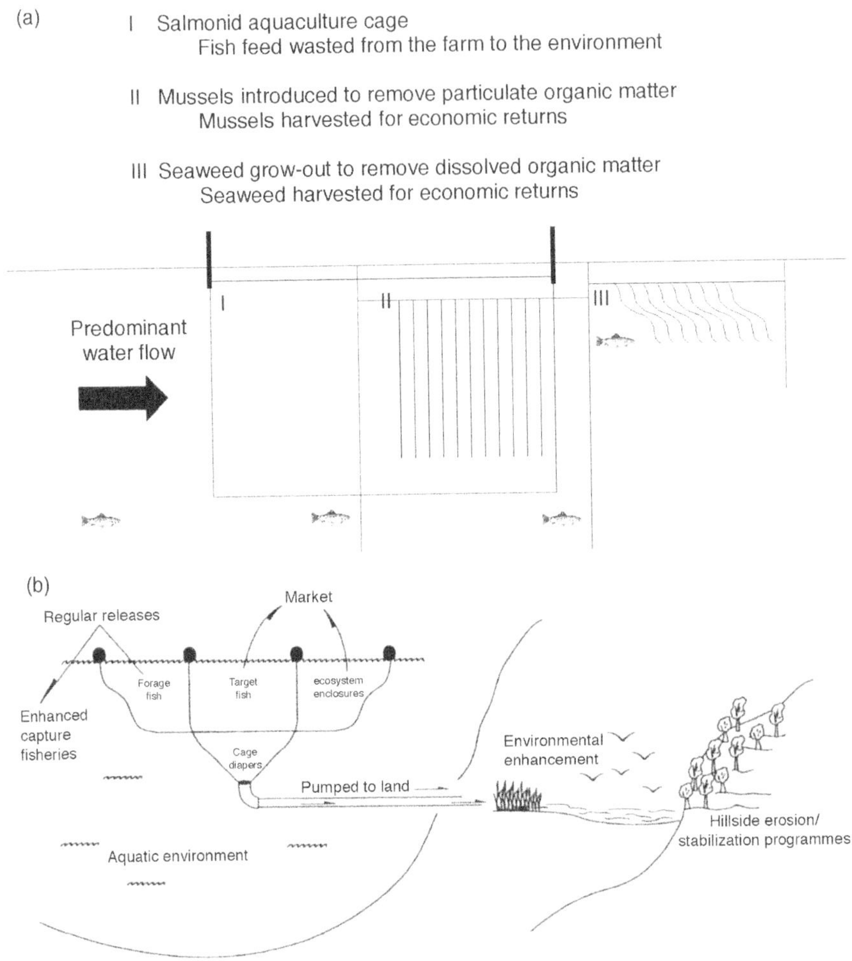

Fig. 8.6. Ecological engineering principles applied to the salmonid aquaculture sector to develop a sustainable aquaculture industry and decrease the ecological footprint of salmonid aquaculture operations (a) and integration of cage aquaculture, waste treatment, sustainable land use and restoration, and enhanced recreational fisheries in mine pit lakes in Minnesota, USA (b) (Costa-Pierce, 1997a).

murky, dark wastelands that were hazards to society. For example, in 1951, Florida newspapers reported '300 homes blackened' and 'two men killed' by 'mangrove root gas' in Miami (*Miami Herald*, 15 November 1951, and the *Miami News*, 28 July 1961). As late as 1969, a short-stature mangrove forest in south Florida was considered 'a form of wasteland' (Lugo and Snedaker, 1974). Until the early 1970s the United States Department of Agriculture Soil Conservation Service considered mangrove areas only for their relative suitability for crops, pastures, woodland, wildlife or other uses. Wetlands were used as acceptable disposal sites for the most contaminated types of refuse and industrial chemicals. As such, vast areas of mangroves were cut, drained, and converted to dry land.

For centuries mangroves in tropical countries were valued not only for their timber but also for food production and soil stabilization. In Southeast Asia, traditional mangrove agroforestry was conducted using the indigenous knowledge that the plants could stabilize dikes, pond banks and transportation pathways (Lugo and Snedaker, 1974; Fitzgerald, 2001). But as populations grew, mangroves began to be looked upon as a vast timber resource to be exploited. Massive deforestation of the mangrove forests of coastal Asia resulted. Macane (1968) published black and white photos of forests of *Rhizophora mucronata* with trunks '30–40 cm in diameter and trees *c.* 35 m tall', taken south of Ranong, Thailand. All of these forests are now gone.

Today, mangrove ecosystems are valued not only as a timber resource but for their ecosystem services in soil building, watershed stabilization, coastal protection, fish habitat and nursery areas, and as habitats for organisms that rural villagers gather as protein sources. When valuation of all of the ecosystem services mangroves provide is accomplished, mangrove forests left standing are more valuable than for almost any alternative use. Ronnback (1999) estimated that the annual market value of capture fisheries supported by mangroves ranges from US\$750 to US\$16,750 ha^{-1}. In Bintuni Bay, Indonesia, the intact mangrove forest is worth US\$4800 ha^{-1}, while the timber is worth US\$3600 ha^{-1}. Not cutting the forest gives local people there US\$10 million a year in ecosystem services, and protects fisheries worth US\$25 million a year (Ruitenbeek, 1992).

Ecosystem services provided by mangroves

Mangrove ecosystems are important, transformative interfaces between land and sea. The mangrove ecosystem imports nutrients from terrestrial systems and exports organic matter as detritus (both dissolved and particulate organic matter) to marine ecosystems. Mangrove litter production has been measured at 896 g dry wt m^{-2} $year^{-1}$ (Lugo and Snedaker, 1974), adding 224 g C m^{-2} $year^{-1}$ to wetland soils and waters. Mangrove leaves are colonized by fungi such as *Phycomycetes*, bacteria and nematodes, etc., upon deposition into water. Mangrove leaf litter provides an important nutrient base for food webs leading to commercially important nearshore fisheries.

Value to fisheries

Mangrove swamps serve as vital nursery grounds for economically important nearshore species such as snappers (Lutjanidae), jacks (Carangidae) and mullet (Mugilidae). Snedaker (1978) estimated that upwards of 90% of nearshore marine species can be found in mangrove areas during one or more parts of their life cycles. Robertson and Blaber (1992) found that from 26 to 197 fish species were reported to use mangrove habitats in the Indo-west Pacific and tropical Atlantic. The success of nearshore fisheries in many tropical regions depends as much for fish recruitment success on the physical environment in mangrove habitat as on the detrital foods produced in that habitat. Robertson and Blaber (1992) state, however, that '. . . the role of mangroves in estuarine dependence by fish remains to be clarified'. The three main hypotheses to explain the high densities of fish species and biomass in mangroves are:

- Turbidity reduces the effectiveness of large predators.
- Mangroves are important feeding sites for fish.
- The structural complexity and increased living space gives shelter from predators.

Robertson and Blaber (1992) stated that 'These three hypotheses . . . are probably all important in explaining the importance of mangrove habitats to fish. The relative significance of each hypothesis . . . will vary depending upon the fishes in question and the particular nature of each mangrove habitat.'

Shoreline protection

Because of the high sedimentation rates in the mangroves, they build land. As such, they are an important pioneer species extending into the coastal zone and connecting marine ecosystems to the edge of the rainforest. Removal of mangroves has been blamed for the increased erosion and severity of storm impacts in Bangladesh and other areas of the world. Mangroves have been used for years to protect seaward-facing areas from erosion, and also for protecting causeways, railroads and embankments in Florida, Hawaii and Sri Lanka. Mangroves have also been used for centuries to stabilize the banks of traditional polyculture ponds in Java (Macane, 1968; Fitzgerald, 2001).

Habitat

There are a variety of habitats in the mangroves (Robertson and Alongi, 1992):

1. The forest canopy. This is essentially a terrestrial environment rich in epiphytes and orchids, which resembles a rainforest with high diversity. Many species of birds are found nesting and caring for young.
2. The soil. Some organisms move away from seawater and up mangrove trees upon inundation. Others bury in the soil when the tides rise. Large numbers of snails abound.
3. Root holes and clefts. A freshwater environment with a large number of insect larvae.
4. Permanent or semi-permanent pools. These contain a variety of shrimp, crabs, fish, snails and frogs normally found in the canals and river branches in the mangroves.

5. Branches and aerial roots of mangroves. Dominated by filter feeders such as barnacles, sea squirts, oysters, mussels, etc. An extremely high biomass is present here.

Use of mangrove ecosystems as repositories for shrimp pond wastes

Owing to the well-known abilities of marshes and mangrove ecosystems to absorb nutrients and tolerate adverse environmental conditions, there has been a great deal of interest in using wetland ecosystems worldwide to treat humanity's wastes. Mangrove ecosystems have been proposed as alternative, low-cost sewage treatment systems (Nedwell, 1975; Clough, 1983) and, more recently, as a possible repository of effluents and sediments from shrimp aquaculture (Rajendran and Kathiresan, 1997).

Mangroves grow in waterlogged, saline, anaerobic soils and have developed remarkable adaptations to deal with these conditions, such as oxygen-transporting prop roots, pneumatophores and resistant, viviparous seeds. The strongly anaerobic environment makes the mangrove ecosystem a unique nutrient regime. Heavy metals are strongly bound to anaerobic sediments, either by adsorption, binding or precipitation as insoluble sulphides. Pesticides are also strongly adsorbed on to mangrove sediments, especially those with high clay and humic acid contents. Nitrate is almost absent and iron exists as Fe^{2+}. The low pH and redox potentials lead to release of phosphate. There is evidence that these characteristics lead to some mangroves being nutrient limited by both N and P.

When soluble P is added to mangrove soils it is rapidly adsorbed, increasing the exchangeable (or labile) phosphorus. However, the capacity of mangrove soils to immobilize phosphorus is related to the number of available exchange sites, and these sites can be saturated completely. A reduced adsorption maximum was found for mangrove soils that had received sewage effluent for 20 years (Clough, 1983). Holford and Patrick (1979) concluded that mangrove soils had a limited short-term capacity to reduce phosphorus in sewage effluent below that required for significant biological activity in receiving waters.

Shrimp pond and sewage effluents contain BODs of 10–70 mg l^{-1} (Rajendran and Kathiresan, 1997). Shrimp pond effluents will increase bacterial activity and increase the anaerobic nature of the sediments. While mangroves are well adapted to anaerobic environments, there is evidence that anaerobic conditions are the major factor controlling plant growth. Additional organic inputs would increase an already stressful situation for the plants and decrease growth. However, addition of soluble nutrients could possibly help ameliorate nutrient stress in mangroves (Clough, 1983). Rajendran and Kathiresan (1997) found that diluted (70%) shrimp pond effluent increased mangrove shoot biomass production, but that full-strength effluent decreased biomass production. However, while mangrove ecosystems could serve as traps for nutrient, pesticide, or heavy-metal-rich sediments, there could be an adverse effect on the mangrove food web.

Reasons for mangrove destruction

Mangroves represent one of earth's most endangered ecosystems. There are many threats to continued survival: clear cutting and land reclamation, pollution, diversion of freshwater, and coastal aquaculture.

Uncontrolled population growth

Demands for resources (food, water and energy, etc.) are driven by the addition of nearly 90 million persons a year to the Earth. Added to this unparalleled increase in population is the unprecedented growth in affluence in East Asia. Population growth is tied to urbanization, ecosystem destruction, food, water, energy, health, public safety, user conflicts and community stability issues. Curbing population growth is the key issue for future environmental and social sustainability of the mangrove ecosystem and the peoples who depend upon mangroves for survival.

In the United States, the net increase of 3 million people per year is blamed for the loss of millions of hectares of farmland to urbanization and a lack of adequate water resources. Immigration is responsible for 60% of US population growth, a portion that will rise to 90% in coming decades if current immigration policies continue (Pimental, 1997). In developing nations, the population/food nexus is even more alarming. Over the next 20 years, of every 10,000 new births, only 50 will be in what have been called rich countries (Swaminathan, 1992). 'A poverty curtain divides the world materially and philosophically. One world is literate, the other largely illiterate; one industrial and urban, the other predominately agrarian and rural; one consumption oriented, and the other struggling for survival' (Swaminathan, 1992). The rich countries today consume about 20 times more resources per capita than the poor countries (Costa-Pierce, 2002). But as economies have strengthened, especially in Southeast Asia, resource consumption rates are rising dramatically. A group of scientists have banded together to ask world leaders to initiate incentives to reduce family size and conserve natural resources in order to maintain an ethical and decent standard of living (Pimental and Dodds, 1997).

Loss to rapidly expanding urban areas

Mangroves are still considered wastelands in many economically depressed countries, which are now rapidly developing their coastal lands (as Europe and North America did before them). In earlier times, mangroves were valuable as sturdy poles in Arabian towns, where the idea of the skyscraper was born. Macane (1968) stated: 'Much of downtown Singapore is built on piles of mangrove trees, chiefly *Rhizophora*'. In the past in Southeast Asia – and today in many areas of coastal Africa, South Asia, and Latin America – mangroves are still cut for firewood and charcoal. Macane (1968) stated that: 'The extensive estuarine forests near Suratthani in peninsular Thailand have been virtually exterminated for this purpose'.

There is a massive amount of rural to urban migration occurring in the developing nations. Most of this movement is from inland rural locales to coastal areas. The rapid expansion of coastal cities throughout the world is exerting major impacts on all coastal ecosystems. The population of coastal Calcutta, India, increases by about 1000 persons a day. China's Vice Minister of Construction has announced a plan to build 600 new cities by 2010, doubling the 633 already in existence. Rapid urbanization of the coastal zone threatens the livelihoods of all remaining peoples who depend on mangroves. In Bangladesh, it has been estimated that one-third of the population is dependent on mangroves for their income (Robertson and Alongi, 1992). Extensive destruction of the mangroves along the coasts of South and Southeast Asia may be one of the reasons for the repeated catastrophic flooding and damage by storms reported in the past decade. There are also reports of extensive mangrove destruction caused by oil pollution, herbicides and war.

Losses to aquaculture development

Coastal aquaculture development (fish and shrimp ponds) has caused a reduction in mangrove areas in Asia and the Americas. The extent of destruction due to aquaculture is, however, debatable and is likely overstated (Boyd and Schmittou, 1999; Chapter 9). Primavera (1991) estimates that Philippines' mangroves declined from 400,000–500,000 ha in the 1920s to 140,000 ha by 1990, with 60% of the destruction due to coastal aquaculture of milkfish and shrimp ponds. In contrast, Larsson *et al.* (1994) reported that there was no evidence that mangroves have been cut down for aquaculture development in Columbia. Binh *et al.* (1997) reported that from 1983 to 1991 one district of Vietnam (the Ngoc Hien district) lost 48% of its mangroves to shrimp pond development.

There are hopeful signs that governments and shrimp farmers are realizing that it is not in their best economic interest to destroy mangroves, and some are legislating against locating aquaculture ponds in mangrove areas. Ochoa (1997) described how one effort involving a non-governmental organization, a local community and a shrimp aquaculture company negotiated a deal with the following conditions:

- All shrimp ponds were to be located 50 m behind the mangroves.
- No alteration of mangrove cover would be allowed.
- No alteration of natural water flows by dams, walls or diversions was permitted.

- Traditional uses and access to mangrove areas would be guaranteed to the local peoples.
- Ecotourism activities and collaborative research were to be encouraged.

Mangrove Restoration Aquaculture

Mangrove restoration can be successful even if the trees and hydric soils have been removed. Lewis and Streever (2000) provided the essential technical background needed to develop mangrove restoration programmes. Lewis *et al.* (1995) emphasized, however, that defining the goals of mangrove restoration – selecting which ecosystem traits and natural functions are to be restored – from the outset is essential since complete restoration to pristine conditions is often unattainable in modern times.

It is important to determine first – before any mangrove restoration aquaculture and planting are conducted – the reasons for mangrove losses, to remove those causes, and to work with natural recovery processes to re-establish mangrove ecosystems (Lewis and Streever, 2000). Erftemeijer and Lewis (1999) reviewed the failure of numerous multimillion dollar mangrove restoration efforts on unstable mudflats, where planting programmes were initiated before the socioeconomic, geomorphometric and nearshore oceanographic reasons for loss had been determined. The most important aspects of mangrove restoration aquaculture are: site soil/sediment characteristics and stability, siltation rates, wave and tidal exposures, depth of tidal inundation, height of the water table, freshwater sources, presence of pests and availability of propagules (Hamilton and Snedaker, 1984; Olsen and Arriaga, 1989; Cintron-Molero, 1992; Field, 1996; Turner and Lewis, 1997). Mangrove ecosystems can self-repair in 15–30 years if the tidal hydrology is restored and adequate seed supplies (propagules) are available from nearby mangrove forests. For example, Soemodihardjo *et al.* (1996) found an area of Tembilaha, Indonesia, that had more than 2500 natural mangrove propagules available per hectare.

Lewis and Marshall (1997) listed five steps to achieve successful mangrove restoration. One needs to:

1. Understand mangrove reproduction patterns, propagule distributions and establishment.
2. Understand tidal hydrology that controls propagule distribution and establishment.
3. Assess the physical environmental modifications needed.
4. Design hydrological restoration.
5. Establish a mangrove planting programme if steps 1–4 will not adequately restore the ecosystem.

If the natural mangrove seed supplies have been removed, mangrove restoration aquaculture programmes can be designed. Planting of ripe

Rhizophora propagules directly by hand is very successful. Direct planting for *Avicennia* and *Laguncularia* is not possible however, since those genera need to be rooted with the cotyledons exposed, and they need to lose their seed coat before becoming established. *Avicennia* and *Laguncularia* are easily produced in nurseries, however, and are planted as 1-year-old trees at 10,000 ha^{-1} (Lewis and Streever, 2000). At 50% survival, a mature planted forest will have 1000 trees ha^{-1}. In 15 years, planted mangroves will be 5 m high with a closed overhead canopy, and will have well-established prop roots and pneumatophore networks (Lewis and Streever, 2000).

If no natural seed sources exist and hydrology must be completely restored, mangrove restoration projects can cost as much as US$200,000 ha^{-1}. However, costs can be as low as US$250 ha^{-1} if hydrological restoration can be achieved without major excavation and with adequate natural seed sources (Lewis and Streever, 2000).

In contrast to the pioneer stages of shrimp aquaculture development (the period from the 1960s through the 1980s), mangrove restoration projects today are important parts of marine animal aquaculture projects worldwide. There are numerous examples of commercially viable silvofisheries which integrate mangrove forestry and mariculture operations (Binh *et al.*, 1997; Fitzgerald, 1997, 2002; Primavera and Agbayani, 1997; Johnston *et al.*, 2000). Stevenson *et al.* (1999) stated that:

> The days when disused shrimp ponds were cited as evidence to denigrate and point accusatory fingers towards the shrimp industry are gone. Instead a real determination to tackle the issue appears to be emerging and all those involved should continue to focus on solving the issue and resist the temptation to lay blame. That is not to say that the reasons for disuse should be ignored; these are very important and should be examined as part of a logical evaluation . . ..

Seagrass Restoration Aquaculture

Seagrass ecosystems are extremely complex environments, providing habitats for numerous trophic levels and serving many important ecological roles. Seagrass beds provide habitats for fish – both juvenile and adult forms – and invertebrate populations (Virnstein *et al.*, 1983; Heck *et al.*, 1995). Tidal seagrass wetlands also stimulate microbial activity enhancing nutrient recycling to the sediments (Blaabjerg *et al.*, 1998; Risgaard-Petersen *et al.*, 1998). Plant stems provide physical structure to the environment, which attenuates wave and tidal action thereby enhancing sediment deposition (Harlin *et al.*, 1982) while decreasing sediment erosion and reducing water column turbidity (Ward *et al.*, 1984).

Seagrass ecosystems are experiencing rapid destruction and deterioration worldwide. Anthropogenic changes to the environment and influences on regional hydrology are having severe impacts on seagrass abundance and distribution (Fourqurean and Robblee, 1999). Destruction of seagrasses also may be associated with natural climate cycles (Marba and Duarte, 1995) and conditions (Hicks *et al.*, 1998).

Successful seagrass restoration requires improvement of water quality to a level appropriate for plant health to ensure growth and survival (Greening *et al.*, 1996; Janicki and Wade, 1996). Numerous restoration and management programmes transplant from healthy natural populations to target regions for restoration or propagate seedlings artificially, which are later transferred to the target restoration area (Fonseca, 1994; Fonseca *et al.*, 1998). Several methods (reviewed by Phillips, 1990) have been developed for transplanting and restoring seagrass beds:

- Horizontal rhizome method (Davis and Short, 1997).
- Staple method (Fonseca *et al.*, 1985).
- Peat pot methods (Fonseca *et al.*, 1990; Harrison, 1990).
- Seedlings (Harrison, 1991; Orth *et al.*, 1994).
- Seed pelletization and seed-impregnated planting meshes (Granger *et al.*, 1996).

Selection of the most suitable method is dependent on species, transplanting site, economics and logistics (Heidelbaugh *et al.*, 2000). Each of these methods is time consuming and in most cases cost prohibitive if large areas of seagrass beds are to be restored.

Researchers at the University of Rhode Island, USA, are currently developing a new method to plant eelgrass (*Zostera marina*) to minimize the level of effort necessary to ensure success. A seagrass seed methodology is being developed whereby numerous seeds may be collected with a modest effort (approximately 50–55 person-hours to collect 500,000 seeds (Granger *et al.*, 2000a,b)) and planted together to decrease the necessary labour force and project costs. The novel technique utilizes a boat-pulled sled that automatically deposits eelgrass seeds below the sediment (Fig. 8.7). Development of such a sled allows enormous hectarage to be seeded in a shorter time and with much less labour than traditional restoration methods. Additionally, seeds are planted directly in the first few centimetres of sediment, thereby reducing the likelihood that they will be washed away or consumed by predators (Granger *et al.*, 2000b; Anonymous, 2001).

Salt Marsh Restoration Aquaculture

Before European colonization, the United States contained an estimated 89.5 million ha of wetlands (Dahl, 1990). Frayer *et al.* (1983) estimated the rate of wetland destruction at 185,400 ha year^{-1} between the mid-1950s and the mid-1970s. In 1986, the United States Emergency Wetlands Restoration Act (Public Law 99-645) was enacted to stem the rate of wetland destruction, conserve remaining areas, and restore damaged wetland ecosystems. From the mid-1970s to the mid-1980s, the estimated rate of wetland loss declined to

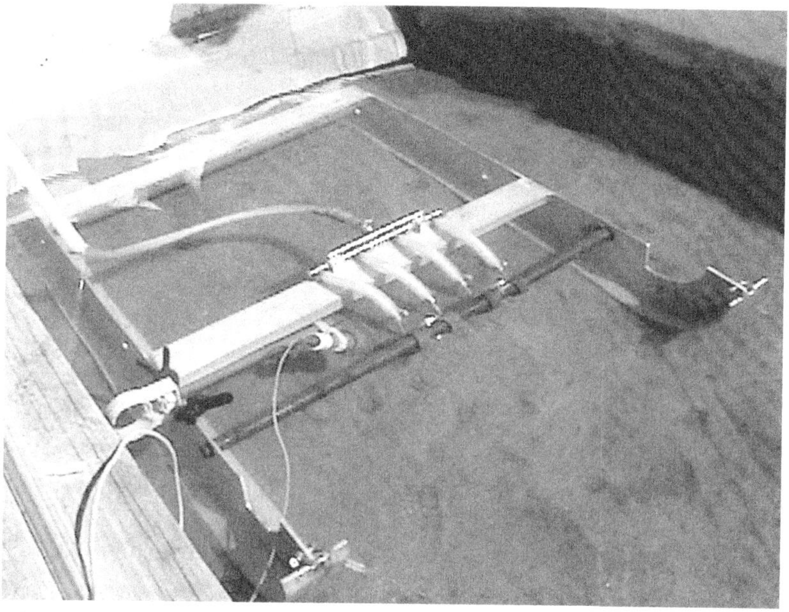

Fig. 8.7. University of Rhode Island, USA, boat-pulled sled to automatically deposit eelgrass seeds below the sediment (courtesy Scott W. Nixon, University of Rhode Island, Kingston, Rhode Island, USA).

117,400 ha year^{-1} (Dahl and Johnson, 1991). It declined further to 23,700 ha year^{-1} between 1986 and 1997 (Dahl, 2000). Today, only 42.7 million ha of wetlands are left in the United States, less than half of the original amount (Dahl, 2000).

Salt marsh restoration aquaculture methods are readily available (Kusler and Kentula, 1990; Thayer, 1992; Matthews and Minello, 1994; Dreyer and Niering, 1995; Zedler, 1996; Broome and Craft, 1998; Copeland, 1999). Similar to mangrove restoration methods and protocols, restoration of tidal hydrology and removal of obstructions to normal freshwater and tidal flows are the most important criteria for restoration success. Establishment of secondary and tertiary creek channels is also essential to restoration success (Weinstein *et al.*, 1997; West and Zedler, 2000). Many salt marshes have been filled, so restoration of the proper elevation of the marsh substrate is critical to establishment of wetland plants. If nearby natural seed sources for salt marsh wetland plants exist, restoration of the proper marsh elevations achieved, and tidal flows and salinities restored, salt marshes will re-establish naturally within 5–15 years (Sinicrope *et al.*, 1990; Dreyer and Niering, 1995; Simenstad and Thom, 1996; Broome *et al.*, 1999).

If natural seed sources of *Spartina* sp. are not available, planting of nursery-produced marsh grasses can be conducted (Broome and Craft, 1998). Plants must be compatible with the salinity, soils and tidal energy at the site. *S. alterniflora* is planted most commonly along the Atlantic and Gulf coasts

(Broome and Craft, 1998), *S. foliosa* in California (Zedler, 1996) and *Carex lyngbyei* in the Pacific Northwest (Simenstad and Thom, 1996). Invasion by *Phragmites* can be eliminated within 5–10 years by regular flushing with water greater than 18 ppt salinity (Rozsa, 1995). Planting materials for propagation originate from natural marshes nearby the restored site to eliminate genetic concerns about invasive strains. Seeds have been collected from natural sites and cultivated in greenhouses (Garbish *et al.*, 1975). Tissue culture methods have also been developed (Li *et al.*, 1995). Personnel at the United States Department of Agriculture Plant and Materials Center (PMC) at Golden Meadow, Louisiana, evaluate, select and release plants for the conservation, protection, restoration and enhancement of coastal wetlands. The objectives of the PMC are to:

- Develop improved plants which will persist in a changing coastal marsh environment.
- Develop culture techniques for the successful use of improved plant materials.
- Release and provide for the commercial increase of improved plant materials.
- Promote the use of tested and proven plant materials to solve coastal wetland conservation problems.
- Serve as a learning centre to stimulate and foster an understanding of the importance of plants in the environment and their role in controlling erosion.

Plant spacing on 60 × 60 cm centres provides plant cover in one growing season if the marsh is protected from tidal and freshwater erosion (Broome *et al.*, 1992). Fertilization at 100 kg N ha^{-1} with ammonium sulphate and 50 kg P ha^{-1} of superphosphate are recommended at transplanting for quick maturity. Restored salt marshes often lack adequate organic matter owing to oxidation and/or removal by dredging or sea level rise (Zedler, 1996). Organic amendments are used after assessing soil conditions and hydrology (Broome *et al.*, 1999).

Oyster Reef and Shellfish Habitats

Restoration of shellfish beds and oyster reefs provides valuable substrate which will lead to the addition of biodiversity to an area, beneficial impacts on water quality, and, of course, food. Bivalves filter large quantities of water, removing particulate matter and nitrogen from the water column. It has been estimated that the 26,400 tonnes of northern quahog clams (*Mercenaria mercenaria*) harvested annually from the Narragansett Bay, Rhode Island, USA, filter 21% of the water during each tidal cycle (Rice, 2001). Newell and Ott

(1999) estimated that at the beginning of the 20th century, oysters in Chesapeake Bay, USA, would have filtered the entire bay in 6 days. By 1985 – due to overharvesting – these estuarine filters would have taken over a year to accomplish purification.

Oyster reef restoration has been practised in the Chesapeake Bay and the Gulf of Mexico for more than 100 years (Berrigan *et al.*, 1991) using simple methods. Oyster larvae attach to shells of other oysters making thick, elevated deposits of aggregated shells (reefs). There are large, 10 km^2 oyster reefs off Louisiana and Mississippi having oyster shell deposits greater than 10 m deep (Berrigan *et al.*, 1991). These reefs are essential habitats for a wide variety of marine organisms that serve as essential forage species for valuable marine fish populations. The principal problems and threats to these reef ecosystems are mining/dredging for shell and building materials, bottom trawling, sedimentation, pollution and altered salinity regimes (removal of freshwater flows).

Suitable oyster growout areas may lack sufficient bottom substrate for productive natural populations to thrive. Oyster restoration aquaculture replenishes bottom substrate with cleaned, unencrusted shell (cultch) for oyster larvae in the water column (meroplankton) to settle (set) and grow out as adult filter feeders. In addition, restoration aquaculture efforts have evolved into community-based initiatives called oyster gardening.

In the 1980s the Virginia Institute of Marine Science, USA, encouraged a group of amateur shellfish growers to culture oysters using gardening techniques. The Chesapeake programme has grown to 2000 participants, spread to neighbouring states – including to a large programme in Maryland, USA, that can be seen at http://www.mdsg.umd.edu/oysters/garden/index.html – and has been newly adopted in Mobile Bay, Alabama, USA. Oyster gardening begins with small seed oysters obtained from an aquaculture hatchery where broodstock are spawned. Eggs are fertilized in the water column, and planktonic larvae settle on to provided cultch (crushed gravel or broken shell), where they are grown to the juvenile stage. Floating cage baskets made of PVC frames with plastic or metal mesh netting are used to hold juvenile oysters. The baskets are suspended just beneath the water surface. Baskets have lids to exclude predators. Mesh sides are cleaned regularly by gardeners to permit good water exchange. At 2 months of age, seed oysters are distributed to gardeners for planting in areas deemed suitable for oyster growout.

An innovative restoration aquaculture project in Maryland, USA, is based on the hypothesis that seagrasses have declined because of declines in oyster populations in Chesapeake Bay. Decline in oyster filtering capacity due to population decline would lead to greater water turbidity and consequent declines in seagrasses. As seagrasses decline, they are less able to trap sediment, which would lead to an even greater reduction in water clarity owing to increased sediment resuspension (Leffler, 2001). Researchers are now working on methods to restore both seagrasses and oysters (Leffler, 2001).

Aquaculture Ponds as Managed Wetlands

Aquaculture ponds are actually managed wetland ecosystems that preserve ecosystem services, the local economy, cultural traditions and community values.

Louisiana has lost millions of hectares of coastal wetlands (Dahl, 2000). In the past 50 years, over 68,000 ha of red swamp crawfish–rice wetlands have been created. From mid-autumn to mid-spring, growers cultivate red swamp crawfish (*Procambrus clarkii*), and in the summer they grow rice in the ponds. Huner (2000a,b) demonstrates that the crawfish–rice aquaculture ecosystem is essential habitat for colonial water birds in Louisiana, arguing that crawfish wetlands are the focus of a stunning recovery of these populations, which were previously considered endangered and threatened. The Lake Martin rookery east of Lafayette, Louisiana, USA, is one example. The rookery is surrounded by over 13,600 ha crawfish wetlands which support robust nesting populations of egrets, herons, ibis and spoonbills, which have rebounded dramatically. Catfish farmers in the Mississippi delta region have similarly been witness to an extraordinary population boom in double-crested cormorants, which have been implicated in economic losses. In contrast, Huner (2000b) stated that problems with water birds in crawfish aquaculture may be unjustified, stating that: 'As yet, no definitive studies have been funded to determine whether or not "perceived" problems are real'.

Bird tourism, recreation, and sport hunting incomes are worth millions of dollars a year in many rural areas. Instead of looking for ways to destroy valuable birds, scientific analyses of the entire aquaculture wetland ecosystem and actual farming losses are needed. This information could serve as the basis of new, more creative, interactions with the public and wildlife managers so that aquatic farmers can both retain profits and enhance nature's goods and services. One means may be to provide government and/or private subsidies to assist aquaculture operators to initiate activities in the restoration and management of expanded populations of colonial water birds on their farms.

References

Aarsnes, J.V., Rudi, H. and Loland, G. (1990) Current forces on cages and net deflection. In: *Engineering for Offshore Fish Farming*. Thomas Telford, London, pp. 137–152.

Alabaster, J. (1982) Survey of fish-farm effluents in some EIFAC countries. In: Aalabaster, J. (ed.) *Report of the EIFAC Workshop on Fish-Farm Effluents*. Food and Agriculture Organization of the United Nations, Rome, pp. 5–20.

Angel, D.L. and Spanier, E. (1999) Artificial reefs to reduce organic enrichment caused by net cage fish farming – preliminary results. In: *Proceedings of the Seventh International Conference on Artificial Reefs*, 7–11 October 1999, Sanremo, pp. 478–485.

Anonymous (1999) Project summary: Salmonid recapture technology for aquaculture operations. Canada/Newfoundland Agreement on Economic Renewal ACERA #1.

Anonymous (2001) Sowing seeds for eelgrass restoration. *Coastlines: Information about Estuaries and Near Coastal Waters* 11(1), 6–7

Bayne, B.L. (1965) Growth and delay of the metamorphosis of the larvae of *Mytilus edulis* (L.). *Ophelia* 2, 1–47.

Becker, K. (1993) Attachment strength and colonization patterns of two macrofouling species on substrata with different surface tension (*in situ* studies). *Marine Biology* 117, 301–309.

Berrigan, M., Candies, T., Cirino, J., Dugas, R., Dyer, C., Gray, J., Herrington, T., Keithly, W., Leard, R., Nelson, J.R. and van Hoose, M. (1991) The oyster fishery of the Gulf of Mexico, United States: a regional management plan. Gulf States Marine Fisheries Commission Number 24.

Binh, C.T., Phillips, M.J. and Demaine, H. (1997) Integrated shrimp–mangrove farming systems in the Mekong delta of Vietnam. *Aquaculture Research* 28, 599–610.

Blaabjerg, V., Mouritsen, K. and Finster, K. (1998) Diel cycles of sulphate reduction rates in sediments of a *Zostera marina* bed (Denmark). *Aquatic Microbial Ecology* 15, 97–102.

Black, E., Gowen, R., Rosenthal, H., Roth, E., Stechy, D. and Taylor, F.J.R. (1997) The costs of eutrophication from salmon farming: implications for policy – A comment. *Journal of Environmental Management* 50, 105–109.

Black, K.D. (2000) *Environmental Impacts of Aquaculture*. CRC Press, Boca Raton, Florida.

Bonar, D.B., Weiner, R.M. and Colwell, R.R. (1986) Microbial–invertebrate interactions and potential for biotechnology. *Microbial Ecology* 12, 101–110.

Bougrova, L.A. and Bugrov, L.Y. (1994) Artificial reefs as fish-cage anchors. *Bulletin of Marine Science* 55, 1122–1136.

Boyd, C.E. and Schmitton, H.R. (1999) Achievement of sustainable aquaculture through environmental management. *Aquaculture Economics and Management* 3, 59–69.

Bray, R.H. and Ebeling, A.W. (1975) Food, activity and habitat of three 'picker-type' microcarnivorous fishes in the kelp forest off Santa Barbara, California. *Fishery Bulletin* 73, 815–829.

Bridger, C.J. and Garber, A.F. (2002) Interactions of aquaculture escapees and the environment. In: Costa-Pierce, B.A. (ed.) *Ecological Aquaculture: the Evolution of the Blue Revolution*. Blackwell Science, Oxford, UK, pp. 77–102.

Bridger, C.J., Booth, R.K., McKinley, R.S. and Scruton, D. (2001a) Site fidelity and dispersal patterns of domestic triploid steelhead trout (*Oncorhynchus mykiss*) released in the wild. *ICES Journal of Marine Science* 58, 510–516.

Bridger, C.J., Booth, R.K., McKinley, R.S., Scruton, D.A. and Lindstrom, R.T. (2001b) Monitoring fish behavior with a remote, combined acoustic/radio biotelemetry system. *Journal of Applied Ichthyology* 17, 126–129.

Broome, S.W. and Craft, C.B. (1998) Tidal salt marsh restoration, creation, and mitigation. In: Barnhisel, R.I., Daniels, W.L. and Darnady, R.G. (eds) *Reclamation of Drastically Disturbed Lands*. American Society of Agronomy, Madison, Wisconsin, pp. 101–135.

Broome, S.W., Rogers, S.M. Jr and Seneca, E.D. (1992) Shoreline erosion control using marsh vegetation and low-cost structures. North Carolina Sea Grant College Program, Raleigh, North Carolina. UNC-SG-92-12.

Broome, S.W., Craft, C.B. and Toomey, W.A. Jr (1999) Soil organic matter effects on infaunal community structure in restored and created tidal marshes. In:

Weinstein, M.P. and Kreeger, D.A. (eds) *Concepts and Controversies in Tidal Marsh Ecology*. Kluwer Academic Publishers, Dordrecht, Netherlands, pp. 49–75.

Brown, A.W., Hoppe, H.-G. and Rosenthal, H. (2000) Changes in bacterial abundance and community structure in cage fish-culture caused by water turbulence during feeding. *Journal of Applied Ichthyology* 16, 27–31.

Brown, J.R., Gowen, R.J. and McLusky, D.S. (1987) The effect of salmon farming on the benthos of a Scottish sea loch. *Journal of Experimental Marine Biology and Ecology* 109, 39–51.

Brown, L. (1993) *Aquaculture for Veterinarians: Fish Husbandry and Medicine*. Pergamon Press, New York.

Brusca, R.C. and Brusca, G.J. (1990) *Invertebrates*. Sinauer Associates, Sunderland, Massachusetts.

Bub, F. (2000) Development of an open-ocean aquaculture demonstration project: Progress report – February 2000. <http://ekman.sr.unh.edu/AQUACULTURE/PROGRESS_FEB00/OOA_Prog_Feb00_Front.html>. Revised 29 February 2000.

Buckley, R.M. and Hueckel, G.J. (1985) Biological processes and ecological development on an artificial reef in Puget Sound, Washington. *Bulletin of Marine Science* 37, 50–69.

Bundell, K. and Maybin, E. (1996) *After the Prawn Rush: the Human and Environmental Costs of Commercial Prawn Farming*. Christian Aid, London.

Buschmann, A.H., Troell, M., Kautsky, N. and Kautsky, L. (1996) Integrated tank cultivation of salmonids and *Gracilaria chilensis* (Gracilariales, Rhodophyta). *Hydrobiologia* 326/327, 75–82.

Butman, C.A. (1987) Larval settlement of soft-sediment invertebrates: the spatial scales of pattern explained by active habitat selection and the emerging role of hydrodynamical processes. *Oceanography and Marine Biology Annual Reviews* 25, 113–165.

Callow, M.E. and Fletcher, R.L. (1994) The influence of low surface energy materials on bioadhesion – a review. *International Biodeterioration and Biodegradation* 1994, 333–348.

Carss, D.N. (1990) Concentrations of wild and escaped fishes immediately adjacent to fish farm cages. *Aquaculture* 90, 29–40.

Cintron-Molero, G. (1992) Restoring mangrove systems. In: Thayer, G.W. (ed.) *Restoring the Nation's Marine Environment*. Maryland Sea Grant Program, College Park, Maryland, pp. 223–277.

Clough, B. (1983) Mangroves and sewage: a re-evaluation. In: Teas, H. (ed.) *Biology and Ecology of Mangroves*. Dr W. Junk Publishers, The Hague, Netherlands, pp. 173–187.

Collins, R.A. (1971) Cage cultured of catfish in reservoir lakes. *Proceedings of the Southeastern Association of the Game and Fish Commissioner* 24, 489–496.

Copeland, B.J. (1999) *Salt Marsh Restoration: Coastal Habitat Enhancement*. North Carolina Sea Grant College Program, Raleigh, North Carolina. UNC-SG-98-08.

Costa-Pierce, B.A. (1997a) Environmental impacts of nutrients discharged from aquaculture: towards the evolution of sustainable, ecological aquaculture systems. In: Baird, D.J., Beveridge, M.C.M., Kelly, L.A. and Muir, J.F. (eds) *Aquaculture and Water Resource Management*. Blackwell Science, Oxford, UK, pp. 81–113.

Costa-Pierce, B.A. (1997b) From farmers to fishers: developing reservoir aquaculture for people displaced by dams. World Bank Technical Paper No. 369 (Fisheries Series), World Bank, Washington, DC.

Costa-Pierce, B.A. (ed.) (2002) *Ecological Aquaculture: the Evolution of the Blue Revolution*. Blackwell Science, Oxford, UK.

Cuba, T.R. and Blake, N.J. (1983) The initial development of a marine fouling assemblage on a natural substrate in a subtropical estuary. *Botanica Marina* 26, 259–264.

Dahl, T.E. (1990) *Wetland Losses in the United States 1780s to 1980s*. Department of Interior, United States Fish and Wildlife Service, Washington, DC.

Dahl, T.E. (2000). *Status and Trends of Wetlands in the Conterminous United States 1986 to 1997*. Department of Interior, United States Fish and Wildlife Service, Washington, DC.

Dahl, T.E. and Johnson, C.E. (1991) *Status and Trends of Wetlands in the Conterminous United States, mid-1970s to mid-1980s*. Department of Interior, United States Fish and Wildlife Service, Washington, DC.

Dahlem, C., Moran, P.J. and Grant, T.R. (1984) Larval settlement of marine sessile invertebrates on surface of different colour and position. *Ocean Science and Engineering* 9, 225–236.

Davis, R.C. and Short, F.T. (1997) Restoring eelgrass, *Zostera marina* L., habitat using a new transplanting technique: the horizontal rhizome method. *Aquatic Botany* 59, 1–15.

Dominguez, L.M., Calero, G.L., Martin, J.M.V. and Robaina, L.R. (2001) A comparative study of sediments under a marine cage farm at Gran Canaria Island (Spain): preliminary results. *Aquaculture* 192, 225–231.

Dooley, J.K. (1972) Fishes associated with the pelagic *Sargassum* complex, with a discussion of the *Sargassum* community. *Contributions in Marine Science* 16, 1–32.

Dreyer, G.D. and Niering, W.A. (eds) (1995) *Tidal Marshes of Long Island Sound: Ecology, History and Restoration. Bulletin 34*. The Connecticut College Arboretum, New London, Connecticut.

Edwards, P. (1998) A systems approach for the promotion of integrated aquaculture. *Aquaculture Economics and Management* 2, 1–12.

Edwards, P. and Demaine, H. (1997) *Rural Aquaculture: Overview and Framework for Country Reviews*. Regional Office for Asia and the Pacific (RAP), Food and Agriculture Organization of the United Nations (FAO), Bangkok, Thailand. RAP Publication 1997/36.

Erftemeijer, P.L.A. and Lewis, R.R. (1999) Planting mangroves on intertidal mudflats: habitat restoration or habitat conversion? Paper presented at the ECOTONE-VIII Seminar 'Enhancing Coastal Ecosystem Restoration for the 21st Century', Ranong & Phuket, 23–28 May 1999.

Evans, L.V. (1981) Marine algae and fouling: a review, with particular reference to ship-fouling. *Botanica Marina* 24, 167–171.

FAO (1995) *Code of Conduct for Responsible Fisheries*. Food and Agriculture Organization of the United Nations, Rome, 41pp.

Ferreira, S. and Seeliger, U. (1985) The colonization process of algal epiphytes on *Ruppia maritima* L. *Botanica Marina* 28, 245–249.

Field, C.D. (ed.) (1996) *Restoration of Mangrove Ecosystems*. International Society for Mangrove Ecosystems, Okinawa, Japan.

Fitzgerald, W.J. (1997) Silvofisheries – an environmentally sensitive integrated mangrove forest and aquaculture system. *Aquaculture Asia* 2, 9–17.

Fitzgerald, W.J. (2002) Aquaculture–silviculture. In: Costa-Pierce, B.A. (ed.) *Ecological Aquaculture: The Evolution of the Blue Revolution*. Blackwell Science, Oxford, UK, pp. 161–262.

Folke, C. and Kautsky, N. (1992) Aquaculture with its environment: prospects for sustainability. *Ocean and Coastal Management* 17, 5–24.

Folke, C., Kautsky, N. and Troell, M. (1994) The cost of eutrophication from salmon farming: implications for policy. *Journal of Environmental Management* 40, 173–182.

Folke, C., Kautsky, N., Berg, H., Jansson, A. and Troell, M. (1998) The ecological footprint concept for sustainable seafood production: a review. *Ecological Applications* 8, S63–S71.

Fonseca, M.S. (1994) *A Guide to Planting Seagrasses in the Gulf of Mexico.* Texas Sea Grant College Program TAMU-SG-94-601, Texas A&M University, College Station, Texas, 24pp.

Fonseca, M.S., Kenworthy, W.J., Thayer, G.W., Heller, D.Y. and Cheap, K.M. (1985) Transplanting of the seagrasses *Zostera marina* and *Halodule wrightii* for sediment stabilization and habitat development of the east coast of the United States. United States Army Corps of Engineers Technical Report EL-85-9.

Fonseca, M.S., Kenworthy, W.J., Colby, D.R., Rittmaster, K.A. and Thayer, G.W. (1990) Comparisons of fauna among natural and transplanted eelgrass *Zostera marina* meadows: criteria for mitigation. *Marine Ecology Progress Series* 65, 251–264.

Fonseca, M.S., Kenworthy, W.J. and Thayer, G.W. (1998) Guidelines for the conservation and restoration of seagrasses in the United States and adjacent waters. National Oceanic and Atmospheric Administration Coastal Ocean Program Decision Analysis Series No. 12, NOAA Coastal Ocean Office, Silver Spring, Maryland.

Fourqurean, J.W. and Robblee, M.B. (1999) Florida Bay: a history of recent ecological changes. *Estuaries* 22(2B), 345–357.

Frayer, W.E., Monahan, T.J., Bowden, D.C. and Graybill, F.A. (1983) *Status and Trends of Wetlands and Deepwater Habitats in the Conterminous United States, 1950s to 1970s.* Colorado State University, Fort Collins, Colorado.

Garbisch, E.W. Jr, Woller, P.B. and McCallum, R.J. (1975) Salt marsh establishment and development. Technical Memorandum No. 52. United States Army Corps of Engineers, Coastal Engineering Research Center, Fort Belvoir, Virginia.

Goering, J.J. and Parker, P.L. (1972) Nitrogen fixation by epiphytes on sea grasses. *Limnology and Oceanography* 17, 320–323.

Goodland, R. and Daly, H. (1996) Environmental sustainability: universal and non-negotiable. *Ecological Applications* 6, 1002–1017.

Gowen, R., Brown, J., Bradbury, N. and McLusky, D. (1988) Investigations into benthic enrichment, hypernutrification and eutrophication associated with mariculture in Scottish coastal waters (1984–1988). Department of Biology, University of Stirling, Stirling, UK.

Gowen, R.J., Smyth, D. and Silvert, W. (1994) Modelling the spatial distribution and loading of organic fish farm waste to the seabed. In: Hargrave, B.T. (ed.) *Modelling Benthic Impacts of Organic Enrichment from Marine Aquaculture.* Canadian Technical Report of Fisheries and Aquatic Sciences 1949, pp. 19–30.

Granger, S., Nixon, S., Traber, M. and Keyes, R. (1996) The application of horticultural techniques in the propagation of eelgrass (*Zostera marina* L.) from seed. In: Kuo, J., Phillips, R.C., Walker, D.I. and Kirkman, H. (eds) *Seagrass Biology. Proceedings of an International Workshop.* Rottnest Island, Australia, p. 377. (Abstract.)

Granger, S.L., Traber, M.S. and Nixon, S.W. (2000a) Propagation of *Zostera marina* L. from seed. In: Shepard, C. (ed.) *Seas at the Millennium: an Environmental Evaluation,* Vol. 3. Pergamon Press, London, pp. 4–5.

Granger, S.L., Traber, M.S. and Nixon, S.W. (2000b) The influence of planting depth and density on germination and development of *Zostera marina* L. seeds. *Biologia Marina Mediterranea* 7, 55–58.

Greening, H.S., Morrison, G., Eckenrod, R.M. and Perry, M.J. (1996) The Tampa Bay resource-based management approach. In: Treat, S.F. (ed.) *Proceedings, Tampa Bay Area Scientific Information Symposium 3*. Tampa, Florida, pp. 349–355.

Grove, T.L. and Edwards, C.A. (1993) Do we need a new developmental paradigm? *Agriculture, Ecosystems and Environment* 46, 135–145.

Hadfield, M.G. (1986) Settlement and recruitment of marine invertebrates: a perspective and some proposals. *Bulletin of Marine Science* 39, 418–425.

Hagler, M. (1997) Shrimp – the devastating delicacy: the explosion of shrimp farming and the negative impacts on people and the environment. A Greenpeace Report.

Hair, C.A., Bell, J.D. and Kingsford, M.J. (1994) Effects of position in the water column, vertical movement and shade on settlement of fish to artificial habitats. *Bulletin of Marine Science* 55, 434–444.

Hall, P. and Holby, O. (1986) Environmental impact of a marine fish cage culture. ICES, C.M. 1986: F46. International Council for the Exploration of the Sea.

Hamilton, L.S. and Snedaker, S.C. (eds) (1984) Restoration and establishment. Section III. In: *Handbook of Mangrove Area Management*. East West Center, Honolulu, Hawaii, pp. 102–108.

Hansen, L.P., Hastein, T., Naevdal, G., Saunders, R.L. and Thorpe, J.E. (eds) (1991) Interactions between cultured and wild Atlantic salmon. *Proceedings of the symposium hosted by the Directorate for Nature Management and Norwegian Institute for Nature Research*, Loen, Norway, April 23–26 1990. *Aquaculture* 98, 1–324.

Harlin, M.M. and Lindbergh, J.M. (1977) Selection of substrata by seaweeds: optimal surface relief. *Marine Biology* 40, 33–40.

Harlin, M., Thorne-Miller, B. and Boothroyd, J. (1982) Seagrass–sediment dynamics of a flood-tidal delta in Rhode Island (USA). *Aquatic Botany* 23, 127–138.

Harrison, P.G. (1990) Variations in success of eelgrass transplants over a five-year period. *Environmental Conservation* 17, 157–163.

Harrison, P.G. (1991) Mechanisms of seed dormancy in an annual population of *Zostera marina* (eelgrass) from Netherlands. *Canadian Journal of Botany* 69, 1972–1976.

Heck, K.L., Able, K., Roman, C. and Fahay, M. (1995) Composition, abundance, biomass and production of macrofauna in a New England estuary: comparison among eelgrass meadows and other nursery habitats. *Estuaries* 18, 379–389.

Heidelbaugh, W.S., Hall, L.M., Kenworthy, W.J., Whitfield, P., Virnstein, R.W., Morris, L.J. and Hanisak, M.D. (2000) Reciprocal transplanting of the threatened seagrass *Halophile johnsonii* (Johnson's seagrass) in the Indian River Lagoon, Florida. In: Bortone, S.A. (ed.) *Seagrasses: Monitoring, Ecology, Physiology, and Management*. Boca Raton, Florida, pp. 197–210.

Hem, S. and Avit, J.L.B. (1994) First results on 'acadja-enclos' as an extensive aquaculture system (West Africa). *Bulletin of Marine Science* 55, 1038–1049.

Hicks, D.W., Onuf, C.P. and Tunnell, J.W. (1998) Response of shoal grass, *Halodule wrightii*, to extreme winter conditions in the Lower Laguna Madre, Texas. *Aquatic Botany* 62, 107–114.

Hoagland, K.D., Roemer, S.C. and Rosowski, J.R. (1982) Colonization and community structure of two periphyton assemblages, with emphasis on the diatoms (Bacillariophyceae). *American Journal of Botany* 69, 188–213.

Hodson, S.L. and Burke, C. (1994) Microfouling of salmon-cage netting: a preliminary investigation. *Biofouling* 8, 93–105.

Hodson, S.L., Lewis, T.E. and Burke, C.M. (1997) Biofouling of fish-cage netting: efficacy and problems of *in situ* cleaning. *Aquaculture* 152, 77–90.

Hodson, S.L., Burke, C.M. and Bissett, A.P. (2000) Biofouling of fish-cage netting: the efficacy of a silicone coating and the effect of netting colour. *Aquaculture* 184, 277–290.

Holford, I. and Patrick, W. (1979) Effects of redox potential and pH on phosphate removal from wastewater during land application. *Progress Water Technology* 11, 215–225.

Hudon, C. and Bourget, E. (1981) Initial colonization of artificial substrate: community development and structure studies by scanning electron microscopy. *Canadian Journal of Fisheries and Aquatic Sciences* 38, 1371–1384.

Huguenin, J.E. (1997) The design, operations and economics of cage culture systems. *Aquacultural Engineering* 16, 167–203.

Huguenin, J.E. and Ansuini, F.J. (1978) A review of the technology and economics of marine fish cage systems. *Aquaculture* 15, 151–170.

Huner, J.V. (2000a) Crawfish and water birds. *American Scientist* 88, 301–303.

Huner, J.V. (2000b) Importance of crawfish in diets of wading birds in Louisiana. *Aquaculture News* 2, 11–12.

Hutchinson, P. (ed.) (1997) Interactions between salmon culture and wild stocks of Atlantic salmon: the scientific and management issues. *Proceedings of the ICES/NASCO Symposium*, Bath, England, 18–22 April 1997. *ICES Journal of Marine Science* 54, 963–1227.

IOA (Institute of Aquaculture), Institute of Freshwater Biology and Institute of Terrestrial Ecology (1990) *Fish Farming and the Scottish Freshwater Environment*. Nature Conservancy Council, Edinburgh.

Janicki, A. and Wade, D. (1996) Estimating critical nitrogen loads for the Tampa Bay estuary: an empirically based approach to setting management targets. Tampa Bay National Estuary Program, Technical Publication 06-96, Coastal Environmental, Inc., St Petersburg, Florida.

Johannessen, P.J., Botnen, H.B. and Tvedten, O.F. (1994) Macrobenthos: before, during and after a fish farm. *Aquaculture and Fisheries Management* 25, 55–66.

Johnston, D., Trong, N.V., Tien, D.V. and Xuan, T.T. (2000) Shrimp yields and harvest characteristics of mixed shrimp–mangrove forestry farms in southern Vietnam: factors affecting production. *Aquaculture* 188, 263–284.

Kaspar, H.F., Hall, G.H. and Holland, A.J. (1988) Effects of sea cage farming on sediment nitrification and dissimilatory nitrate reductions. *Aquaculture* 70, 333–344.

Kautsky, N. and Folke, C. (1991) Integrating open system aquaculture: ecological engineering for increased production and environmental improvement through nutrient recycling. In: Etnier, C. and Guterstam, B. (eds) *Ecological Engineering for Wastewater Treatment*. Bokskogen Publisher, Gothenburg, Sweden, pp. 320–324.

Kent, E.B. (1980) Effect of food level on reproduction and metamorphosis in the suctorian protozoan *Tokophrya lemnarum* Stein. *Microbial Ecology* 6, 153–159.

Kent, M.L. (1992) *Diseases of Seawater Netpen-reared Salmonid Fishes in the Pacific Northwest*. Canadian Special Publication of Fisheries and Aquatic Sciences No. 116.

Kilambi, R., Hoffman, C., Brown, A., Adams, J. and Wickizer, W. (1976) *Effects of Cage Culture Fish Production upon the Biotic and Abiotic Environment of Crystal Lake, Arkansas*. Department of Zoology, University of Arkansas, Fayetteville, Arkansas.

Klima, E.F. and Wickham, D.A. (1971) Attraction of coastal pelagic fishes with artificial structures. *Transactions of the American Fisheries Society* 100, 86–99.

Krost, P., Chrzan, T., Schomann, H. and Rosenthal, H. (1994) Effects of a floating fish farm in Kiel Fjord on the sediment. *Journal of Applied Ichthyology* 10, 353–361.

Kupka-Hansen P., Pittman, K. and Ervik, A. (1991) Organic waste from marine fish farms – effects on the seabed. In: *Marine Aquaculture and the Environment*. Nordic Council of Ministers, Copenhagen, pp. 105–119.

Kusler, J.A. and Kentula, M.E. (1990) *Wetland Creation and Restoration: The Status of the Science*. Island Press, Washington, DC.

Larsson, J., Folke, C. and Kautsky, N. (1994) Ecological limitations and appropriation of ecosystem support by shrimp farming in Columbia. *Environmental Management* 18, 663–676.

Laur, D. and Ebeling, A.W. (1983) Predator–prey relationship in a guild of surfperches. *Environmental Biology of Fishes* 8, 217–229.

Leffler, M. (2001) Oyster reefs: key to restoring bay grasses? Maryland Marine Notes Online 18, 1–11.

Lewis, R.R. and Marshall, M.J. (1997) Principles of successful restoration of shrimp aquaculture ponds back to mangrove forests. In: Programa/resumes de Marcuba '97, September 15–20, 1997. Palacio de Convenciones de La Habana, Cuba, p. 126. [Abstract.]

Lewis, R.R. and Streever, B. (2000) Restoration of mangrove habitat. WRP Technical Notes Collection (ERDC TN-WRP-VN-RS-3.2), United States Army Engineer Research and Development Center, Vicksburg, Mississippi.

Lewis, R.R., Kusler, J.A. and Erwin, K.L. (1995) Lessons learned from five decades of wetland restoration and creation in North America. In: Montes, C., Oliver, G., Molina, F. and Cobos, J. (eds) Bases ecologicas para la Restauracion de Humedales en la Cuenca Mediterranea. *Proceedings of a meeting held at the University of La Rabida, Spain*. 7–11 June 1993. Junta de Andaluca, Spain, pp. 107–122.

Li, X., Seliskar, D.M., Moga, J.A. and Gallagher, J.C. (1995) Plant regeneration from callus cultures of salt marsh hay, *Spartina patens*, and its cellular based salt tolerance. *Aquatic Botany* 51, 103–113.

Little, B.J. (1984) Succession in microfouling. In: Costlow, J.D. and Tipper, R.C. (eds) *Marine Biodeterioration: An Interdisciplinary Study*. E. & F.N. Spon Ltd, London, pp. 63–67.

Love, M.S. and Ebeling, A.W. (1978) Food and habitat of three switch feeding fishes in the kelp forests off Santa Barbara, California. *Fishery Bulletin* 76, 257–271.

Loverich, G. and Forster, J. (2000) Advances in offshore cage design using spar buoys. *Marine Technology Society Journal* 34, 18–28.

Loverich, G.F. and Gace, L. (1998) The effect of currents and waves on several classes of offshore sea cages. In: Helsley, C.E. (ed.) *Open Ocean Aquaculture '97, Charting the Future of Ocean Farming. Proceedings of an International Conference*. 23–25 April 1997. Maui, Hawaii. University of Hawaii Sea Grant College Program CP-98-08, Honolulu, Hawaii, pp. 131–144.

Loyacano, H.A. and Smith, D.C. (1976) Attraction of native fish to catfish culture cages in reservoirs. *Proceedings of the Southeastern Association of Game and Fish Commissioners* 29, 63–73.

Lugo, A.E. and Snedaker, S.C. (1974) The ecology of mangroves. *Annual Review of Ecology and Systematics* 5, 39–64.

Lumb, C.M. (1989) Self-pollution by Scottish salmon farms? *Marine Pollution Bulletin* 20, 375–379.

Lynch, J.M., Fletcher, M. and Latham, M.J. (1979) Biological interactions. In: Lynch, J.M. and Poole, N.J. (eds) *Microbial Ecology: a Conceptual Approach.* Blackwell Scientific, Oxford, UK, pp. 171–187.

Macane, W. (1968) A general account of the fauna and flora of mangrove swamps and forests in the Indo-West-Pacific region. *Advances in Marine Biology* 6, 73–270.

Maki, J.S., Rittschof, D., Costlow, J.D. and Mitchell, R. (1988) Inhibition of attachment of larval barnacles, *Balanus amphitrite*, by bacterial films. *Marine Biology* 97, 199–206.

Maki, J.S., Rittschof, D., Samuelsson, M.-O., Szewzyk, U., Yule, A.B., Kjelleberg, S., Costlow, J.D. and Mitchell, R. (1990) Effect of marine bacteria and their exopolymers on the attachment of barnacle cypris larvae. *Bulletin of Marine Science* 46, 499–511.

Mallet, A. and Myrand, B. (1995) The culture of the blue mussel in Atlantic Canada. In: Boghen, A.D. (ed.) *Cold-water Aquaculture in Atlantic Canada.* The Canadian Institute for Research on Regional Development, The Tribune Press Ltd, Sackville, New Brunswick, Canada, pp. 255–296.

Manahan, D. (1990) Adaptations by invertebrate larvae for nutrient acquisition from seawater. *American Zoologist* 30, 147–160.

Manahan, D., Wright, S.H., Stevens, G.C. and Rice, M.A. (1982) Transport of dissolved amino acids by the blue mussel, *Mytilus edulis*: demonstration of net uptake from natural seawater. *Science* 215, 1253–1255.

Marba, N. and Duarte, C. (1995) Coupling of seagrass (*Cymodocea nodosa*) patch dynamics to subaqueous dune migration. *Journal of Ecology* 83, 381.

Marshall, K.C., Stout, R. and Mitchell, R. (1971) Mechanisms of the initial events in the sorption of marine bacteria to surfaces. *Journal of General Microbiology* 68, 337–348.

Matthews, G.A. and Minello, T.J. (1994) *Technology and Success in Restoration, Creation and Enhancement of* Spartina alterniflora *Marshes in the United States.* NOAA Coastal Ocean Program Decision Analysis Series No. 2, Washington, DC, 2 volumes.

Mattsson, J. and Linden, O. (1983) Benthic macrofauna succession under mussels, *Mytilus edulis* L. (Bivalvia), cultured on hanging long-lines. *Sarsia* 68, 97–102.

McGhie, T.K., Crawford, C.M., Mitchell, I.M. and O'Brien, D. (2000) The degradation of fish-cage waste in sediments during fallowing. *Aquaculture* 187, 351–366.

Mearns, A.J. and Word, J.Q. (1982) Forecasting effects of sewage solids on marine benthic communities. In: Mayer, G.F. (ed.) *Ecological Stress and the New York Bight: Science and Management.* Columbia South Carolina Estuarine Research Federation, Columbia, South Carolina, pp. 495–512.

Mitchell, C.T. and Hunter, J.R. (1970) Fishes associated with drifting kelp, *Macrocystis pyrifers*, off the coast of southern California and northern Baja California. *California Fish and Game* 56, 288–297.

Mitsch, W.J. and Jorgensen, S.E. (eds) (1989) *Ecological Engineering: An Introduction to Ecotechnology.* Wiley Interscience, New York.

Morrisey, D.J., Gibbs, M.M., Pickmere, S.E. and Cole, R.G. (2000) Predicting impacts and recovery of marine-farm sites in Stewart Island, New Zealand, from the Findlay–Watling model. *Aquaculture* 185, 257–271.

Myrseth, B. (2000) Automation of feeding management in cage culture. In: Liao, I.C. and Lin, C.K. (eds) *Cage Aquaculture in Asia. Proceedings of the First International*

Symposium on Cage Aquaculture in Asia. Asian Fisheries Society, Manila, Philippiness and World Aquaculture Society – Southeast Asian Chapter, Bangkok, Thailand, pp. 151–155.

Naylor, R.L., Goldberg, R.J., Mooney, H., Beveridge, M., Clay, J., Folke, C., Kautsky, N., Lubchenco, J., Primavera, J. and Williams, M. (1998) Nature's subsidies to shrimp and salmon farming. *Science* 282, 883–884.

Naylor, R.L., Goldberg, R.J., Primavera, J.H., Kautsky, N., Beveridge, M.C.M., Clay, J., Folke, C., Lubchenco, J., Mooney, H. and Troell, M. (2000) Effect of aquaculture on world fish supplies. *Nature* 405, 1017–1024.

Nedwell, D. (1975) Inorganic nitrogen metabolism in a eutrophicated tropical estuary. *Water Research* 9, 221–231.

Neori, A., Shpigel, M. and Ben-Ezra, D. (2000) A sustainable integrated system for culture of fish, seaweed and abalone. *Aquaculture* 186, 279–291.

Newell, R.I.E. and Ott, J. (1999) Macrobenthic communities and eutrophication. In: Malone, T.C., Malej, A., Harding, L.W. Jr and Smodlaka, N. (eds) *Ecosystems at the Land–Sea Margin: Drainage Basin to Coastal Sea.* Coastal and Estuarine Studies, American Geophysical Union, San Francisco, California, pp. 265–293.

Nickels, J.S., Parker, J.H., Bobbie, R.J., Martz, R.F., Lott, D.F., Benson, P.H. and White, D.C. (1981) Effect of cleaning with flow-driven brushes on the biomass and community composition of the marine microfouling film on aluminum and titanium surfaces. *International Biodeterioration Bulletin* 17, 87–94.

Ochoa, E. (1997) Majagual: the tallest mangroves in the world. *Intercoast Special Mangrove Edition* No. 1 (March), 17.

Odum, E.P. (1985) Trends expected in stressed ecosystems. *BioScience* 35, 419–422.

Olsen, S. and Arriaga, L. (1989) *Establishing a Sustainable Shrimp Mariculture Industry in Ecuador.* The University of Rhode Island, Coastal Resources Center, United States Agency for International Development, Narragansett, Rhode Island.

O'Neill, S.M., Sutterlin, A.M. and Aggett, D. (1983) The effects of size-selective feeding by starfish, *Asterias vulgaris*, on the production of mussels, *Mytilus edulis*, cultured on nets. *Aquaculture* 38, 211–220.

Orth, R.J., Luckenbach, M. and Moore, K.A. (1994) Seed dispersal in a marine macrophyte: implications for colonization and restoration. *Ecology* 75, 1927–1939.

Phillips, R.C. (1990) Transplant methods. In: Phillips, R.C. and McRoy, C.P. (eds) *Seagrass Research Methods*, UNESCO, Paris, pp. 51–54.

Pimental, D. (1997) *Techniques for Reducing Pesticides: Environmental and Economic Benefits*. John Wiley & Sons, Chichester, UK.

Pimental, D. and Dodds, H. (1997) Human population and resource use: a letter to world leaders from scientists. http://climate.konza.ksu.edu/~popres/

Porter, C.B., Krost, P., Gordin, H. and Angel, D.L. (1996) Preliminary assessment of grey mullet (*Mugil cephalus*) as a forager of organically enriched sediments below marine fish farms. *Bamidgeh* 48, 47–55.

Primavera, J.H. (1991) Intensive prawn farming in the Philippines: ecological, social and economic implications. *Ambio* 20, 28–33.

Primavera, J.H. and Agbayani, R.F. (1997) Comparative strategies in community based mangrove rehabilitation programs in the Philippines. In: Hong, P.N., Ishwaran, N., San, H.T., Tri, N.H. and Tuan, M.S. (eds) *Proceedings of the ECOTONE V Regional Seminar: Community Participation in Conservation, Sustainable Use and Rehabilitation of Mangroves in Southeast Asia*, 8–12 January 1996. Ho Chi Minh City, Vietnam, pp. 229–243.

Prince, E.D. and Gotshall, D.W. (1976) Food of the copper rockfish, *Sebastes caurinus* Richardson, associated with an artificial reef in south Humboldt Bay, California. *California Fish and Game* 62, 274–285.

Pullin, R.S.V., Rosenthal, H. and Maclean, J.L. (eds) (1993) *Environment and Aquaculture in Developing Countries*. International Council for Living Aquatic Resources Management, Manila, Philippines. *Conference Proceedings* 31.

Rajendran, N. and Kathiresan, K. (1997) Effect of effluent from a shrimp pond on shoot biomass of mangrove seedlings. *Aquaculture Research* 27, 745–747.

Rees, W.E. and Wackernagel, M. (1994) Ecological footprints and appropriated carrying capacity. In: Jansson, A.M., Hammer, M., Folke, C. and Costanza, R. (eds) *Investing in Natural Capital: The Ecological Economics Approach to Sustainability*. Island Press, Washington, DC, pp. 362–390.

Reiswig, H.M. (1985) *In situ* feeding in two shallow-water hexactinellid sponges. In: Rutzler, K. (ed.) *New Perspectives in Sponge Biology*. Smithsonian Institution Press, Washington, DC, pp. 504–510.

Relini, M., Relini Orsi, L. and Relini, G. (1994) An offshore buoy as a FAD in the Mediterranean. *Bulletin of Marine Science* 55, 1099–1105.

Rice, M.A. (2001) Environmental impacts of shellfish aquaculture: filter feeding to control eutrophication. In: Tlusty, M., Bengston, D., Halvorson, H., Oktay, S., Pearce, J. and Rheault, R. (eds) *Marine Aquaculture and the Environment*. Cape Cod Press, Falmouth, Massachusetts, pp. 77–86.

Risgaard-Petersen, N., Dalsgaard, T., Rysgaard, S., Christensen, P.B., Borum, J., McGlathery, K. and Nielsen, L.P. (1998) Nitrogen balance of a temperate eelgrass *Zostera marina* bed. *Marine Ecology Progress Series* 174, 281–291.

Ritz, D.A., Lewis, M.E. and Shen, M. (1989) Response to organic enrichment of infaunal macrobenthic communities under salmonid sea cages. *Marine Biology* 103, 211–214.

Robertson, A. and Alongi, D. (1992) *Tropical Mangrove Ecosystems*. American Geophysical Union, Washington, DC.

Robertson, A.I. and Blaber, S.J.M. (1992) Plankton, epibenthos and fish communities. In: Robertson, A.I. and Alongi, D.M. (eds) *Tropical Mangrove Ecosystems*. Coastal and Estuarine Studies No. 41. American Geophysical Union, Washington, DC, pp. 173–224.

Ronnback, P. (1999) The ecological basis for economic value of seafood production supported by mangrove ecosystems. *Ecological Economics* 29, 235–252.

Roscoe, D.T. and Walker, G. (1995) Observations on the adhesion of the calcareous tube of *Pomatoceros lamarckii* and the holdfast of *Laminaria digitata*. *Biofouling* 9, 39–50.

Rozsa, R. (1995) Tidal wetland restoration in Connecticut. In: Dreyer, G.D. and Niering, W.A. (eds) *Tidal Marshes of Long Island Sound: Ecology, History and Restoration. Bulletin 34*. The Connecticut College Arboretum, New London, Connecticut, pp. 51–65.

Ruitenbeek, H. (1992) Mangrove management: an economic analysis of management options with a focus on Bintuni Bay, Irian Jaya, Indonesia. *Environmental Reports* No. 8. EMDI, Gabriola Island, Canada.

Ruokolahti, C. (1988) Effects of fish farming on growth and chlorophyll *a* content of *Cladophora*. *Marine Pollution Bulletin* 19, 166–169.

Ryder, R.A. and Kerr, S.R. (1989) Environmental priorities: placing habitat in hierarchic perspective. In: Levings, C.D., Holtby, L.B. and Henderson, M.A. (eds)

Proceedings of the National Workshop on Effects of Habitat Alteration on Salmonid Stocks. Canadian Special Publication of Fisheries and Aquatic Sciences 105, pp. 2–12.

Shpigel, M., Gasith, A. and Kimmel, E. (1997) A biomechanical filter for treating fish-pond effluents. *Aquaculture* 152, 103–117.

Silvert, W. and Sowles, J.W. (1996) Modelling environmental impacts of marine finfish aquaculture. *Journal of Applied Ichthyology* 12, 75–81.

Simenstad, C.A. and Thom, R.M. (1996) Functional equivalency trajectories of the restored Gog-Le-Hi-Te estuarine wetland. *Ecological Applications* 6, 38–56.

Sinicrope, T., Hine, P., Warren, R. and Niering, W. (1990) Restoration of an impounded salt marsh in New England. *Estuaries* 13, 25–30.

Snedaker, S. (1978) Mangroves: their value and perpetuation. *Nature and Resources* 14, 6–13.

Soemodihardjo, S., Wiroatmodja, P., Mulia, F. and Harahap, M.K. (1996) Mangroves in Indonesia – a case study of Tembilahan, Sumatra. In: Field, C. (ed.) *Restoration of Mangrove Ecosystems.* International Society for Mangrove Ecosystems, Okinawa, Japan, pp. 97–110.

Spanier, E., Tom, M. and Pisanty, S. (1985) Enhancement of fish recruitment by artificial enrichment of man-made reefs in the southeastern Mediterranean. *Bulletin of Marine Science* 37, 356–363.

Stephens, J.S. and Zerba, K.E. (1981) Factors affecting fish diversity on a temperate reef. *Environmental Biology of Fishes* 6, 111–121.

Stevenson, N.J., Lewis, R.R. and Burbridge, P.R. (1999) Disused shrimp ponds and mangrove rehabilitation. In: Streever, W. (ed.) *An International Perspective on Wetland Rehabilitation*, Island Press, Washington, DC, pp. 277–297.

Stirling, H.P. and Okumus, I. (1995) Growth and production of mussels (*Mytilus edulis* L.) suspended at salmon cages and shellfish farms in two Scottish sea lochs. *Aquaculture* 134, 193–210.

Swaminathan, M.S. (1992) Cultivating food for a developing world. *Environmental Science and Technology* 26, 1104–1107.

Thayer, G.W. (ed.) (1992) *Restoring the Nation's Marine Environment.* Maryland Sea Grant Publication, College Park, Maryland.

Turner, R.E. and Lewis, R.R. (1997) Hydrologic restoration of coastal wetlands. *Wetlands Ecology and Management* 4,65–72.

Virnstein, R.W., Mikkelsen, P.S., Cairns, K.D. and Capone, M.A. (1983) Seagrass beds versus sand bottoms: the trophic importance of their associated benthic invertebrates. *Florida Scientist* 46, 363–381.

Wahl, M. (1989) Marine epibiosis. I. Fouling and antifouling: some basic aspects. *Marine Ecology Progress Series* 58, 175–189.

Ward, L.G., Kemp, W.M. and Boynton, W.R. (1984) The influence of waves and seagrass communities on suspended sediment dynamics in an estuarine embayment. *Marine Geology* 59, 85–103.

WCED (1987) *Our Common Future.* World Commission on Environment and Development. Oxford University Press, Oxford, UK.

Weinstein, M.P., Balletto, J.H., Teal, J.M. and Ludwig, D.F. (1997) Success criteria and adaptive management for a large-scale wetland restoration project. *Wetlands Ecology and Management* 4, 111–127.

West, J.M. and Zedler, J.B. (2000) Marsh–creek connectivity: fish use of a tidal salt marsh in southern California. *Estuaries* 23, 699–710.

Weston, D.P. (1986) *The Environmental Effects of Floating Mariculture in Puget Sound.* Special Report, Washington Department of Fisheries and Ecology, Seattle, Washington.

Weston, D.P. (1990) Quantitative examination of macrobenthic community changes along an organic enrichment gradient. *Marine Ecology Progress Series* 61, 233–244.

Wickham, D.A., Watson, J.W. Jr and Ogren, L.H. (1973) The efficacy of midwater artificial structures for attracting pelagic sport fish. *Transactions of the American Fisheries Society* 103, 563–572.

Willoughby, S. (1999) *Manual of Salmonid Farming.* Fishing News Books, London.

Workman, I.K., Landry, A.M. Jr, Watson, J.W. Jr and Blackwell, J.W. (1985) A midwater fish attraction device study conducted from Hydrolab. *Bulletin of Marine Science* 37, 377–386.

Wu, R.S.S., Lam, K.S., MacKay, D.W., Lau, T.C. and Yam, V. (1994) Impact of marine fish farming on water quality and bottom sediment: a case study of the sub-tropical environment. *Marine Environmental Research* 38, 115–145.

Zedler, J.B. (1996) *Tidal Wetland Restoration: a Scientific Perspective and Southern California Focus.* California Sea Grant College Publication T-038, La Jolla, California.

Mangroves and Coastal Aquaculture

9

Claude E. Boyd

Department of Fisheries and Allied Aquacultures, Swingle Hall, Auburn University, Auburn, AL 36849, USA

Abstract

There is considerable concern over destruction of mangroves by shrimp farming and other types of aquaculture, and it is true that mangroves have been destroyed for installation of aquaculture ponds. In most countries, destruction of mangroves results from people seeking wood for building or for fuel. Also, mangroves may be destroyed for development unrelated to shrimp or other types of aquaculture. Worldwide, probably less than 10% of mangrove loss may be attributed to aquaculture, but in some countries, aquaculture is responsible for 20% or more of historic loss of mangroves.

Mangrove areas are not ideal places for shrimp farming or other types of aquaculture. This is because of the following: low elevation, poor drainage characteristics, location in complex and sensitive ecosystems where it is difficult to prevent spread of aquatic animal diseases and where bird predation of culture species may be great, and the tendency of soils to have high organic matter concentrations and to be highly acidic from sulphide-sulphur oxidation. Thus, shrimp farms and other aquaculture activities in mangrove areas tend to be less productive than those in non-mangrove areas. Mangrove areas afford shelter from storms and provide biological treatment of pollution entering waters from various human activities. Thus, their destruction is harmful to aquaculture. It is no longer popular to locate aquaculture farms in mangroves and in most countries governmental regulations make it illegal to do so.

Areas behind mangrove forests often are excellent sites for aquaculture farms. Governments implementing regulations to exclude aquaculture activities from mangrove areas should allow shrimp and fish farmers to route water supply and discharge canals through mangrove areas and to locate pump stations in mangrove areas if necessary. It is possible to have aquaculture ponds and mangroves situated harmoniously in the same area, and management practices to encourage this objective are provided.

Introduction

Pond aquaculture of marine and brackish water species has become a significant activity in many tropical and subtropical nations. Coastal aquaculture contributes heavily to the world production of several species. For example, about 30% of shrimp are produced from aquaculture. Aquaculture is also important to local and national economies. However, the expansion of coastal aquaculture has caused some negative environmental impacts, and it has attracted the attention of environmental advocate groups. These groups have expressed concern over the contribution of coastal aquaculture to conversion of agricultural land, loss of mangroves and other sensitive ecological areas, water pollution, decrease in biodiversity, social inequities and other adverse impacts (Primavera, 1993; Naylor *et al.*, 1998). Most aquacultural scientists believe that the social and environmental performance of aquaculture can be improved, but they also feel that the concerns of the environmental groups are often embellished (Boyd and Schmittou, 1999). Thus, there is currently a great effort on the part of the aquaculture industry to improve their image and environmental and social performance through presentation of factual information on aquaculture and by adoption of more responsible production techniques and operating methods.

The purpose of this chapter is to discuss the influence of aquaculture on mangroves and to outline practices that should be followed for protecting mangroves in aquaculture areas.

Mangrove Ecosystems

According to Mitsch and Gosselink (1993), a mangrove forest is an association of halophytic trees, shrubs, palms, ferns and other plants growing in brackish to saline tidal waters on mudflats, riverbanks and coastlines in tropical and subtropical regions. This vegetation has the common characteristic of living in the zone inundated by the highest tides and exposed by the lowest tides. All mangrove species also share a common characteristic of salt tolerance. About 70 species of plants are recognized as mangrove species, and the greatest diversity of mangrove species is in Southeast Asia and Australia (Spalding *et al.*, 1997). Mangrove forests can be considered mangrove ecosystems, for they are inhabited by a wide variety of animals coming mostly from the sea but some are from the land. Mangrove forests contain high biodiversity, serve as nursery grounds for many aquatic species, and tend to stabilize coastlines from storm surges (Massaut, 1999).

The most recent estimate of the world's mangrove forests is about 181,000 km^2 (Spalding *et al.*, 1997). However, two other fairly recent estimates were 168,810 km^2 and 198,818 km^2 (Table 9.1). The average of the three estimates, 182,920 km^2, is very close to the most recent estimate, so

Table 9.1. Estimates of mangrove areas (km^2) together with percentages of global totals (Spalding *et al.*, 1997).

Region	Mangrove area (IUCN, 1983)	Mangrove area (Fisher and Spalding, 1993)	Mangrove area (Spalding *et al.*, 1977)
South and SE Asia	51,766 (30.7%)	76,226 (38.3%)	75,173 (41.5%)
Australasia	16,980 (10.0%)	15,145 (7.6%)	18,789 (10.4%)
The Americas	67,446 (40.0%)	51,286 (25.8%)	49,096 (27.1%)
West Africa	27,100 (16.0%)	49,500 (24.9%)	27,995 (15.5%)
E. Africa and the Middle East	5,508 (3.3%)	6,661 (3.4%)	10,024 (5.5%)
Total area	168,810	198,818	181,077

about 180,000 km^2 is an easily remembered estimate of the world's mangrove areas. Most of the mangrove area is in South and Southeast Asia and in the Americas (Table 9.1). The amount of mangrove has been declining during the past century, and possibly 50% of the world's mangrove area has been lost during the past 100 years (Massaut, 1999).

Importance of Mangroves

Functions

Mangroves form dense stands along estuarine streams and coastlines. They have the physical effects of stabilizing soil against erosion, acting to reduce the energy of waves and storm surges, shielding coastal areas from wind, and filtering runoff entering coastal waters from rivers. They have a remarkable ability to protect coastlines from the ravages of wind, waves and floods.

Mangrove ecosystems are the most productive ecosystems found along tropical and subtropical shorelines. They provide habitat for many bird species, and nutrients contained in bird faeces are a major source of nutrients in mangroves. Particulate organic matter originating from leaf fall is also an important source of nutrient for a wide variety of aquatic species. Thus, mangroves are excellent nursery grounds for a wide variety of marine and estuarine species because they provide shelter and food (Massaut, 1999).

Mangrove ecosystems tend to act as a large sedimentation area and biofilter. Water entering from the land and passing through mangroves is purified through the processes of sedimentation, filtration, microbial activity, adsorption of bottom soils, etc. Thus, mangroves serve to protect coastal water quality.

Attributes

Mangroves are a source of great biological diversity and thus very important to the ecology of coastal areas. Because of this diversity, mangroves are of aesthetic value, wilderness value and educational value. Many coastal communities depend heavily upon resources from mangrove ecosystems. Thus, mangroves represent wilderness areas of significant public and scientific interest and, in some nations, tracts of mangrove are being set aside as national forests.

Uses

The most direct use of mangroves is for fuelwood and construction material. Certain types, e.g. *Rhizophora*, are particularly sought for making charcoal. In construction, mangroves are used to fabricate the frames of small buildings, for piers, for scaffolding and props, etc. Agricultural use of mangroves usually has involved conversion of mangrove areas to rice fields. Mangroves have been used as fodder or green forage for livestock, and livestock is sometimes allowed to graze in mangrove forests. Mangroves have been used for various industrial purposes such as production of paper, synthetic fibres, dye for cloth, and tannins for leather processing. The use of mangroves for paper has resulted in major losses of mangrove reserves (Ong, 1982; Spalding *et al.*, 1997). Production of wood chips for pulp and paper has been particularly destructive of mangrove in Southeast Asia (Ong, 1982; Macintosh and Phillips, 1992). Mangroves also may be cleared to construct salt ponds (Paw and Chua, 1991). Fish, reptiles, shrimp, crabs and other animals may be captured in mangrove areas and used for human food. Mangrove bees are a source of honey and wax for local communities. Mangroves also have been consumed in urban sprawl around many major cities of the world.

The support of coastal fisheries by mangroves is especially important. As mentioned above, mangroves supply nutrients to coastal waters and serve as nursery grounds. As a result, many highly productive fishing grounds are adjacent to mangrove ecosystems. In Malaysia, about 32% of fish landings are associated with mangroves (Jothy, 1984), and in the Philippines, 72% of municipal fish catches are associated with mangroves (Paw and Chua, 1991). It has been reported that 80% of the marine fishery in Florida depends in some way on mangroves (Mangrove Working Group, 1998).

Mangrove forests also have been converted to fish and shrimp ponds. However, healthy mangroves should be maintained in subtropical and tropical areas for aquaculture. Areas adjacent to mangroves are excellent sites for cockle beds (FAO/NACA, 1995). Fish and shrimp larvae for stocking coastal aquaculture ponds often are captured in mangrove ecosystems or adjacent areas. Mangroves also protect coastal aquaculture areas from storms and floods, and improvements in coastal water quality resulting from natural purification in mangroves benefits aquaculture. Removal of mangroves can be detrimental to aquaculture. For example, Hong and San (1993) reported the following adverse effects of mangrove destruction on coastal aquaculture: coastal erosion, salinity intrusion, decline in shrimp postlarvae abundance, decrease in mud crab (*Scylla serrata*) abundance, acidification of pond soils and waters, and declining shrimp harvest from ponds.

Social issues

The loss of mangroves can have adverse economic and social impacts (Bailey, 1988), since local communities depend upon them for food, construction material, fuelwood and employment. Removal or restriction of access to mangroves results in declining socioeconomic conditions for traditional fishermen and others who hunt and gather goods from these ecosystems. It is not surprising that intrusion of non-traditional activities into mangroves may result in social conflict.

The Mangrove Controversy

Mangrove ecosystems have many beneficial uses. Some of the uses by local communities can be managed in a sustainable manner, while other uses are destructive to mangrove ecosystems. The most destructive use involves clear cutting the mangrove for wood products or clearing them in order to use the land for other purposes. Coastal aquaculture has been responsible for some of the loss of mangroves through land use conversion, but it is only one of several activities that have destroyed mangroves. The loss of mangroves to coastal aquaculture is less than often perceived. Examination of historical data showed that the largest loss of mangrove occurred before the recent boom in coastal aquaculture and shrimp farming in particular (Hambrey, 1996).

Data provided by the World Resources Institute (1996) showed that individual nations in Africa and South and Southeast Asia had experienced losses of original mangrove from 0% in the Congo to 85% in India (Table 9.2). The unweighted average was 61% loss in South and Southeast Asia and 55% loss in Africa. This is an interesting finding because there is very little coastal aquaculture in Africa but South and Southeast Asia have the majority of the world's coastal fish and shrimp farms. If aquaculture had been the major factor causing loss of mangrove, one would expect the loss in South and Southeast Asia to have greatly exceeded the loss in Africa.

A survey of former land use at shrimp farms in the world's leading shrimp-farming nation, Thailand, revealed that only 13.7% of farms were located in former mangrove areas (Macintosh and Phillips, 1992). In another survey of use in 1993 of pre-1961 mangrove area in Thailand (Table 9.3), 17.5% of shrimp farms were assigned to former mangrove areas. A much greater loss of mangroves was attributed to agriculture, salt farms, mining and infrastructure development than to shrimp farming. In Indonesia, also a major shrimp-farming nation, the total area of shrimp farming is only 5% of the existing mangrove reserves of 4 million hectares. Conversion of mangrove to aquaculture ponds was a major factor in reducing Philippine mangrove reserves from 448,000 ha in 1968 to 110,000 ha in 1988, but most of the ponds were for milkfish culture rather than shrimp production. A

Table 9.2. Estimated loss of original mangrove area in different regions (based on country, data available from WRI (1996)).

Region	Country	Loss of original (%)
South and Southeast Asia	Bangladesh	73
	Brunei	17
	India	85
	Malaysia	32
	Myanmar	58
	Pakistan	78
	Singapore	76
	Thailand	87
	Vietnam	62
	Unweighted average	61
Africa	Angola	50
	Congo	00
	Djibouti	70
	Equatorial Guinea	60
	Gabon	50
	Guinea	60
	Guinea-Bissau	70
	Kenya	70
	Madagascar	40
	Mozambique	60
	Somalia	70
	Tanzania	60
	Zaire	50
	Unweighted average	55

Table 9.3. Utilization of pre-1961 mangrove areas in 1993 (Menasveta, 1997).

Land use type	Area (ha)	Per cent
Shrimp farms	64,992	17.5
Community use	4,961	1.3
Agriculture, salt farms, mining, infrastructure	133,813	35.9
Mangrove remaining	168,683	45.3
Total	372,449	100

comprehensive survey of land use for shrimp farming in Asia (Asian Development Bank and Network of Aquaculture Centers in Asia Pacific, 1997), which included about 5000 farms in 12 nations, showed that 41.9% of extensive farms were in former mangrove areas, but only about 19% of semi-intensive and intensive farms were in such areas (Table 9.4).

In Ecuador, there were about 204,000 ha of mangrove forest in 1969 before shrimp farming began (Table 9.5). By 1992, 120,000 ha of shrimp farms had been constructed, but there were still 162,000 ha of mangroves – a loss of 42,000 ha, but this loss represented only 35% of the shrimp farm area (CLIRSEN, 1997). The study estimated that only 15% to 20% of the mangrove loss was actually caused by shrimp farm construction.

It is estimated that less than 10% of the world's mangrove resources have been converted to shrimp ponds (Boyd and Clay, 1998). Furthermore, much of the mangrove area converted to shrimp or other aquaculture farms

Table 9.4. Estimated land use type for shrimp farming in 12 Asian countries based on a survey of around 5000 farms conducted during 1995 (ADB/NACA, 1997). The study was conducted in Bangladesh, Cambodia, China, India, Indonesia, Korea, Malaysia, The Philippines, Sri Lanka, Taiwan, Thailand and Vietnam.

	Intensive farms		Semi-intensive farms		Extensive farms	
Land use type prior to shrimp farming	Area (ha)	Per cent	Area (ha)	Per cent	Area (ha)	Per cent
Intertidal land						
Ex-mangrove	14,142	19.0	24,786	18.6	359,118	41.9
Non-forested wetland	8,669	11.6	25,206	19.0	136,121	15.9
Salt pans	7,496	10.0	4,242	3.2	20,649	2.4
Other intertidal	6,470	8.7	14,603	11.0	195,948	22.9
Supratidal land						
Rice farm	22,515	30.2	19,397	14.6	122,087	14.3
Other agriculture	8,432	11.3	4,603	3.5	8,215	0.9
Non-agriculture	7,397	9.9	36,278	27.3	25,601	3.0
Estimated total farm area	75,121	100	129,115	100	867,739	100

Table 9.5. Statistics associated with loss of mangroves from 1969 to 1992 and increase in shrimp farming area in Ecuador, 1970–1992 (CLIRSEN, 1997).

Mangrove forest in 1969	204,000 ha
Mangrove forest in 1992	162,000 ha
Loss	42,000 ha
Shrimp farms constructed (1970–1992)	120,000 ha
Maximum possible use of mangrove	35%
Estimated actual use	15–20%

was secondary scrub and fringe areas and not primary mangrove forest. Many shrimp farms were constructed on land behind mangroves or on islands in estuaries where salt flats are often surrounded by mangrove trees. Environmentalists often see such areas from aeroplanes and assume that the area currently occupied by shrimp ponds was mangroves, when that is not the case. Available data do not agree with claims that shrimp farming has been a major factor in the decline of world mangrove forests (Hambrey, 1996).

Soils in mangrove forests often are highly acidic because they contain large amounts of sulphur and organic matter. The soil typically has a carbon : nitrogen ratio above 20 (Sonnenholzner and Boyd, 2000), and there may be a high percentage of clay (Table 9.6). Such soil is not desirable for shrimp pond bottoms. Moreover, mangroves are in low-lying areas where water exchange to tidal flow in the creeks and channels of the swamp is incomplete. Pond effluents often are not transported away completely after discharge. As a result, cross-contamination of intake and discharge water occurs on farms located in mangrove areas. This can cause deterioration of water supplies on farms and diseases may spread from farm to farm. Crabs and other possible carriers of shrimp diseases are abundant in mangrove areas, and they cannot be excluded from ponds. Today, common practice is to construct shrimp and other aquaculture ponds behind mangrove areas, and intake and discharge canals often must be routed through mangrove.

Nevertheless, mangrove may still be used for aquaculture. Although large farms are seldom established in mangroves, small farmers in some countries still develop fish or shrimp ponds in mangroves, for it is the only land available to them. They usually do not have the right to use the land, and the operations may be illegal. Nevertheless, in some countries, the government does not seriously attempt to curtail such activities. Larger and better-organized aquacultural producers sometimes are blamed for the adverse environmental impacts caused by the small producers. Governments need to develop better policies and regulations about mangrove use and enforce those regulations fairly and consistently.

Recent research has shown that mangrove forests can enhance the removal of solids and nutrients from shrimp farm effluents (Robertson and Phillips, 1995). Discharges from shrimp farms can be routed through mangrove areas to effect water treatment. It is thought that this practice can have the combined benefits of stimulating mangrove productivity, enhancing

Table 9.6. Typical analysis of mangrove soil.

pH	4.3
Total carbon	15.2%
Total nitrogen	0.41%
Carbon/nitrogen ratio	37
Total sulphur	1.18%
Lime requirement (by H_2O_2 oxidation)	47 tonnes (upper 15 cm)

coastal fisheries, minimizing pollution of the coastal environment, and providing a higher quality water supply for shrimp farms. It is not difficult to produce mangrove seedlings for transplantation, and the coastal aquaculture industry could initiate mangrove reforestation projects to enhance water quality.

In summary, shrimp farming and other kinds of aquaculture have encroached on mangroves in some nations, but the impact of aquaculture on mangrove loss is rather small compared to the loss of mangroves as the result of other human activities. Aquaculturists have learned that mangrove swamps are not the best sites for shrimp and other aquaculture farms. Most governments are learning of the benefits of mangroves and regulating the use of mangrove areas. Controlled and sustainable use of the mangrove resources of the world is desirable, and it is evident that coastal development will not be permitted to consume large tracts of mangrove in the future. However, mangrove resources should be used and managed rather than preserved (Boyd and Tucker, 1998), and aquaculture methods that do not have adverse effects on mangroves have been developed in recent years.

Aquaculture Practices and Mangrove Protection

In many nations, environmental laws and regulations may be weak or not enforced. Thus, environmental protection measures seldom are incorporated in development or production activities. In order to demonstrate environmental and social responsibility, many aquaculture organizations are developing codes of conduct with codes of practices for responsible aquaculture (Boyd *et al.*, 2001). These codes contain practices for protection of mangrove and other ecologically important habitats.

The Global Aquaculture Alliance (GAA) was formed by shrimp and fish aquaculture companies and supporting firms to promote environmentally responsible aquaculture. The GAA has a code of conduct and a manual on codes of practices for responsible aquaculture (Boyd, 1999). The GAA has developed a technique for preparing a compliance plan for preventing negative environmental impacts from shrimp aquaculture, and a third-party inspection programme will be used to verify that farms are in compliance with the environmental plan. A major feature of the GAA programme is mangrove conservation. The GAA sponsored a working group of shrimp farming and mangrove experts in Bangkok, Thailand, in 1998 for the purpose of developing guidelines and best management practices for mangrove protection. The GAA adopted most of the recommendations of the mangrove working group (Mangrove Working Group, 1998). The GAA requires that its members adhere to a policy of no net loss of mangrove as a result of shrimp farming activities. The principles in the GAA mangrove code are as follows:

- Construction of new farms or expansions of existing farms should not interfere with mangrove ecosystems.

- Farms should be operated in a manner to prevent damage to mangroves.
- A monitoring programme should be established to verify that mangroves are not damaged by shrimp farming.
- Abandoned shrimp farms on public land should be replanted to mangroves if the land was a former mangrove ecosystem.
- Traditional uses of mangroves by local communities should not be prevented.

The GAA guidelines for siting shrimp farms require that new shrimp farm development be located outside of mangroves and other wetlands of significant ecological value. Roads, canals and other infrastructure should not block flow into and out of wetlands. Particular care should be given to preventing impoundment of freshwater in brackish water areas or vice versa. When shrimp farms are constructed behind mangroves, mangrove trees must be removed to construct pumping stations, canals, roads and other infrastructure. In order to comply with the principle of no net loss of mangroves, shrimp farm developers must replant mangroves in other places. The suggested method is to pay into a mangrove reforestation and conservation programme an amount necessary to replant three times more mangroves than were destroyed. This assumes that mangrove replacement is done by appropriate techniques and in suitable habitat.

Construction practices should be planned and organized to protect mangroves and provide a sustainable shrimp farm. A survey of site soils should be made. Areas of highly acidic or highly organic soils revealed in the survey should be excluded from the farm layout, or suitable soil mitigation procedures used (Hajek and Boyd, 1994). The construction should be done by standard engineering practices and all regulations related to construction should be obeyed. Construction activities should be planned to minimize disturbances to mangroves. Records should be maintained on mangrove removal for canals, roads and other infrastructure, and a plan for mitigating the loss should be implemented. Natural water flow should not be altered to cause changes in sedimentation patterns or salinity. Construction of earthwork should be done by cut and fill procedures so that spoil piles and barrow pits are not created on the periphery of the farm. Potential acid-sulphate soils should not be exposed to the land surface. Construction should be done in the dry season or methods for preventing runoff from the site should be installed, for runoff from construction sites is high in suspended solids.

Farms also should be operated in a manner that will not harm neighbouring mangrove ecosystems. Farm garbage should be incinerated or put in a landfill. It should not be dumped into mangroves as is sometimes done. Although studies show that mangroves and other wetlands may be efficient in treating shrimp farm effluents, this practice should not be initiated where effluents could cause enough sedimentation to harm the mangroves. If mangrove areas are to be used for treating effluents, a monitoring programme should be implemented to verify the health of the mangrove ecosystem. Water exchange should be reduced as much as practical on shrimp farms, and effluent should not be discharged into stagnant areas of mangroves. Dead shrimp should be disposed of in a sanitary manner and not placed in mangroves. When drugs, antibiotics, or other chemicals are applied to shrimp ponds, pond water should not be discharged into mangrove areas until the chemicals have degraded to non-toxic forms. Care should be taken to prevent pollution of mangroves with fuel or other petroleum products. Pumping stations for shrimp farms are often near mangroves. Fuel storage tanks and pump motors should be fitted with secondary containment devices to prevent fuel spills.

Farms should monitor neighbouring mangrove areas to ascertain that negative impacts are not occurring. Factors to consider in mangrove assessments are changes in the area of mangroves, changes in species diversity, presence of dead or dying trees, freshwater impoundment, saline water intrusion, sedimentation, hydrological changes and use of mangroves by local people.

Conclusions

Coastal aquaculture has not been the major factor causing a decline in world mangrove reserves, but it has contributed to the overall loss of mangroves. Mangrove ecosystems are not the best sites for aquaculture ponds, and their use for this purpose should be discouraged by the aquaculture industry and regulated by governments. The shrimp farming industry is developing better practices for protecting mangrove, and other coastal aquaculture industries should do likewise. Healthy mangrove ecosystems are beneficial to coastal aquaculture, and mangrove protection by aquaculturists represents a 'win–win' situation. It is possible to have sustainable aquaculture farms and sustainable mangrove ecosystems in the same vicinity through more enlightened management methods and dedication by aquaculturists to environmental stewardship.

References

Asian Development Bank and Network of Aquaculture Centers in Asia Pacific (ADB/NACA) (1997) *Final Report on the Regional Study and Workshop on Aquaculture Sustainability and the Environment (RETA 5534)*. Asian Development Bank and Network of Aquaculture Centres in Asia-Pacific. NACA, Bangkok, Thailand, 21pp.

Bailey, C. (1988) The social consequences of tropical shrimp mariculture development. *Ocean and Shoreline Management* 11, 31–44.

Boyd, C.E. (1999) *Codes of Practice for Responsible Shrimp Farming*. Global Aquaculture Alliance, St Louis, Missouri, 49pp.

Boyd, C.E. (2001) *The Responsible Aquaculture Program*. Global Aquaculture Alliance, St Louis, Missouri.

Boyd, C.E. and Clay, J.W. (1998) Shrimp aquaculture and the environment. *Scientific American* 278(June), 58–65.

Boyd, C.E. and Schmittou, H.R. (1999) Achievement of sustainable aquaculture through environmental management. *International Journal of Aquaculture Economics and Management* 2, 59–69.

Boyd, C.E. and Tucker, C.S. (1998) *Pond Aquaculture Water Quality Management*. Kluwer Academic, Boston, Massachusetts, 700pp.

Boyd, C.E., Hargreaves, J.A. and Clay, J. (2001) Codes of conduct for marine shrimp aquaculture. In: Browdy, C.L. and Jory, D.E. (eds) *The New Wave. Proceedings of the Special Session on Sustainable Shrimp Culture, Aquaculture 2001*. The World Aquaculture Society, Baton Rouge, Louisiana.

CLIRSEN (Centro de Levantamiento Integrado por Recursos Sensores) (1997) Estudio multitemporal de camaroneras, manglares y salinas, 1995. *Aquacultura del Ecuador* 18(1), 27–35.

FAO/NACA (1995) Regional study and workshop on the environmental assessment and management of aquaculture development. Food and Agricultural Organization of the United Nations and Network of Aquaculture Centres in Asia-Pacific, Bangkok, Thailand.

Fisher, P. and Spalding, M.D. (1993) *Protected Areas with Mangrove Habitat*. Draft Report. World Conservation Monitoring Centre, Cambridge, UK, 60pp.

Hajek, B.F. and Boyd, C.E. (1994) Rating soil and water information for aquaculture. *Aquacultural Engineering* 13, 115–128.

Hambrey, J. (1996) Comparative economics of land use options in mangrove. *Aquaculture Asia* 1(2), 10–14.

Hong, P.N. and San, H.T. (1993) *Mangroves in Vietnam*. International Union for the Conservation of Nature, Regional Wetlands Office, Bangkok, Thailand.

IUCN (1983) *Global Status of Mangrove Ecosystems*. Commission on Ecology Papers No. 3 (Saenger, P., Hegerl, E.J. and Davie, J.D.S. eds) International Union for Conservation of Nature and Natural Resources, Gland, Switzerland, 88pp.

Jothy, A.A. (1984) Capture fisheries and the mangrove ecosystem. In: Ong, J.E. and Gong, W.K. (eds) *Productivity of the Mangrove Ecosystem: Management Implications*. Universiti Sains Malaysia, Penang, Malaysia, pp. 129–141.

Macintosh, D.J. and Phillips, M.J. (1992) Environmental issues in shrimp farming. In: de Saram, H. and Singh, T. (eds) *Shrimp '92, Proceedings of the 3rd Global Conference on the Shrimp Industry*. Infofish, Kuala Lumpur, Malaysia, pp. 118–145.

Mangrove Working Group (1998) *Coastal Shrimp Aquaculture and Mangrove Forests. Part 1: A Background Report*. Global Aquaculture Alliance, St Louis, Missouri, 66pp.

Massaut, L. (1999) *Mangrove Management and Shrimp Aquaculture*. Department of Fisheries and Allied Aquacultures, and International Center for Aquaculture and Aquatic Environments, Auburn University, Alabama, 45pp.

Menasveta, P. (1997) Intensive and efficient shrimp culture system – the Thai way – and save mangroves. *Aquaculture Asia* 2(1), 38–44.

Mitsch, W.J. and Gosselink, J.G. (1993) *Wetlands*, 2nd edn. Van Nostrand Reinhold, New York, 722pp.

Naylor, R.L., Goldburg, R.J., Mooney, H., Beveridge, M., Clay, J., Folke, C., Kautsky, N., Lubchenco, J., Primavera, J. and Williams, M. (1998) Nature's subsidies to shrimp and salmon farming. *Science* 282, 883–884.

Ong, J.E. (1982) Mangroves and aquaculture in Malaysia. *Ambio* 11, 252–257.

Paw, J.N. and Chua, T.E. (1991) An assessment of the ecological and economic impact of mangrove conversion in Southeast Asia. In: Chou, L.M., Chua, T.E., Khoo, H.W., Lim, P.E., Paw, J.N., Silvestre, G.T., Valencia, M.J., White, A.T. and Wong, P.K. (eds) *Towards an Integrated Management of Tropical Coastal Resources. ICLARM Conference Proceedings 22*, pp. 201–212.

Primavera, H.J. (1993) A critical review of shrimp pond culture in the Philippines. *Reviews in Fisheries Science* 1(2), 151–201.

Robertson, A.I. and Phillips, M.J. (1995) Mangroves as filters of shrimp pond effluents: predictions and biogeochemical considerations. *Hydrobiologia* 295, 311–321.

Sonnenholzner, S. and Boyd, C.E. (2000) Chemical and physical properties of shrimp pond bottom soils in Ecuador. *Journal of the World Aquaculture Society* 31(3), 358–375.

Spalding, M., Blasco, F. and Field, C. (1997) *World Mangrove Atlas*. The International Society for Mangrove Ecosystems, Okinawa, Japan, 178pp.

World Resources Institute (1996) *World Resources 1996*. A guide to the global environment. Database diskette and users guide. World Resources Institute, Washington, DC.

Environmental Effects Associated with Marine Netpen Waste with Emphasis on Salmon Farming in the Pacific Northwest

10

Kenneth M. Brooks,[1] Conrad Mahnken[2] and Colin Nash[2]

[1]Aquatic Environmental Sciences, 644 Old Eaglemount Road, Port Townsend, WA 98368, USA; [2]National Marine Fisheries Service, Northwest Fisheries Science Center, Manchester Laboratory, 7305 Beach Drive East, Manchester, WA 98366, USA

Abstract

There is considerable evidence available in the international scientific literature to evaluate any potential risk of netpen salmon farming on the marine environment. Environmental impacts have been studied in great detail for some 20 years, mostly in temperate environments. The results are well documented, and a common denominator is that the potential for environmental impact depends primarily on the site of each individual farm.

Scientific evidence in the literature indicates that the potential changes in the sediments below operating netpen farms bear the most risk for the environment. Continuous monitoring of the sediments under and around farm sites for many years has produced an extensive database of chemical and biological information, and specific parameters are now being used to predict the environmental effects. Key parameters include, *inter alia*, sediment grain size, total volatile solids or total organic carbon, redox potential, free sulphide concentrations, and ultimately, invertebrate community assessment. Modelling programmes are also beginning to provide insight into the environmental response to farm waste, but these are not yet adequate to make reasonable quantitative predictions.

Long-term monitoring of the sediments has also revealed that chemical and biological recovery of the substrate under and around farm sites occurs naturally without human intervention or mitigation. *In situ* data show that physicochemical recovery can occur within weeks or months at some sites, and within 2 or 3 years at others.

This chapter reviews the available scientific literature on environmental effects of netpen farming with emphasis on studies conducted in the state of Washington, USA, and the province of British Columbia, Canada – the primary locations of salmon farming in the eastern north Pacific Ocean. The review addresses:

1. The physicochemical and biological impacts of biodeposits (fish faeces and uneaten feed) from farm operations on the environment beneath netpens.
2. The impact on benthic and epifaunal communities of the accumulation of heavy metals in the sediments below netpens.
3. The impact on non-target organisms of the use of therapeutic compounds (both pharmaceuticals and pesticides) at netpen farms.
4. The physiological effect of low dissolved oxygen levels on other biota in the water column.
5. The toxic effect of hydrogen sulphide and ammonia from the biodeposits below netpen farms on other biota in the water column.
6. The toxic effect of algal blooms enhanced by the dissolved inorganic wastes in the water column around netpen farms.
7. Changes in the macrofaunal community caused by the accumulation of organic wastes in sediments below netpen farms.

Introduction

This chapter reviews current information on the effects of the many activities associated with netpen aquaculture on the environment. Where possible the review attempts to deal with those effects in quantitative terms, and the measures that are currently used to reduce them. Following this introduction, the chapter contains seven sections. The first section reviews the potential effects of the organic wastes emanating from netpen salmon farming. The origins of such effects are uneaten or waste feed, faeces from the fish, and biofouling organisms on the netpen structures. The second section reviews the potential effects of inorganic wastes, specifically nitrogen and phosphorus, and heavy metals on benthic and pelagic biota. The third section deals with pathogenic organisms which might be in the vicinity of fish farms, and the risks to human health from wastes which might contain such pathogens. The next section deals with the therapeutic compounds commonly used to control parasites and diseases. The fifth section reviews the biological and chemical changes in the sediments and in the water column both beneath netpen farms and downstream. Where possible, quantitative information is provided and then applied collectively in terms of an operating farm with reference points. The sixth section reviews information on the chemical and biological recovery of sediments under salmon farms. The final section reviews alternatives being used for management of all the aforementioned environmental effects. These include monitoring experiences with a number of indicators and models, followed by a review of other government policies and methodologies.

Types of Solid Organic Wastes Associated with Salmon Netpen Farms

Waste feed

Early diets for farmed Atlantic salmon contained 45–50% protein, 16–22% lipids and 17% carbohydrates. Technology now permits the production of high-energy salmon diets containing about 30–35% lipids, 40% protein and the minimum level of digestible carbohydrate (about 10%) necessary to bind the pellets. These high-energy diets more closely resemble the composition of the natural prey of salmon than did previous formulations. More recent salmon feeds were reported by Einen *et al.* (1995) and Rosenthal *et al.* (1995) to contain 7% nitrogen and 1% phosphorus. Mann (EWOS Canada Ltd, personal communication) estimated that current salmon diets contain 38–39% crude protein, 6.5% nitrogen and 1% phosphorus. Half of the lipid content in current diets is fish oils and the other half plant oils, such as flax and linseed oils, which are high in ω-3 fatty acids.

The amount of waste feed depends on feeding efficiency, which is principally influenced by feed composition, feeding methodology, water currents at the site and netpen configuration. Beveridge *et al.* (1991) stated that up to 30% of feed was lost. Rosenthal *et al.* (1995) reported higher losses (up to 35%) for wet feeds – which might contain more than 30% moisture – than dry feeds. Weston (1986) suggested that less than 5% of dry feed was lost at Puget Sound salmon farms. This is consistent with research by Gowen and Bradbury (1987), who reported that losses were least (1–5%) with dry feeds, which contained less than 10% moisture. Findlay and Watling (1994) reported maximum feed loss rates of 5–11%, with an average feed wastage of < 5%. Dry and semi-moist feeds are now used exclusively in the Pacific Northwest USA and current feed loss rates are estimated at between 3% and 5% (J. Mann, EWOS Canada Ltd, personal communication).

The amount of feed loss is also dependent on feeding methods and strategies. Cross (1990) reported that feed wastage at a commercial salmon farm in Sooke Inlet, British Columbia, Canada, was 3.6% delivered by hand and 8.8% delivered by automatic feeders (probably owing to the abrasion of feed pellets in some, though not all, automatic feeders).

New technologies, such as feedback cones and underwater video or acoustical devices described by Mayer and McLean (1995), are now commonly used to monitor feeding behaviour in efforts to minimize losses of uneaten feed from netpens. Sutherland *et al.* (2000) conducted a study at a salmon farm in Canada to quantify suspended particulates during peak feeding times and to make point-in-time estimates of organic loading. Based on stable carbon isotope analysis, they concluded that very little feed was not consumed by the fish at the farm under study. In a series of video reports to the British Columbia, Canada, Ministry of Environment documenting the environmental conditions on the perimeter of several salmon farms in the Province, Brooks (2000a–f) recorded no observable wasted feed pellets.

Thus, the indication is that 5% or less of the dry feed delivered to cultured salmon in netpens is lost to the environment. This result has been attributed to improved feedback technologies and the practice of feeding the fish to satiation once or twice daily. Improvements in feed delivery systems to minimize pellet disintegration will probably reduce losses to well below 5%, much less than the 20–30% used in many aquaculture models as discussed below.

Fish faeces

Weston (1986) estimated that 25–33% of feed consumed by the fish was ejected as faeces. Modern diets are approximately 87–88% digestible (J. Mann, EWOS Canada Ltd, personal communication). The remaining ash consists primarily of calcium and inorganic phosphate, and represents 8.0–8.5% of the feed. This implies that approximately 12.5% of the weight of ingested feed will be ejected in faeces. About 4% of the feed ingested is estimated as being ejected as labile organic material in the faeces. If 5% of the feed is uneaten (Findlay and Watling, 1994), and faeces contribute organic matter equivalent to 4% of the feed weight, then approximately 8.8% of the labile organic carbon delivered in feed is discharged from netpens in particulate form, contributing to the biochemical oxygen demand (BOD) of the sediments.

Feed conversion ratios (FCRs) of farmed fish are frequently quoted in literature. They are typically measured as the ratio of the dry weight of feed provided to the wet weight of salmon produced. They are considered an essential metric by the aquaculture industry for assessing producer programme efficiency. The FCR is affected by genetic and environmental factors; feed quality and composition, including palatability and nutrient balance; together with fish health; and feeding methods.

Two types of FCR are typically defined by industry:

- The economic feed conversion ratio (EFCR) is defined as the amount of feed supplied to a farm divided by the round dressed weight of fish produced for market.
- The biological feed conversion ratio (BFCR). This metric is biologically and environmentally more meaningful but more difficult to determine. It is equal to the feed actually consumed by the fish (feed provided *less* the uneaten portion) divided by the total fish biomass produced on the farm, including escapes and mortalities.

Enell and Ackefors (1992) reported that marine FCRs in Norway declined from 2.25 in 1974 to an average of 1.2–1.3 in 1992. The authors calculated that improvement resulted in a 23% decrease in nitrogen loading and a 50% decrease in phosphorus loading associated with farm operations. Rosenthal *et al.* (1995) estimated FCRs for Atlantic salmon to be 1.2, while Levings (1997) estimated an even lower FCR of 1.17 for operations in British Columbia, Canada. Additional improvement in FCRs will be difficult to achieve.

Fish carcasses as wastes

Winsby *et al.* (1996) reviewed and analysed the mortality of fish at netpen salmon farms in British Columbia, Canada, in 1994. Their data suggested approximately 2000 tonnes of salmon died at farms that year, or approximately 9% of the total production of 22,000 tonnes.

No inappropriate disposal of salmon carcasses has been documented in the literature. Losses of fish on netpen salmon farms are restricted to individual fish, which may have been attacked and killed by a predator, and numbers of fish which died as a consequence of an algal bloom or disease. Best management practices (BMPs) for netpen salmon farms require physical removal of any carcasses on a daily basis, and therefore they do not contribute to biological loading on the environment.

Biofouling organisms as wastes

Biological fouling is a significant factor in coastal environments, and large masses of mussels, barnacles, ascidians and bryozoans can weigh down nets and restrict water flow through a netpen complex. Heavily fouled nets can also compromise netpen structural integrity.

Weston (1986) concluded that biofouling organisms on netpens, and the debris released as a result of net cleaning, were not significant sources of organic input to sediments beneath salmon farms. Winsby *et al.* (1996) discussed the mechanics of removing fouling organisms from nets associated with salmon farms. Quantitative information on biofouling of netpens is not currently available.

Brooks (1994a) defined a neutral impact zone (NIZ) as that distance from the perimeter of a salmon farm at which there was neither an apparent increase nor a decrease in the abundance and diversity of the benthic infaunal community when compared with a local control site. In annual observations at a poorly flushed farm site, he noted that the NIZ was influenced by several factors, including, for example, deposits of debris following the pressure-washing of nets *in situ*. He observed a 30-cm-deep layer of mussel debris in sediments from the perimeter of the farm stretching a distance of 6 m downstream. The downstream location of the NIZ increased from 12 m in 1993 to approximately 22 m in 1994, after *in situ* cleaning. He concluded that it was not possible to establish a cause and effect relationship, but the presence of the mussel shells undoubtedly had an effect on the benthos.

Measurement of organic wastes

Brown *et al.* (1987) compared the areal extent of benthic impacts associated with organic wastes from fish farms in Sweden with that from sewage treatment plants and pulp mill effluents. They found reducing (anaerobic) sediments covering 0.6 km^2 around a poorly flushed salmon farm located in shallow water. Significant changes in the benthic community were observed within 15 m of the perimeter of the farm. In comparison, they cited Stanley *et al.* (1980) in noting that a pulp mill in Loch Eil, Scotland, had created reducing conditions in sediments extending over an area of 5 km^2. Pearson (1986) observed reducing sediments covering 23 km^2 associated with a sewage disposal site at Garroch Head, Firth of Clyde, Scotland. The impact associated with that single sewage discharge covered an area 38 times as large as that impacted by the salmon farm.

Ellis (1996) suggested that waste feed and faeces from salmon farms in British Columbia, Canada, were equivalent to the human sewage from a city of 500,000 people, and Folke *et al.* (1994) compared the waste from 100 tonnes of salmon a human settlement of 850 to 3200 persons. Ackefors and Enell (1990) criticized the assumptions upon which such comparisons were made. Their argument was based on differences in the form of nitrogen released from sewage treatment plants and fish farms and differences in the ratio of carbon, nitrogen and phosphorus discharged from the two activities.

Taylor *et al.* (1998) found that organic enrichment adjacent to the Macaulay and Clover Point outfalls in the city of Victoria, British Columbia, Canada, was similar to that expected at a productive salmon farm. Adverse effects on benthic infauna associated with organic enrichment by sewage treatment plants were generally restricted to distances less than 100 m from the diffusers. The authors concluded that the magnitude and extent of the observed effects ($<$ 400 m from the outfalls) indicated little cause for concern for human health. The effects at the outfalls extended four times further from the source than is allowed at salmon farms complying with the British Columbia Draft Waste Management Policy.

Ackefors and Enell (1994) estimated the total organic output from salmon farms to be of the order of 2.5 tonnes wet weight per tonne of fish produced. Gowen *et al.* (1991) cited three studies assessing the flux of carbon through salmon netpens. In all three cases the harvested fish retained 21–23% of the carbon in the feed and it was estimated that 75–80% of the carbon was lost to the environment mostly as CO_2. Merican and Phillips (1985) estimated that 35.6% of the carbon, 21.8% of the nitrogen and 65.9% of the phosphorus were lost to the environment in solid form. Other estimates of the total suspended solids output from intensive netpen culture by Kadowaki *et al.* (1980), Warrer-Hansen (1982), Enell and Lof (1983) and Merican and Phillips (1985) range from 5 to 50 g m^{-2} day^{-1} of suspended solids. Those publications do not reflect recent improvements in fish feed and feeding technologies.

Gowen and Bradbury (1987) estimated organic waste sedimentation rates of 27.4 g m^{-2} day^{-1} under Irish salmon farms, and an average of 8.2 g m^{-2} day^{-1} immediately adjacent to the perimeter of the netpen. Gowen *et al.* (1988) measured average rates of 82.2 g dry weight m^{-2} day^{-1} on the perimeter of a netpen in Washington, and Cross (1990) estimated an average overall sedimentation rate of 42.7 g total volatile solids (TVS) m^{-2} day^{-1} with a maximum of 94.5 g TVS m^{-2} day^{-1} at seven salmon farms in British Columbia, Canada. More recent work by Findlay and Watling (1994) in Maine, USA, measured sedimentation rates on the perimeter of salmon farms at between 1.0 and 1.6 g C m^{-2} day^{-1}, and Hargrave (1994) summarized sedimentation rates from less than 1 g to over 100 g C m^{-2} day^{-1} from salmon cage operations described by a number of authors.

Brooks, using his published data from many original sources (Brooks, 2000a–f), derived a theoretical estimate of contemporary TVS loading near fish farms. Given a feed with 11% moisture content and FCR of 1.2, the feed provided 1.07 kg dry feed per kilogram of fish produced. This was, in turn, equal to 0.094 kg of labile volatile solids per wet weight kilogram of fish produced. Thus, he estimated that a salmon farm producing 1500 tonnes of salmon during a 16–20-month production cycle would discharge 141 tonnes of organic waste on a dry weight basis. Furthermore, assuming a fish density of 10 kg m^{-3} in cages 15 m deep and a growout cycle of 18 months, the annual sediment load on average would be:

$$\left[\frac{\left(10\,\text{kg fish m}^{-3}\times 15\,\text{m deep}\times 0.094\,\text{kg TVS kg}^{-1}\,\text{fish}\right)}{(548\ \text{days})}\right]$$

which is equal to 25.7 g TVS m^{-2} day^{-1}. The load would, in reality, be lower at the beginning of the growout cycle and increase with increasing fish biomass.

Brooks (2000e) analysed sediments collected in canisters deployed 5 m above the bottom at varying distances from two farms in British Columbia and at reference stations. The mean loading of volatile solids on the perimeter of these farms was 39.2 g TVS m^{-2} day^{-1}. The mean deposition of volatile material at the control stations was 6.3 g TVS m^{-2} day^{-1} and the contribution by the farm was approximately 32.9 g TVS m^{-2} day^{-1}. These studies were completed near peak salmon biomass and the observed values would therefore be greater than the theoretical average of 25.7 g TVS m^{-2} day^{-1} calculated above. None the less, these observed and theoretical values are reasonably close.

In summary, sedimentation rates on the perimeter of salmon netpens have remained fairly constant in the range 15.1–100 g TVS m^{-2} day^{-1} despite the typical increase in farm size from 200–300 tonnes in the 1980s and early 1990s to the 1500 tonnes of recent years. A recent study by Brooks (2000e) found a TVS loading of 32.9 g m^{-2} day^{-1} on the perimeter of a salmon farm at peak production. This value is reasonably close to a theoretical average of 25.7 g TVS m^{-2} day^{-1} calculated for an entire 18-month production cycle.

Inorganic Wastes Associated with Salmon Netpen Farms

Dissolved nitrogen and phosphorus

Salmon excrete 75–90% of their ammonia and ammonium waste across gill epithelia (Gormican, 1989) or in concentrated urea (Persson, 1988; Gowen *et al.*, 1991). Brett and Zala (1975) reported a constant urea excretion rate by sockeye salmon of 2.2 mg N kg^{-1} h^{-1}. Nitrogen and phosphorus are also dissolved from waste feed and faeces during and after descent to bottom sediments. All these dissolved forms of nitrogen are readily available for uptake by phytoplankton. Silvert (1994a) suggested that 66–85% of phosphorus in feed is lost in a dissolved form to the environment at salmon farms.

Winsby *et al.* (1996) reported significant variation in observable increases in soluble nitrogen and phosphorus levels in the water column at salmon farms. Johnsen and Wandsvik (1991) and Johnsen *et al.* (1993) estimated that 20.5–30.0 g of nitrogen and 6.7 g of phosphorus are released per kilogram of Atlantic salmon produced when fed modern high-energy diets containing 30% lipid. Levings (1997) used these estimates to conclude that 844 tonnes of nitrogen and 188.6 tonnes of phosphorus are released to marine environments in British Columbia each year by salmon farms. The values do not include nitrogen and phosphorus associated with uneaten feed.

Statistically significant increases in soluble nutrients at salmon farms have infrequently been observed in Puget Sound (Rensel, 1989; Brooks, 1994a,b, 1995a,b). Aquatic Lands Leases (ALLs) for salmon farms in Washington, USA, have required monitoring of NO_3^-, NO_2^- and total ammonia (NH_3 + NH_4^+) in water samples taken within an hour of slack tide at stations located 30 m upstream, and 6 m and 30 m downstream at all permitted farms at a depth equal to one-half the depth of the containment nets. In general, the variability between replicate samples taken at the 6 m downstream station was as great, or greater, than any observed increase in nitrogen between upstream and downstream stations. No significant increases in nitrogen were observed at any of the 30 m downstream stations.

The highest observed level of toxic un-ionized ammonia (NH_3) was 0.0004 mg l^{-1}. This is lower (by a factor of 87.5) than the EPA (US Environmental Protection Agency) chronic exposure (4 day) concentration limit of 0.035 mg l^{-1} at pH = 8 and T = 15°C when sensitive salmonid species are present.

Rensel (1989) studied dissolved nitrogen levels at two poorly flushed farms in Washington. He compared dissolved nitrogen and un-ionized ammonia concentrations within the salmon pens with upstream and downstream levels during early ebb tides. Upstream dissolved nitrogen levels of 0.0003 mg l^{-1} were increased to 0.0023 mg l^{-1} at the centre of the netpen complex, but decreased to background levels at downstream stations. He also observed maximum un-ionized ammonia levels equivalent to 6% of the EPA criteria in the centre of these netpen complexes during slack tide.

Weston (1986) reported ambient levels of dissolved inorganic nitrogen (DIN) in Puget Sound at 0.3–1.9 mg l^{-1}, indicating high variability. The greatest increase in DIN reported by Brooks (1991, 1992, 1993a, 1994a, 1995a) was 5.29 μmol l^{-1} (0.09 mg l^{-1}), or 8% of the mean value reported by Weston (1986).

The literature indicates that the concentration of dissolved inorganic nitrogen added to marine water at salmon farms is very low on the perimeter of netpen farms, and essentially immeasurable at distances greater than 9 m from the farm perimeter.

Heavy metal accumulation in sediments

Zinc

Zinc is an essential element, important for insulin structure and function and as a co-factor of carbonic anhydrase. Historically, it has been added to salmon feeds in trace amounts equal to 30 to 100 mg kg^{-1} of feed (see Chow and Schell, 1978; Anderson, 1998).

Long *et al.* (1995) provided an effects range-low (ER-L) of 150 μg zinc g^{-1} dry sediment weight, an effects range-moderate (ER-M) of 410 μg g^{-1}, and an overall apparent effects threshold (AET) of 260 μg g^{-1}. The Washington State (USA) sediment quality criterion for zinc is 270 μg Zn g^{-1} dry sediment (WAC, 1991). Other available benchmarks include the mean of the threshold effects and the probable effects levels (TEL + PEL)/2 or 197.5 μg Zn g^{-1} dry sediment. Information on the development of the TEL and the PEL can be found in MacDonald (1994).

Brooks (2000b) summarized 193 analyses for zinc in sediments collected near 27 salmon farms in British Columbia, Canada. Nineteen samples from eight farms exceeded the AET of 260 μg Zn g^{-1} dry sediment. All of the high zinc samples also contained significantly elevated sediment sulphide and TVS concentrations. There was a statistically significant correlation between zinc and both TVS and total sediment sulphide levels.

In response to the observed high sediment zinc concentration at some farms, fish feed manufacturing companies in British Columbia have reduced the amount of zinc in feed to the minimum necessary to maintain salmon health. They have also changed the form of zinc from zinc sulphate to a methionine analogue, which is more bioavailable.

Di Toro *et al.* (1992) described the relationship between sediment acid volatile sulphides, metal concentrations and toxicity to infauna. Acid volatile sulphides (AVS) are a reactive pool of solid-phase sulphide that is available to bind metals and render them biologically unavailable and non-toxic to biota. The AVS method is useful in predicting when sediments with elevated metal concentrations are potentially toxic. Metal toxicity is considered additive, and each must be added to obtain a sum of toxic units. This methodology has been validated for copper and zinc (EPA, 1994). None of the zinc concentrations observed by Brooks (2000a) was toxic, as they were all associated with sediments containing high concentrations of sulphide.

Brooks (2000a) reported that initially high sediment concentrations of zinc under one salmon farm declined to background levels during a post-production fallow period. At the peak of production, sediment zinc concentrations were elevated to 200 μg g^{-1} on the perimeter of the farm. They declined exponentially and reached a background concentration of *c*.25 μg g^{-1} at a distance of between 30 and 75 m from the netpen perimeter on the downstream transect. Sediment zinc declined during chemical remediation and was at background concentrations after 6 months of fallowing. It was hypothesized that zinc was bound by sulphides in the sediments. Sediment sulphides decrease with decreasing BOD during chemical remediation. When the sediments become aerobic, the sulphide is oxidized back to sulphate, releasing the zinc, which is diluted in the overlying water column. This hypothesis is consistent with chemical theory and with evidence collected earlier by Brooks (2000b).

Copper

Levels of copper may be elevated in the environment around netpen farms which use preventive treatments for biofouling. Several antifouling paints and solutions containing copper are approved for use in the marine environment, and are therefore used on salmon farms.

Anti-predator nets and fish containment nets are increasingly treated with a copper antifouling solution to inhibit the settlement of organisms and biofouling. The practice benefits the environment by reducing carbon inputs to the benthos. However, if cleaning is accomplished *in situ* the displaced organisms can exacerbate organic loading to the benthos under the farm. Best management practices (BMPs) should require that nets first be removed and then washed by hand or machine on a barge or at an upland facility.

Copper is a micronutrient. At moderately low levels the cupric ion is toxic to marine organisms – particularly the larval stages of invertebrates. Until 1995 the EPA marine chronic water quality criterion for copper was 2.5 μg l^{-1} (EPA, 1986). Based on new information that level is now being increased to 3.1 μg l^{-1} dissolved copper (EPA, 1995).

Lewis and Metaxas (1991) examined copper concentrations immediately adjacent to newly installed copper-treated nets at netpen salmon farms in British Columbia, Canada. They measured ambient copper concentrations of 0.38 μg l^{-1} in July and 0.37 μg l^{-1} in August. The concentration inside the pen was 0.54 μg l^{-1} in July after a freshly treated net was installed, and 0.54 μg l^{-1} 1 month later. The small addition of copper in the water from the treated net (0.16–0.17 μg l^{-1}) was not biologically significant except to organisms that tried to settle on the net.

Peterson *et al.* (1991) compared copper levels in muscle and liver tissue from chinook salmon grown in pens treated with Americoat 675®, a copper-based antifoulant, with those in a pen with untreated nets. No statistically significant differences in the copper levels in like-sized fish from these two farms were observed, suggesting that the copper released from the treated nets was not significantly concentrated by chinook salmon.

Brooks (2000d) conducted *in vitro* studies on the leaching of copper from Flexgard XI®, the most commonly used antifouling product on the west coast of North America. Initial losses of 155 μg Cu cm^{-2} day^{-1} declined exponentially during the period of the study. Brooks (2000d) used the data to develop a spreadsheet model that predicts copper concentration in the water as a function of the maximum current speeds observed at a site, and the netpen configuration and orientation of the complex to the currents. His model predicted that containment nets treated with Flexgard XI® would not exceed the EPA copper water quality criteria when fewer than 24 cages were installed in two rows oriented parallel to currents flowing with a maximum speed greater than 20 cm s^{-1}. The model predicted that unless the configuration of netpens or their orientation with the currents was changed, the use of nets treated with Flexgard XI® would result in exceeding the chronic water quality copper criterion at a small percentage of existing farms. Assumptions used in his model were conservative, and probably predicted higher copper concentrations than would actually be observed in the field.

Brooks (2000d) compared sediment copper concentrations at salmon farms in British Columbia using Flexgard XI®-treated nets with those at farms using untreated nets, and reference stations. Farmers in British Columbia typically treat their nets at the beginning of each production period. The mean concentration of copper in the sediments of 117 farm stations using treated nets was 48.24 ± 27.00 μg Cu g^{-1}. This level was not significantly different from the mean concentration of 12.01 ± 2.77 μg Cu g^{-1} measured at the reference stations, or the mean concentration of 26.3 μg Cu g^{-1} observed at farms not using copper-treated nets.

Brooks (2000d) found a great deal of variability in sediment copper concentration at farms using copper-treated nets. The concentration of copper in two of the 117 samples collected at 14 farms exceeded the State of Washington, USA, sediment quality criterion of 390 μg Cu g^{-1}. Thirteen of the samples exceeded the mean of the TEL and PEL used as a regulatory benchmark in British Columbia, Canada. All samples exceeding the lower benchmark were collected at five of the 14 farms using copper-treated nets. The producers indicated that nets on those five farms were washed on barges during fallow periods. The fouling debris cleaned from the nets was not retained but washed over the side. Consequently, the concentrations observed at five of the 14 farms were not directly associated with the copper treatment itself, but with an ancillary activity. Other anecdotal information revealed that the copper-containing latex paint was abraded from the nets during washing and that significant quantities of the latex chips (with copper embedded) were then washed over the side of the barge with the fouling organisms. Brooks

(2000d) concluded that all copper-treated nets should be removed after fish harvesting, washed and retreated at upland stations. Furthermore, all debris should be buried at an approved landfill site.

Pathogenic Organisms in the Vicinity of Netpen Salmon Farms

Faecal coliform bacteria

The National Shellfish Sanitation Programme (NSSP) certifies commercial shellfish beds in the USA and their harvest is governed by some very specific regulations (NSSP, 1997); for example, harvesting shellfish is forbidden within 1 mile (*c.*1500 m) of any outfall from a sewage treatment plant. This is because of public health concerns associated with toxicants released in industrial and residential waste, and because many human pathogens are associated with treated human sewage. Shellfish sanitation is not adversely affected by nutrients.

Viruses are generally taxon specific, and viruses pathogenic to fish, such as infectious pancreatic necrosis (IPN), viral haemorrhagic septicaemia (VHS), and infectious haematopoietic necrosis (IHN) have no documented effect on human beings. However, faecal coliform (FC) bacteria persist in sediments high in total organic carbon (TOC) for varying periods. These bacteria are specific to warm-blooded animals (mammals and birds) and are not a normal part of the microflora found in fish intestines. However, mammals and birds are strongly attracted to fish farms, increasing the potential for enhanced faecal coliform levels in the sediments near salmon farms. There is no potential for an increase in faecal coliform bacteria associated with cultured fish.

Brooks (2000a) analysed 33 water samples from the vicinity of an operating salmon farm during every quarter of the year. The MPN for all stations was less than the NSSP requirements for an Approved Shellfish Harvest Classification (14 FC/100 ml), and all stations met NSSP requirements. He also examined shellfish tissues for FC bacteria. NSSP has established an allowable upper limit of 230 FC/100 g of tissue for product entering interstate or international commerce. Average clam tissue levels at the closest station located 200 m from the farm were 130 FC/100 g tissue. At 500 m the average level dropped to 50 FC/100 g, and at the reference station it was 20 FC/100 g. All of the shellfish samples tested met the NSSP requirement for shellfish tissues in commerce. The sources of observed bacteria were not determined, but potential sources include farm workers themselves and, more probably, the birds and mammals which congregate around salmon farms. Water and shellfish tissues were consistently of high quality and met all bacteriological requirements imposed by NSSP.

Other pathogenic organisms in farm wastes

Ellis (1996) postulated that waste feed and faeces might enhance populations of a variety of ubiquitous marine bacteria pathogenic to humans. There is no direct evidence in the scientific literature that salmon farm wastes enhance pathogenic marine bacteria. In an extensive review of the epidemiological records for shellfish in the waters of Washington State, USA, spanning 20 years and ending in 1993, Brooks (unpublished data) did not find one documented case of *Vibrio vulnificus*-induced disease. This, he concluded, was because thermophilic bacteria, like *V. vulnificus*, required high water temperatures in addition to a rich source of organic material to thrive. Elevated ambient water temperatures would likely be a requirement of most bacteria that are pathogenic to homeotherms. Bacteria that flourish in warm-blooded animals are unlikely to proliferate in cold Pacific Northwest waters under salmon farms. Also, as salmon are poikilotherms, FC and other disease-causing bacteria that flourish in warm-blooded animals would probably not multiply in the gut of salmon. Thus, there appears to be no basis for assuming that the faeces of caged salmon would contain more than ambient concentrations of bacteria pathogenic to humans.

Calderwood *et al.* (unpublished) examined the kidney, liver, spleen, heart and muscle tissues of wild and farmed salmon for the presence of viruses and 16 bacterial species, including several later hypothesized as risks by Ellis (1996). They compared 68 adult wild steelhead trout and 50 wild adult coho with cultured chinook salmon.

V. vulnificus, a potentially serious human pathogen in immunocompromised individuals, was not detected in the cultured fish. However, 44% of wild fish returning from sea were positive for this bacterium. Other *Vibrio* species, including the potential human pathogen, *V. parahaemolyticus*, were not found in the farmed salmon, but were found with a prevalence of 9% and 44% in the wild fish. *Acinetobacter calcoaceticus* var. *anitratus*, which is a common bacterium found in water and soil, and one that has been associated with pneumonia, meningitis and septicaemia in humans, was observed with an average prevalence of 14% in cultured fish, and five times higher (76%) in wild fish. *Aeromonas hydrophila* was observed in over half of the wild fish. Both *A. hydrophila* and *A. salmonicida* are common fish pathogens (Roberts, 1978) but neither was isolated from tissues of cultured chinook salmon.

This early work by Calderwood *et al.* (unpublished) suggests that wild fish are far more likely to be a source of disease-causing bacteria than farmed fish. Their data do not support any hypothesis that environmental conditions on farms with healthy chinook salmon are enhancing populations of pathogenic bacteria. Furthermore, in their search for human diseases, epidemiologists most frequently examine the population having the greatest exposure to the suspected aetiologic agent. On salmon farms, the population most exposed to the fish are the farm workers and processors. If farms are a significant source of human pathogens, then farm workers and fish processors should show some history of such diseases. There are no epidemiological records that show evidence of any infectious outbreaks of disease.

There is no evidence that salmon farms create conditions leading to a proliferation of pathogenic bacteria. Furthermore, from a perspective of human health there appears to be no basis for suggesting fish farm wastes are comparable with human sewage from either large cites or even small towns.

The effects of therapeutic compounds

The control of sea lice is important to the health of farmed salmon and to reduce the potential for salmon farms to act as vectors for the infestation of wild stocks of salmon and sea trout. A review of the available treatments suggests that great care must be exercised in the use of these therapeutic compounds. They are all non-specific, at least within the Class Crustacea, and several are broad-spectrum biocides with potential to affect many phyla adversely. However, field studies have not found significant widespread adverse effects to either pelagic or benthic resources associated with the authorized use of these pharmaceuticals or pesticides.

The majority of therapeutic compounds used on salmon farms are for the control of sea lice. Sea lice, particularly *Caligus elongatus*, *Lepeophtheirus salmonis* and *Ergasilus labracis*, have caused extensive losses of fish, particularly at farms in the northeastern Atlantic. They have not presented significant problems to producers in the Pacific Northwest, and salmon produced in Washington, USA, have not been treated for lice for the last 15 years (A. Mogster, Northwest Seafarms, personal communication). Both ivermectin and emamectin have been used infrequently to control sea lice in British Columbia, Canada.

Costello (1993) and Roth *et al.* (1993) described the physical, chemical and biological methods used to control sea lice on fish farms in other areas. Current practices rely primarily on the administration of chemotherapeutic compounds in food or as a bath. The following treatments have been authorized for use.

Ivermectin

Ivermectin (22, 23-dihydroavermectin B1) has been widely used in agriculture to control parasites, and has been reported effective in controlling sea lice on caged salmon (Johnson and Margolis, 1993; Smith *et al.*, 1993). It is administered as a coating on feed at a rate of 0.025 mg kg^{-1} of fish at 10°C twice weekly for 4 weeks. The dose is increased by 10% for every 1°C decrease in ambient temperature. In Scotland the maximum number of weekly treatments is three per year (SEPA, 2000). The compound is reported to concentrate in *Mytilus edulis* by a relatively low factor (SEPA, 1998a). Grant and Briggs (1998a) stated that it did not appear to accumulate or concentrate in the food chain.

Dissolved concentrations of ivermectin are lethal to a number of marine organisms, ranging from a 96 h LC_{50} of 0.022 μg l^{-1} for *Mysidopsis bahia* (Davies *et al.*, 1997) to > 10,000 μg l^{-1} for nematodes (Grant and Briggs, 1998b). Most of the 96 h LC_{50} values are less than 1000 μg l^{-1}. Collier and Pinn (1998) and Grant and Briggs (1998b) found that crustaceans and polychaetes are more susceptible to ivermectin than are molluscs. Studies in Scotland did not detect ivermectin in the water column at a Scottish farm undergoing treatment (ERT, 1997, 1998).

Burridge and Haya (1993) found that ivermectin-coated waste feed affected non-target species, such as shrimp (*Crangon septemspinosa*), at concentrations of 8.5 μg g^{-1} food. Ivermectin toxicity was demonstrated in laboratory sediments at concentrations ranging from a 10 day LC_{50} of 23 μg kg^{-1} dry sediment for *Arenicola marina* by Thain *et al.* (1997), to 180 μg kg^{-1} for *Asterias rubens* by Davies *et al.* (1998). Black *et al.* (1997) documented significant mortality of polychaetes at ivermectin accumulations > 78.4 mg m^{-2} (SEPA, 1998a). This is equivalent to a concentration of approximately 2.54 mg kg^{-1} if the ivermectin is mixed into the top 2 cm of sediments having a density of 1.6 g cm^{-3}. Of particular interest was the adverse effect on the organic-carbon-tolerant opportunist *Capitella capitata*, the abundance of which was significantly reduced at ivermectin concentrations above the calculated value of 2.54 mg kg^{-1}. Black *et al.* (1997) did not discuss the increased chemical and biological remediation times that might result from a significant reduction in the abundance of *C. capitata*.

The fate of ivermectin not absorbed by Atlantic salmon appears to be sedimentation followed by slow degradation. Collier and Pinn (1998) noted that the breakdown of ivermectin in marine sediments was dependent on light and temperature. Davies *et al.* (1998) determined a half-life of ivermectin in marine sediments > 100 days.

ERT (1997, 1998) detected ivermectin at concentrations ranging from 5 to 11 μg kg^{-1} (wet sediment weight) in only three of 54 sediment samples at a farm undergoing treatment. Ivermectin was also detected in sediment traps deployed on the perimeter of farms undergoing treatment at 42 g kg^{-1}. Ivermectin was detected in two of 108 mussel samples at concentrations of 5 and < 5 μg kg^{-1} around farms undergoing treatment. The active ingredient was not detected in wild shrimp (*Nephrops norvegicus*) collected in the vicinity of the treated farm. In Canada, ivermectin was not found in American lobsters (*Homarus americanus*) around salmon farms (DFO, 1996); however, ivermectin was detected in sediments at distances up to 50 m from a farm in the Bay of Fundy where the compound was used.

These results suggest that ivermectin is most likely to be detected in sediments and not in the water column. In general, significant sediment concentrations of ivermectin have not been observed at distances beyond 10–20 m. The half-life of ivermectin is approximately 3 months (DFO, 1996). Permission to use ivermectin on farms in Scotland has been withdrawn by the Scottish Environmental Protection Agency (SEPA).

Emamectin benzoate

SEPA reviewed the proposed use in Scotland of the pharmaceutical emamectin benzoate under the proprietary name Slice® (SEPA, 1999a). Emamectin has low water solubility and is expected to accumulate in sediments. However, based on laboratory bioassays, SEPA stated that emamectin was about one-tenth as toxic as ivermectin, at least for the genus *Crangon*. Field studies failed to detect emamectin in water, and maximum observed sediment concentrations were just above the level of detection (1.0 μg kg^{-1} wet weight). The sediment half-life of emamectin was 175 days. No adverse effects on infaunal communities were observed following treatment with emamectin benzoate. Currently Slice® is in use in Chile, Ireland and Norway, and was approved for use in Scotland in January 2000 (SEPA, 2000).

Calicide

Calicide (teflubenzuron) has been licensed for the control of sea lice in Scotland (SEPA, 2000), and is being licensed in Canada, Chile, Ireland and Norway. This therapeutic compound is administered as a 2 g kg^{-1} coating on feed. It is a chitinase inhibitor effective against juvenile sea lice on salmon at a dose of 10 mg kg^{-1} day^{-1} for 7 consecutive days. It has reduced effectiveness after lice become adults and stop moulting.

SEPA (1999b) noted that calicide had a half-life of 115 days in sediments, and could be detected at distances up to 1000 m downstream from farms being treated. However, no adverse effects were detected in benthic communities and SEPA concluded that any residue was not bioavailable.

For teflubenzuron, SEPA (1999b) permitted an allowable seawater quality standard of 6.0 ng l^{-1} for the annual average, and a maximum allowable concentration of 30 ng l^{-1}. In Scotland, there is an 'allowable effects area' of 100 m from salmon farms. A sediment quality standard of 2.0 μg calicide kg^{-1} dry sediment in a 5-cm-deep core has been established outside this area. Sediment concentrations at all distances within 25 m of a treated farm must be maintained at less than 10 mg kg^{-1} dry sediment (5 cm core).

Calicide is approved only for interrupting the life cycle of sea lice. It is not approved for treating adult lice infestations.

Cypermethrin

Cypermethrin® (dichlorvos) is a pesticide being used in investigative programmes, and under some form of temporary registration in Canada, Ireland, Norway, the UK and the USA. It is administered in a 5 μg l^{-1} bath for 60 min within a confined and covered area.

Dichlorvos is toxic to crustaceans, with a LC_{50} of 0.006 μg l^{-1} and a no-observed-effect concentration (NOEC) of 0.003 μg l^{-1} for *Mysidopsis bahia*. It is adsorbed by sediments, where it degrades with a half-life of 35 days in high TVS sediments and 80 days in low TVS environments. It has a 10-day NOEC of 1000 μg kg^{-1} in *Arenicola*, and 64 μg kg^{-1} in amphipods of the genus

Corophium. Concentrations greater than about 10 ng l^{-1} are acutely toxic to some crustaceans, and modelling suggests that this value can be exceeded in the immediate vicinity of pens being treated with this product. SEPA (1998b) concluded that toxic effects to non-target species could occur within a few hundred metres of a treated farm and that these effects might last for several hours.

High mortality of shrimp and lobsters has been observed when they are exposed to a bath of dichlorvos, but the effect has not been observed outside netpens. Based on an absence of demonstrated deleterious effects on non-target animals, SEPA (1998b) recommended authorization, under permit, of Excis® (containing dichlorvos) for an initial 2-year period.

Azamethiphos

Azamethiphos (or Salmosan®) is a pesticide administered in a bath at 0.1 mg l^{-1}. The active ingredient degrades with a half-life of approximately 11 days at neutral pH. Larval lobsters were the most sensitive organism tested with a 96 h LC_{50} of 0.52 μg l^{-1}. A toxicity threshold to lobster larvae was estimated by SEPA (1997) at 0.078 μg l^{-1}. Azamethiphos is very water soluble and is not expected to accumulate in sediments. Compared with dichlorvos, azamethiphos is considered more toxic to crustaceans.

Farm Sediments

Monitoring environmental effects on sediments

Infaunal community analysis, as demonstrated by Pearson and Rosenberg (1978), Mahnken (1993) and Brooks (2000a), is ultimately the most direct and sensitive methodology for assessing the biological response to organic loading from salmon farming. However, benthic communities are not stable, as shown by Mills (1969) and Eagle (1975), and their structure is influenced by many natural processes unrelated to human influence. These processes include seasonal factors (Crisp, 1964; Arntz and Rumohr, 1982; Brooks 2000a) and physicochemical factors (Striplin Environmental Associates, 1996). Skalski and McKenzie (1982) pointed out that this variability typically requires large numbers of samples to achieve reasonable test powers.

The taxonomy required in support of infaunal analysis is expensive and time consuming (GESAMP, 1996). That cost, when coupled with high internal variability, detracts from infaunal analysis as a routine method for evaluating environmental effects as part of regulatory programmes. Therefore, physicochemical end-points, including TVS, TOC, redox potential (*E*h) and total sediment sulphides (S^{-2}), are being used increasingly as rapid and inexpensive surrogates for assessing biological response. GESAMP (1996) concluded that the visually determined depth of the reduction–oxidation discontinuity was of low value because it was semi-quantitative. They did not consider emerging physicochemical end-points, such as sulphide analysis using ion-specific probes (Wildish *et al.*, 1999) or TVS (Hargrave *et al.*, 1995; Brooks, 2000e,f).

Brooks (2000c,e) discussed the relative merits of TVS, TOC, S^{-2} and *E*h for evaluating the environmental response to salmon farms. Organic carbon, whether measured as TOC or TVS, had only a moderate correlation with biological effects, particularly infauna. It was hypothesized that sediment carbon can come in forms that are refractory to microbial catabolism (e.g. woody debris), resulting in low BOD. Drift macroalgae and eelgrass form an intermediate class of sediment carbon broken down within a few months to a year. Fish faeces, on the other hand, are labile and create high BOD, more frequently leading to anaerobic conditions than woody debris or macroalgae. Therefore, 15% TVS may not exceed the assimilative capacity of the sediment if it is in the form of woody debris, whereas 15% TVS associated with salmon farm waste in the same sediment would be more likely to create anaerobic conditions.

*E*h was identified by GESAMP (1996) and Wildish *et al.* (1999) as a valuable end-point for evaluating sediment chemistry near salmon farms because it is rapid, low cost, and permits extensive spatial surveys. Brooks (2000f) found low sediment redox (20.68 ± 22.29 mV) and low sulphide concentrations (29.83 ± 11.59 μmol) in the same samples from reference stations located in areas with greater than 80–90% silts and clays. Low *E*h and high sulphide concentrations in sediments were not always well correlated, and low *E*h could be the result of physical processes, especially in fine-grained sediments. This limited its usefulness in evaluating the effects of carbon input from salmon farms.

Wang and Chapman (1999) described the response of laboratory bioassay test animals to sediment sulphides. However, despite attempts by Brown *et al.* (1987), Henderson and Ross (1995) and Hargrave *et al.* (1997), the literature does not provide a good quantitative description of the response of natural infaunal communities to sediment sulphide concentrations or redox potential.

Benthic infaunal and epifaunal analysis appears to be the most sensitive indicator of environmental health in sediments around salmon farms. However, benthic communities are not stable across environments and depend on the physical environment – including water depths, current speeds, sediment grain size distribution and the availability of organic carbon (Striplin Environmental Associates, 1996). In addition, benthic communities vary by season, as influenced by food input, water temperature, etc. (Arntz and Rumohr, 1982).

Biological changes in the water column and sediments

The environmental changes associated with salmon farms are superimposed on natural changes. These potential effects have been examined in numerous studies during the last two decades. Despite significant site-specific variability there is a consistent thread binding this literature.

Water column changes

Possible changes in the water column associated with the intensive culture of fish could be associated with intoxication due to hydrogen sulphide and ammonia production in underlying sediments, decreases in dissolved oxygen associated with salmon respiration and/or the oxidation of sedimented waste, and eutrophication associated with nitrogen released across gill epithelia and in urine and faeces. The magnitude and consequences of environmental changes associated with these factors is dependent on environmental parameters such as water depth, current speeds, background nutrient availability, salinity, rainfall, wind, etc., which in aggregate constitute the local environment.

Nitrogen and phosphorus loading to the water column

Marine environments along the west coast of North America are especially productive because cold, upwelling, nutrient-rich water replaces surface waters driven offshore by prevailing northwesterly winds. In addition, the relatively high geographic latitude of British Columbia in Canada and Washington, USA, results in reduced light penetration in water compared with more southerly latitudes. Lastly, moisture-laden onshore winds create significant cloud cover throughout much of the year. These factors combine to limit light availability significantly in most temperate marine environments, except during summer months. Furthermore, in most marine environments, nitrogen is the limiting nutrient and not phosphorus.

In the Pacific Northwest, wind-driven vertical-mixing drives a significant proportion of the phytoplankton crop below the compensation depth where cell respiration equals photosynthesis and where they no longer multiply. Where water freely circulates, flood tides replenish nutrients from water upwelling offshore. When coupled with the atmospheric and geographical factors that reduce light availability, the result is that primary productivity in the Pacific Northwest is generally light limited, not nutrient limited. This is especially true during winter months.

There are sheltered, poorly flushed, shallow embayments where salinity and temperature-induced stratification result in a stable water column that allows phytoplankton to remain above the compensation depth. When these conditions occur in the spring or summer, significant blooms can occur following several days or weeks of clear, sunny weather. These blooms eventually wane because winds increase vertical mixing, cloud cover reduces the available light, or nutrients are depleted in the surface water. In the last situation, nutrient input from a salmon farm could further stimulate plant growth, exacerbating the problem. In addition, shallow bays having significant freshwater input and minimal flushing are not considered good sites for netpen growout operations. However, they might be deemed appropriate as smolt introduction sites.

Another point to consider is that nitrogenous compounds are released from fish farms into currents that generally average more than 4–12 cm s^{-1}.

At temperatures of 10–15°C, it takes 1 to 2 days for an algal cell to divide, even if all of its photosynthetic needs are met (Brooks, 2000g). An algal bloom may result in cell densities increasing from a few thousand cells per millilitre to perhaps a million. This requires eight or nine cell generations, or a minimum of 8–16 days. In open bodies of water, moving with a net speed of even 2 cm s^{-1}, a phytoplankton population would move 14 km from the location at which nutrients were added during creation of a bloom. Therefore, it appears reasonable to conclude that, within a single algal cell division (1–2 days), the water passing through the farm would have travelled at least 1.7 km. It is difficult to conclude that the nutrient additions from the farm, generally undetectable at 30 m downstream, would have any effect at all on primary production even if the water body was nutrient limited.

Pease (1977), Rensel (1988, 1989) and Parametrix Inc. (Parametrix, 1990) documented small increases in dissolved nitrogen within and on the perimeter of salmon farms. However, all of these publications were in agreement that the quantity of dissolved nitrogen added by even several farms would have no measurable effect on phytoplankton production. Gowen *et al.* (1988) studied a Scottish loch, with very restricted water exchange to the open sea, and a large salmon farm. The authors concluded that the farm had no measurable effect on phytoplankton density.

Weston (1986) quantitatively assessed the effects of five hypothetical farms located in a small embayment with poor flushing. His analysis suggested that the nitrogen added by five farms could not be expected to adversely affect the phytoplankton abundance in the embayment. He recommended that nutrient-sensitive embayments should be identified and carefully managed.

Banse *et al.* (1990), Parsons *et al.* (1990), Pridmore and Rutherford (1992), Taylor (1993) and Taylor and Horner (1994) examined phytoplankton production and blooms of noxious phytoplankton in the Pacific Northwest. They concluded that nitrogen levels and phytoplankton production at salmon farms were determined by ambient conditions. Furthermore, they found that salmon farms had little or no effect on ambient levels of either nutrients or phytoplankton density.

The literature is consistent with the previous general discussion and strongly supports a thesis that, with the exception of a few shallow, very poorly flushed embayments, the potential for netpen enhancement of phytoplankton populations is remote or non-existent. Based on similar arguments and 10 years of monitoring dissolved nutrients at salmon farms, the state of Washington, USA, eliminated any requirement for water column monitoring in compliance with NPDES (National Pollution Discharge Elimination System) permits issued to all salmon farms in 1996.

Hydrogen sulphide gas production in sediments

When the assimilative capacity of the benthos is exceeded, oxygen is depleted and sulphur-reducing bacteria continue to degrade organic carbon. During the process either ammonia or hydrogen sulphide gas may be produced. These toxic gases are not unique to fish farms and other sources of anthropogenic carbon, and are frequently found in natural environments where organic debris accumulates.

Hargrave *et al.* (1997) examined a suite of physicochemical parameters under 11 salmon farms and 11 reference stations located > 50 m from netpens in the Bay of Fundy, Canada. Sediment concentrations of hydrogen sulphide were found to be significantly different under netpens when compared with reference sediments. Total sulphide concentrations in surface sediments at all cage sites were > 180 μmol l^{-1}, while values at all but one reference location were < 200 μmol l^{-1}. Sulphide concentrations > 2000 μmol l^{-1} were indicative of high organic loading under some netpens and were generally associated with negative *Eh* potentials.

Ammonia and hydrogen sulphide are lighter than water and when significant quantities of those gases accumulate in sediments, they can escape and rise to the surface. As the bubbles rise, the soluble H_2S dissolves in the water column. Samuelsen *et al.* (1988) analysed gas released from sediments underlying poorly flushed salmon farms. They found that 98% of the gas was in the form of CH_4 and CO_2. Less than 1.9% of the gas at the sediment–water interface was sulphide (S^{-2}). Furthermore, they found that, after rising 3 m in the water column, the S^{-2} was reduced to 0.05% of the total gas. The resulting concentration would be 25.5 μg S^{-2} l^{-1} seawater (25 ppb). Water quality standards are based on the undissociated sulphide (H_2S), which is *c.* 10% of S^{-2} at pH = 8.4. Applying this factor predicts an undissociated H_2S level of 2.55 in the 0.5 cm diameter column through which the gas bubble passes. This is approximately equal to the 2 μg l^{-1} chronic water quality criteria established by the EPA (1986) for freshwater and marine environments. In reality, oxidation, diffusion and mechanical mixing significantly reduce concentrations further by a factor of 100 or more.

Samuelsen *et al.* (1988) found that the fraction of the less soluble CH_4 did not appreciably change during transit of the bubbles through the water column. The low concentration of H_2S in the bubbles at the sediment–water interface, and the low water concentrations predicted during ascent, suggests that very large gas emissions would be required before sufficient H_2S could be dissolved in the water column to create toxic conditions.

Dissolved oxygen

Weston (1986) reviewed the effects of salmon culture on ambient dissolved oxygen (DO) levels and concluded that farms could decrease these levels by 0.3 ppm. Brooks (1991, 1992, 1993a, 1994a,b, 1995a,b) observed decreases of as much as 2 ppm in water passing through a large, poorly flushed farm in Puget Sound, Washington, USA. Significant reductions in DO were not observed by Brooks (1994a,b) at farms in well-flushed passages. In no case were DO levels within 6 m of the downstream farm perimeter depressed below 6 ppm, a minimum level for optimum culture of salmonids. Winsby *et al.* (1996) reported a range of results from the literature. His discussion suggested that depressed oxygen levels are associated with the water column immediately overlying anaerobic sediments. Cross (1993) concluded that salmon farms in British Columbia, Canada, have minimal effects on ambient DO levels.

Depressed oxygen levels (3–6 ppm) are infrequently encountered at salmon farms along the Pacific coast of North America. These depressions result from the upwelling of cold, nutrient-rich but oxygen-deficient water to the surface. Conditions favouring depressed DO are most frequently encountered in the Pacific Northwest during the summer and autumn, when northwest trade winds increase oceanic upwelling. Deep fjords, like Hood Canal in Washington State, can also experience depressed concentrations of DO when winds bring anoxic water to the surface. Feeding is suspended and compressors are used to increase DO when these naturally occurring masses of water with low DO levels flow into salmon farms. Although the phenomenon is imposed on the farm, not caused by the farm, the frequency of occurrence of these oxygen-deficient water masses should be assessed in siting a farm. In addition, it could be considered good management on the part of operators to measure DO in bottom water under their farms in an attempt to predict periods of depressed surface oxygen.

Physicochemical changes in the sediment near salmon farms

The chemical and biological effects associated with fish farms have been documented and reviewed by Pease (1977), Braaten *et al.* (1983), Earll *et al.* (1984), Ervik *et al.* (1985), Ackefors (1986), Weston (1986), Aure *et al.* (1988), Rosenthal *et al.* (1988, 1995), Weston and Gowen (1988), Hansen *et al.* (1990), Parametrix (1990), Gowen *et al.* (1991), Johannessen *et al.* (1994), Winsby *et al.* (1996), Mazzola *et al.* (1999) and Morrisey *et al.* (2000). It is possible to model rates of organic loading from netpen operations described by Weston and Gowen (1988), Einen *et al.* (1995), Silvert and Sowles (1996), Ervik *et al.* (1997) and Findlay (unpublished). The fate and transport of those wastes is a far more complex problem. However, the effects of farm wastes on the benthos in a variety of environments have been well documented. Brooks (1992, 1993a, 1994a,b, 1995a,b) studied sediment chemistry and benthic infaunal response at two farms which represented two very different environments in Puget Sound, Washington, USA. In terms of negative environmental effects associated with intensive netpen fish culture, organic loading to the sediments was most significant. Goyette and Brooks (1999) observed statistically significant changes in the composition of the benthic infaunal community in British Columbia, Canada, associated with < 1% change in sediment organic carbon content across the 500 m study area. In general, the literature suggests a lack of appreciation of the sensitivity of the benthos to small additions of organic carbon, particularly labile forms like fish faeces.

Hargrave *et al.* (1995) documented sediment total sulphide concentrations under salmon farms in the Bay of Fundy, Canada that were ≤ 6600 μmol S^{-2}. In contrast, Brooks (2000c,f) observed significantly higher sediment sulphide concentrations (< 16,000 μmol) on the perimeter of salmon farms in British Columbia, Canada, and Wildish *et al.* (1999) reported sediment concentrations up to 36,000 μmol S^{-2} in Bay of Fundy sediments under operating farms.

Effects of farm wastes on benthic infauna

The biological response of infauna to the sediment physicochemical changes occurring as a result of organic loading from salmon farms has been assessed by Hargrave (1994), Henderson and Ross (1995), and Hargrave *et al.* (1997). The toxicity of sulphide to infauna is documented for a few species (Bagarinao 1993; Wang and Chapman 1999), but despite the efforts of Henderson and Ross (1995), quantitative relationships between infauna and physicochemical end-points remain elusive. Brooks (2000a) observed a significant enhancement in infauna during the early stages of production and at the end of a fallow period. However, at the peak biomass there was a significant reduction in the number of invertebrates observed at downstream stations located between 20 m and about 70 m from the farm perimeter. Near-field invertebrate numbers were supplemented by allochthonous input from the fouling community on farm nets. In addition, a significant portion of the invertebrate community associated with near-field sediments during periods of high organic farm input were the TOC-tolerant species *Capitella capitata* and *Ophryotrocha* cf. *vivipara*.

Species richness – the number of species observed in biological samples – is frequently a more sensitive indicator of environmental stress than abundance. Brooks (2000a) observed significant reductions in the number of taxa within 45 m of a farm during peak production. It did not appear that significant effects extended beyond 75 m during the production period. Biological remediation began as soon as harvest was initiated and was essentially complete within 4 months of fallowing. A slight enhancement in taxa richness was evident 5 months following the completion of harvest.

Polychaete abundance was enhanced as sediment organic carbon built up at the beginning of the production cycle. Abundance declined within 80 m of the farm perimeter during peak biomass when farm waste exceeded the assimilative capacity of the sediments, which became anaerobic. Polychaete abundance began increasing again during the winter, approximately 6 months after harvest began. Polychaetes proliferated with the improving benthic conditions and exceeded reference abundance during the last 6 months of the study. The enhanced area ultimately extended from the farm perimeter to a distance of at least 75 m.

Brooks (2000a) found that crustaceans were adversely affected at near-field stations earlier than polychaetes. A steady increase in the number of crustacean taxa was observed as soon as the fish biomass began decreasing during harvest. The salmon farm had little effect on the overall abundance of crustaceans. Molluscs were an abundant and diverse part of the infaunal invertebrate community in reference sediments from the study area.

Statistically significant decreases in the numbers of molluscs were not observed in the study. However, the number of molluscan taxa was significantly reduced at all farm stations during the production cycle. An increase in the number of molluscan taxa was evident at the end of the study, but the number of taxa observed within 50–75 m of the netpen perimeter had not recovered to reference conditions. Brooks (in preparation) has determined the biological response to varying concentrations of sediment TVS, sulphides and *Eh* as a function of farm production, overlying currents and sediment grain size distribution. Preliminary results indicate that reference sulphide concentrations were generally low (10–100 μmol) but could be as high as 250–300 μmol. Reductions in the number of taxa from > 20 species to 12–14 species, with significantly increased abundance and biomass of infauna, were noted with sulphide concentrations in the range of 300–2000 μmol. Sediments containing > 2000 μmol S^{-2} had a reduced infaunal community dominated by *C. capitata*, and those containing greater than 6000 μmol were generally sparse. The findings were consistent with the reports of Wildish *et al.* (1999), Poole *et al.* (1978) and Pearson and Rosenberg (1978).

Recovery and Remediation of Sediments

Chemical and biological recovery of sediments under salmon farms has been documented by, *inter alia*, Ritz *et al.* (1989), Anderson (1992), Mahnken (1993), Brooks (1993b, 2000a), Lu and Wu (1998), Karakassis *et al.* (1999) and Crema *et al.* (2000).

Chemical remediation

Brooks (2000a) defined chemical remediation as the reduction of accumulated organic carbon with a concomitant decrease in hydrogen sulphide and an increase in sediment oxygen concentrations under and adjacent to salmon farms to a level at which aerobic organisms can recruit into the area. At one farm, sediment concentrations of volatile solids declined rapidly as soon as harvest was started in June and they were close to control values when the harvest was completed the following April. By the end of the 10-month harvest, significant differences in TVS were not observed among all reference area and farm stations located at 5, 10 and 15 m from the netpen perimeter. Chemical remediation resulted in increased levels of oxygen in sediment pore water and decreased levels of H_2S and/or ammonia. H_2S was evaluated organoleptically. High levels of sediment H_2S were evident to 20 m during peak production. Moderate levels of H_2S were observed as far as 37 m on the downstream transect. H_2S was detectable at low levels to distances less than 50 m from the netpen perimeter at the peak of production.

Biological remediation

Biological remediation was defined by Brooks (2000a) as the restructuring of the infaunal community to include those taxa representing at least 1% of the total invertebrate abundance observed at a local reference station. Recruitment of rare species (those representing < 1% of the reference area abundance) into the remediation area is not considered necessary for biological remediation to be considered complete.

Brooks (2000a) observed the beginning of biological remediation during a 10-month harvest period. Biological remediation appeared to be nearly complete 5 months following harvest. Several infaunal series were apparent in the data. Farm inputs (fish biomass and 30-day feeding rate) were positively correlated with several sediment physicochemical variables including per cent fines, total volatile solids and the presence of hydrogen sulphide. There was also a significant and positive correlation between the opportunistic polychaetes *C. capitata* and *O. vivipara* and farm inputs.

Striplin Environmental Associates (1996) found that Shannon's diversity index varied between 1.09 and 1.53 at Puget Sound reference stations where the per cent fines (silt and clay) was less than 20% and water depths were < 45 m. Brooks (2000a) found relatively high values of Shannon's index (2.6 ± 0.4) at a reference station. The values suggest that the undisturbed infaunal community was diverse and evenly distributed. Low values of Shannon's index suggest a community dominated by a few species. This condition was evident to 50 m and possibly as far as 100 m from the perimeter of the netpens on the downcurrent transect during the peak of the production period. Shannon's index increased steadily following the initiation of harvest and an enhancement of the invertebrate community was suggested by this index at the end of the 5-month fallow period.

Anderson (1992) aggregated the taxa identified in three replicate Ponar 0.05 m^2 grab samples at each station on fallow farms in British Columbia, Canada, and found Shannon's index (H') values varying from 0.108 to 1.465. The lowest values were found at stations with high TOC loading dominated by *C. capitata*, *Nephtys cornuta* and the gastropod *Mitrella gouldi*. The higher values were generally associated with undisturbed reference stations. He used principal components analysis (PCA) to observe that the factors fallow time, biomass, per cent fines, total taxa richness and H' were positively correlated with each other and negatively correlated with TVS, coarse sand, gravel and sulphide. He concluded that high values of TVS and sulphide were indicative of unhealthy ecological conditions. Anderson (1992) observed recovery times that varied between several months at sites with low initial impact to an estimated 2 years for severely impacted sites, during which physical, chemical and microbial processes acted on sediments to make them hospitable to macrofaunal colonization. This refractory period was referred to as chemical remediation by Brooks (2000a). Once macrofaunal colonization began, Anderson (1992) observed an increased rate of recovery.

In Washington, USA, Brooks (1993b) found recovering sediments dominated by *N. cornuta*, *Glycera* and *Lumbrineris*, with few *C. capitata* in the vicinity of a fallow salmon farm. He hypothesized that following cessation of production in October, an initial period of chemical remediation was followed by a proliferation of opportunistic *C. capitata*. As the organic load was dispersed and catabolized by microbes, and the oxidation–reduction potential increased, predatory polychaetes in the genera *Nephtys*, *Glycera* and *Lumbrineris* flourished by preying on the large standing crop of smaller prey species.

Mahnken (1993) studied the succession of invertebrates at a depositional environment in Puget Sound, Washington, USA, for 2 years during a fallow period at a salmon farm and observed two distinct stanzas of biological recovery. The first was a 3-month period of rapid recovery in abundance and species diversity followed by a 3- to 25-month period when community recovery proceeded more slowly. Species which were numerically dominant in samples from the reference station showed rapid colonization. Rare species were slow to recruit to the area. He identified four successional series including:

- A pre-successional series comprised of species tolerant of sediments having high TOC.
- A pioneering group of early colonizers containing several species of organic-tolerant opportunists.
- An intermediate group of colonizers associated with reduced numbers of deposit-feeding opportunists.
- A group of late colonizers consisting of a group of more conservative and persistent species.

A fifth group of rare species identified at the reference station were still absent from the farm sediments at the end of the 25-month study. Mahnken (1993) observed that succession was most clearly defined in the Order Polychaeta. Biological recovery was initiated by *C. capitata* and followed in successive series characterized in turn by *Armandia brevis*, *Phyllodoce maculata*, *Pectinaria granulata*, *Platynereis bicanaliculata* and finally by more generalist species like *Leitoscoloplos pugettensis*. He concluded that the sequence was best described as a response to changing organic content in the sediments, resulting from biogenic reworking by changing guilds of benthic organisms.

Brooks (2000a) conducted an exhaustive study of a salmon farm in British Columbia, Canada, during production and fallow periods over a period of 2 years. This was the first study documenting the relationship between salmon farm biomass, fish feed inputs, and the physicochemical and biological response of the benthos. The study design focused on quarterly samples collected on the downstream transect from the netpen perimeter to a distance of 75 m, and at a local reference station located approximately 1200 m from the farm. A maximum Atlantic salmon biomass of 1199 tonnes was raised at the farm under study. His data clearly depicted the accumulation of volatile organic material under the farm and out to distances of *c.*40 m from the netpen perimeter during the peak of production. The physicochemical data

(TVS, TOC, hydrogen sulphide, zinc and depth of the reduction–oxidation potential (RPD) discontinuity) were well correlated and internally consistent. Organic carbon accumulations (TOC or TVS) were sensitive indicators of biological effects.

In all of these cases chemical and biological recovery of the benthos occurred within weeks or months at some sites, and within 2 to 3 years at others. The benthic recoveries have occurred naturally with no need for intervention or mitigation.

Managing the Environmental Effects Associated with Salmon Farms

Monitoring experiences

From 1987 until 1996 the Department of Natural Resources, Washington, USA, required monitoring of sediment chemistry (carbon, nitrogen, redox potential and sediment grain size), water chemistry (dissolved oxygen, pH, nitrate, nitrite, total ammonia and un-ionized ammonia), and the benthic community (quantitative infaunal surveys and qualitative SCUBA surveys) as a condition in ALLs for salmon farms. This monitoring experience provided an extensive database upon which to evaluate the effectiveness of each measured parameter in predicting environmental effects. Lessons learned from these studies include:

1. Sediment grain size and water depth were primary factors determining the structure of an undisturbed infaunal community.
2. Absence of any anthropogenic inputs; i.e. in reference areas, the TOC content of undisturbed sediments was significantly correlated with the proportion of fines (silt and clay) contained in superficial sediments (< 2 cm depth). Depositional areas associated with slow current speeds and gyres accumulate both fine sediments and particulate organic materials at higher rates than high-energy areas.
3. The redox potential and health of the infaunal community associated with a particular sediment grain size distribution appears well correlated with the level of TOC in the sediments (Striplin Environmental Associates, 1996; Goyette and Brooks, 1999).
4. Invertebrate community analysis has been a traditional and direct way of evaluating the effects of organic loading (see Pearson and Rosenberg, 1978; Lunz and Kendall, 1982; Weston, 1990; Brooks, 1993a,b, 1994a,b, 1995a,b; Henderson and Ross, 1995). In Washington, USA, invertebrate infaunal community analysis is used as a primary end-point for evaluating benthic sediment quality (WAC 173-204-320).
5. Salmon farms located in well-flushed (> 50 to 100 cm s^{-1} maximum current speed) environments frequently increase both the abundance and taxa

richness of infaunal communities, even at high levels of salmon production (Brooks, 1994b, 1995b).

6. Salmon farms located in poorly flushed (< 10 cm s^{-1} maximum current speed) environments can result in the deposition of significant amounts of carbon to the benthos – even when located in water as deep as 30 m at mean lower water. Adverse effects are generally restricted to an area within 15–22 m of the perimeter of farms located in poorly circulated environments. Increases in both the level of TOC and the distance at which adverse effects are observed are sensitive to farm management practices (Brooks, 1994a). However, the negative effects can be managed so that they remain within 33–100 m of the farm perimeter, even during intensive production of fish.

7. Indicator invertebrate taxa have been identified at several farms studied in the ALL Program of the Washington Department of Natural Resources. These indicator taxa and groups of taxa appear temporally consistent but are specific to different environments (Brooks, 1995a,b). Other authors (Weston, 1990; Tsutsumi *et al.*, 1991; Mahnken, 1993; Henderson and Ross, 1995) have identified similar suites of indicator species in response to organic loading.

Based on the monitoring reports, it appears that TOC can be used as a screening tool to evaluate benthic health indirectly. This is not unlike the use of bioassays as a screening tool in evaluating the effects of toxic industrial and municipal effluents in fine-grained sediments. The use of TOC (or TVS) as a screening tool has the advantage of being fast. Analyses can be completed in a few days, whereas infaunal community analyses take months. In addition, the lower cost of TOC/TVS analysis allows more frequent monitoring. When combined, these factors allow TOC/TVS to be used as a real-time parameter useful to farm managers.

Sediment total sulphides and oxidation–reduction potential measured with ion-specific probes immediately following sample collection are emerging as more biologically relevant physicochemical end-points in ongoing studies in British Columbia, Canada.

Management by modelling salmon farm wastes

There is significant interest in modelling salmon farm waste as a management tool for regulatory agencies. Some of these models are qualitative (Sowles *et al.*, 1994) and others attempt to quantify the dispersal and accumulation of particulate organic matter in sediments (Fox, 1990; Gowen *et al.*, 1994; Silvert, 1994b). It appears that the more basic the model inputs, the more room there is for error. Silvert (1994a) used a simple carbon budget to model salmon farm waste and concluded that 40% of the feed was not consumed by the fish. There is no evidence in the literature substantiating feed loss rates that high. None of the models has been tested to compare predictions with observed carbon deposition rates or sediment physicochemical responses to salmon farm waste.

Findlay and Watling (1994) modelled sediment organic carbon decay rates and developed non-linear regression equations relating oxygen delivery (mmol m^{-2} h^{-1}) and maximum oxidizable organic matter (g carbon m^{-2} day^{-1}) to sediments as a function of current speed (cm s^{-1}). They concluded that the maximum carbon flux not exceeding the assimilative capacity of the sediment is highly dependent on the minimum 2-hour average bottom current speed.

Silvert and Sowles (1996) developed several algorithms considered useful in modelling the environmental response to salmon farming. They concluded that models exist which can help assess impacts and make reasonable management decisions, but this is not substantiated in the existing literature.

Models provide some insight into the environmental response to salmon farm waste. However, they are not adequate for making reasonable quantitative predictions regarding the degree or spatial extent of salmon farm waste.

Risk management through NPDES permit standards

In 1996 Washington, USA, developed sediment management standards for marine netpens (WAC 173-204). The Washington State rule is based on the following assumptions:

1. Salmon farming provides significant benefit to the State and its people.
2. The negative benthic effects associated with netpen operations in poorly flushed environments will remediate naturally following cessation of operation or initiation of a fallow period.
3. The spatial extent of these effects can be managed. The sediment rule for netpens authorizes a sediment impact zone (SIZ) extending 33 m from the perimeter of the farm structure. This distance was chosen because it corresponds to the SIZ provided for other industrial discharges. From a biological point of view, it would seem more appropriate to develop site-specific SIZs that reflect the biological productivity of the site's benthos and the presence of adjacent valuable resources. In that context, SIZ widths could extend considerably further from the perimeter of a farm, perhaps to a distance of 100 m or more.
4. TOC can be used as a screening tool in evaluating the health of the benthos. TOC triggers have been defined as a function of the proportion of silt and clay in the sediment matrix (Table 10.1). If sediments located 33 m from the perimeter of the netpen structures at salmon farms exceed the trigger values, then an evaluation of the health of the infaunal community is required.
5. Biennial monitoring of sediment TOC is required at seven stations at each permitted farm. Four of the stations are located at a distance of 30 m from the perimeter on each side of the farm. Three replicate sediment samples are collected at each station. No further monitoring is required if sediment TOC is not statistically elevated (*t*-test) above the TOC trigger corresponding to the observed per cent fines at each 30 m station. If the measured TOC is

Table 10.1. TOC triggers used in Washington, USA, as a function of silt and clay percentage in the sediments.

Percentage of silt and clay in the sediments	TOC trigger (%)
0–20	0.5
20–50	1.7
50–80	3.2
80–100	2.6

significantly higher than the corresponding trigger, then repeat sampling is required in the summer of the next year together with the collection of five benthic infaunal samples at each station failing the TOC trigger, and at a suitable control. Benthic infaunal analysis is required for any station at which elevated TOC is observed during the second round of sampling.

6. Each farm is required to manage its production such that there are no significant negative effects on benthic resources beyond the boundary of this 33 m SIZ. WAC 173-204 states that biological resources in sediments are considered adversely impacted if the mean numbers of crustaceans, molluscs or polychaetes in the test sediment at the boundary of the SIZ are reduced to significantly less than 50% of the number of animals belonging to the same taxa living in an undisturbed reference sediment. Evaluation is based on a one-tailed *t*-test at $\alpha = 0.05$ for five replicate 0.1 m^2 samples. Infauna are seldom found regularly distributed in sediments. For that reason, if three samples were collected from the same general area, the individual samples would likely contain very different numbers of any one of these taxa. The reason that the rule relies on a 50% reduction in the number of any taxa is not that it is acceptable to kill 50% of the crustaceans, molluscs or polychaetes outside the boundary of the sediment impact zone, but to acknowledge that the collection of a reasonable number of random samples may produce two means which vary by as much as 50%, even though the sediments are not impacted at all, or share the same level of impact.

7. Benthic conditions at each of the four orthogonal 30 m SIZ stations must be photographically documented every 2 years and whenever sediment samples cannot be collected and analysed in conformance with the requirements stipulated in the Puget Sound estuary protocols (PSEP, 1986).

8. Well-managed salmon farms recognize the benefits of vaccination and best management practices (BMPs) in controlling disease. While not examined in this review, records of antibiotic use in Washington indicated a sharp decline at permitted farms between 1992 and 1996. Similarly, Kontali (1996) reported that the use of vaccines in Norway had resulted in reductions in the use of antibiotics from a high of 592 mg kg^{-1} salmon produced in 1987 to 5, 9 and 3 mg kg^{-1} in 1994, 1995 and 1996, respectively.

Based on the current low use, Washington, USA, regulations (WAC 173-204-412) do not require routine monitoring for bacterial resistance at marine netpen sites. All farms are required to maintain an operational log that specifies the date and nature of application of all disease-control chemicals used. In addition, farms are required annually to report the amount of each therapeutic chemical used on each farm.

Based on the absence of adverse effects observed during 10 years of monitoring the water column adjacent to salmon farms in Washington State, the WDOE eliminated all requirements for nutrient and dissolved oxygen monitoring in the water column from the NPDES permits.

This approach by the State of Washington regulatory agencies is appealing for several reasons:

- It recognizes the value of netpen culture while requiring that any negative impacts be restricted to the immediate vicinity of the farm.
- It invokes a realistically achievable performance standard that can (must) be met through proper management practices.
- It is a relatively inexpensive approach as long as TOC levels at the boundary of the SIZ remain below trigger levels. This provides a real incentive to maintain carbon levels below specified triggers.
- The immediate end-point (TOC) can be measured quickly and is useful as a real-time management tool.
- The performance standard encourages future siting in environments that either have fine-grained sediments that support high TOC levels, or in high current areas where TOC will not accumulate.

On the other hand there have been some problems in implementing the State's NPDES permit system. For example, sediment samples collected from coarse bottoms cannot be analysed for TOC because the matrix must be ground to a fine consistency. Sediments from erosional environments are generally composed of coarse gravel and cobble.

The consequences of failing the benthic biological criteria are significant, requiring reductions in the number of fish raised, or the amount of food provided, or actually fallowing the farm for a period. This is a significant penalty when there is a half chance that failure is simply a matter of chance with $\alpha = 0.05$.

Risk management practices in British Columbia, Canada

Following the exhaustive review of the scientific literature (EAO, 1997), the Provincial Government of British Columbia has been developing a performance-based waste management policy (WMP) to ensure that adverse benthic effects associated with salmon farming are managed. The following are essential elements likely to emerge:

1. Only single year-classes of fish will be grown at British Columbia salmon farms. The purpose of this restriction, as practised also in Norway and Chile since the mid-1990s, is to reduce the potential for disease transfer between year-classes and to provide for a fallow period between production cycles sufficient to ensure that sediments chemically remediate to within 30 m of the farm perimeter prior to restocking.
2. Prior to restocking a new year-class, the farm must remain fallow until the level of volatile residue in sediments at a distance of 33 m from the netpen complex perimeter returns to baseline or local reference station values. Farm management must certify to the Ministry of Environment that this condition has been met before restocking fish.
3. At no time will adverse benthic effects be allowed at distances ≥ 100 m from the perimeter of the netpens. This performance standard will be evaluated annually during the months of August–November. This distance is under review.
4. Based on the problems encountered with the analysis of TOC in Washington, USA, British Columbia is using TVS as a primary screening tool. Samples are collected at a distance of either 30 m (for the pre-stocking certification) or 100 m (for the annual monitoring) from the midpoint on each of the sides of the netpen's perimeter. If no local reference station is available, then farm samples will be compared against TVS triggers which represent the upper 90th percentile of historical TVS levels observed at reference stations throughout British Columbia.
5. Hargrave *et al.* (1997) examined the biological and physicochemical attributes at 11 salmon farms and 11 reference stations in the Western Isles region of the Bay of Fundy. They found that organic carbon, sediment sulphides and redox potential were effective end-points for evaluating the benthic effects associated with salmon farming. The results of that study have been incorporated in the WMP by requiring quantitative evaluation of sediments for total sulphides (S^{-2}) and oxidation–reduction potential. Protocols developed by Hargrave *et al.* (1995) were adopted for these analyses. Specific performance standards for sulphides and *E*h will be developed pending the outcome of a series of focused studies designed to determine the biological response to varying concentrations of these end-points.

The British Columbia Provincial Government stated a desire to establish a final performance-based salmon farm management programme in the autumn of 2001. In addition, the DFO, which has the responsibility to enforce the Fisheries Act, is participating at the technical level in developing the British Columbia programme.

Conclusions

Biodeposits from salmon farms settle on to sediments near the netpens and can have definite effects on physicochemistry and the abundance and diversity of the benthic biota. Changes can be anticipated in total volatile solids and sulphur chemistry in the sediments in the immediate vicinity of operational netpens, together with decreased redox potential. Accumulation of volatile organic material under farms can extend to distances of 145–205 m from the netpen perimeter during peak production. The magnitude of the change in any of these parameters is correlated with the degree of flushing in and around each farm site.

Biodeposits can enrich benthic communities but the actual effects depend on the hydrodynamics of each particular site. At poorly circulated sites accumulations of organic wastes can exceed the aerobic assimilative capacity of sediments, leading to reduced oxygen tension and significant changes in the benthic community. Under extreme conditions sediments can become anoxic and depauperate. However, under any circumstances these effects are ephemeral and conditions have returned to normal within a period of weeks to years during fallow periods in all cases studied.

The accumulation of organic wastes in the sediments can change infaunal community abundance and diversity. But prolonged case studies reveal significant differences between poorly flushed and well-flushed sites. At poorly flushed sites benthic effects are highly dependent on farm management practices. Very high salmon production levels and other activities, such as cleaning nets *in situ*, result in significant changes in both abundance and diversity of infauna to distances as great as 30 m from the netpen's perimeter. At reduced production levels, and in the absence of *in situ* net cleaning, the impacts are restricted to as little 15 m, or less, downstream from thc netpens. At well-flushed sites the abundance and diversity of infaunal organisms is positively correlated with total organic carbon, suggesting that the farm stimulates the infaunal community throughout the area.

Both copper, from marine antifouling compounds used on netpens, and zinc, from fish feeds, can be toxic in their ionic forms to marine organisms. Levels of copper are elevated around some netpen farms which use government-approved antifouling paints on structures or, more likely, treat their nets with approved commercial compounds containing copper. The detected additions of copper in the water following the installation of newly treated nets are biologically insignificant, except to organisms which settle on the nets. Zinc is an essential trace element for salmon nutrition, and it is added to feeds as part of the mineral supplement. Sediment concentrations of zinc are typically increased near salmon farms.

The degree of risk is dependent on several factors. Firstly, the concentration of sulphide in the sediment is important, as typically elevated concentrations near salmon farms reduce the bioavailability of both copper and zinc, thus making the observed concentrations non-toxic. Long-term studies have demonstrated that the metal concentrations return to background during the period of chemical remediation, and there is no evidence of a long-term build-up of these metals under salmon farms. Secondly, the formulation of the feed is relevant, as the majority of feed manufacturers now use reduced amounts of a more bioavailable proteinated form of zinc, or a methionine analogue. Management practices play a role, as the potential rate of accumulation of copper in sediments can be significantly reduced by washing the nets at upland facilities and properly disposing of the waste in an approved landfill.

On European salmon farms therapeutic compounds are used for the control of sea lice, both for the health of the fish and to reduce their potential as vectors. The commonly used compounds are all non-specific within the Class Crustacea, and several are broad-spectrum biocides with potential to affect many phyla adversely. The degree of risk is greatly reduced by government regulation for the use of specific therapeutic compounds following extensive research *in vitro* and *in situ* on their effects on marine organisms. Case studies show that some of these compounds can be detected in sediments close to the perimeter of netpen farms, but the levels resulting from their authorized use do not show significant widespread adverse effects on either pelagic or benthic resources

Fish stocked intensively in contained areas are known to have a high oxygen demand. Decades of monitoring in Washington, USA, have found a maximum oxygen reduction of 2 mg l^{-1} in water passing through salmon netpens where large biomasses of fish were being fed. In most cases the reduction in dissolved oxygen has been ≤ 0.5 mg l^{-1}. Salmon are more sensitive than most other species to depressed oxygen levels and 6.0 mg l^{-1} is considered a minimum concentration for optimum health. Therefore, if there was a localized effect associated with netpen culture, the farmed salmon would be the first organisms affected. At coastal (oceanic) sites, farmed salmon are infrequently subjected to low dissolved oxygen concentrations when oxygen-deficient up-welled water naturally intrudes into the growing area. These are oceanographic events which have nothing to do with the culture of fish or shellfish.

The accumulation of any highly labile organic material in sediment produces ammonia and hydrogen sulphide once the oxygen is depleted. These gases most frequently cycle between oxidized and reduced states within superficial sediment layers where they modify the infaunal community. They are infrequently released into the water column. Although there is evidence from *in situ* studies that total sulphide concentrations in surface sediments in areas of high organic loading can exceed 20,000 μmol l^{-1}, there is little soluble hydrogen sulphide in the water column even under poorly flushed sites. Less than 1.9% of the gases at the sediment–water interface are sulphide, and this can be reduced to 0.05% at a distance 3 m above the sediment. The majority of these gases are methane and carbon dioxide. In a well-sited farm, concentrations of hydrogen sulphide gas rising through the water column are rapidly reduced by oxidation, diffusion and mechanical mixing. For these reasons it is unlikely that toxic conditions caused by hydrogen sulphide will ever occur unless there are extremely large emissions at the sediment–water interface in shallow water.

Enhancement of a harmful algal bloom by the inorganic nutrients discharged from salmon farms is feasible but highly unlikely to occur in the Pacific Northwest. First, apart from the summer months, the natural atmospheric and geographical parameters of the region reduce light availability for photosynthesis, and the waters are vertically well mixed, which reduces the time phytoplankton spend in the euphotic zone. Second, the physical characteristics of locations permitted for salmon farming are not conducive to the accumulation of nutrients, even when the water body is nutrient limited. Decades of monitoring have shown minimal increases in inorganic nutrient concentrations downstream from even the few sites having restricted water exchange. Small increases observed at 6 m downstream during slack tide have been statistically insignificant at a distance of 30 m downstream. Nutrient-limited embayments in Washington, USA, have been identified and salmon aquaculture activities in these locations are discouraged and carefully managed when allowed.

Wild salmonids carry genera of marine bacteria, such as *Vibrio*, *Acinetobacter* and *Aeromonas*, some species of which are pathogenic to humans. The concern is that fish faeces and waste feed might enhance populations of these pathogens. There is no evidence in the literature, or in the epidemiological records of Washington State, of any documented case in which the handling or consumption of farmed salmon has led to infectious disease in consumers or farm workers. There are many differences in the physical and chemical composition of salmon farm waste compared with human sewage discharge, and the former does not disperse over large areas but remains localized where it is metabolized by naturally occurring marine bacteria and invertebrates. There is no credible evidence supporting a hypothesis that salmon farming increases the risk of infectious disease in humans or in wild populations of animals.

Public health concerns for the safety of fish and shellfish in the vicinity of discharges of industrial and residential waste are real, and vigilance is maintained by stringent regulations and monitoring programmes. The accumulation of wastes from netpen farms is perceived as another source of human and environmental pathogens. However, there is little evidence substantiating this hypothesis. Viruses pathogenic to fish have no documented effect on human beings because they are taxa-specific. Faecal coliform bacteria are unlikely to persist in netpen sediments rich in total organic carbon as they are specific to warm-blooded animals. Sources of faecal coliform bacteria near salmon farms are more likely to be mammals (such as seals and sea-lions) or birds. *In situ* monitoring at some well-flushed netpen farms revealed slightly more faecal coliform bacteria in water and shellfish tissues at stations closest to the farm perimeter. The sources of observed bacteria were not determined. However, all water and shellfish tissues examined were consistently of high quality and met all bacteriological requirements imposed by the US National Shellfish Sanitation Programme.

Responsible permitting of each site is also playing an important management role. The National Pollution Discharge Elimination System (NPDES) permit has been effective in regulating the degree of allowable effect, but its impact must now be supplemented with the strict adherence by site operators to a well-defined set of industry best management practices which are based on good scientific information. These BMPs can be specific to a particular farm, or they can be overarching for the entire industry.

References

Ackefors, H. (1986) The impact on the environment by cage farming in open water. *Journal of Aquaculture in the Tropics* 1, 25–33.

Ackefors, H. and Enell, M. (1990) Discharge of nutrients from Swedish fish farming to adjacent sea areas. *Ambio* 19(1), 28–35.

Ackefors, H. and Enell, M. (1994) The release of nutrients and organic matter from aquaculture systems in Nordic countries. *Journal of Applied Ichthyology* 10, 225–241.

Anderson, E. (1992) *Benthic Recovery Following Salmon Farming: Study Site Selection and Initial Surveys.* Report to the Water Quality Branch, Ministry of Environment, Lands and Parks, Province of British Columbia, Canada, 170pp.

Anderson, S. (1998) *Dietary Supplementation of Salmon Diets with Zinc: Alternative Sources and Their Effects on the Environment.* Report prepared by Steward Anderson, Aquaculture Industry Coordinator, Hoffmann La-Roche Ltd, 9pp.

Arntz, W.E. and Rumohr, H. (1982) An experimental study of macrobenthic colonization and succession, and the importance of seasonal variation in temperate latitudes. *Journal of Experimental Marine Biology and Ecology* 64, 17–45.

Aure, J., Ervik, A.S., Johannesen, P.J. and Ordemann, T. (1988) *The Environmental Effects of Seawater Fish Farms.* Fisken Havet ISSN 0071-5638 [English abstract].

Bagarinao, T.U. (1993) Sulfide as a toxicant in aquatic habitats. *Aquaculture* 15, 2–4.

Banse, K., Horner, R. and Postel, J. (1990) Fish farms innocent. *Seattle Post Intelligencer* 4 August 1990.

Beveridge, M.C.M., Phillips, M.J. and Clarke, R.M. (1991) A quantitative and qualitative assessment of wastes from aquatic animal production. In: Brune, D. and Tomasso, J.R. (eds) *Aquaculture and Water Quality. Advances in World Aquaculture,* World Aquaculture Society, Baton Rouge, Louisiana, pp. 506–533.

Black, K.D., Fleming, S., Nickell, S.D. and Pereira, P.M.F. (1997) The effects of ivermectin, used to control sea lice on caged farmed salmonids, on infaunal polychaetes. *Journal of Marine Science* 54, 276–279.

Braaten, B., Aure, J., Ervik, A. and Boge, E. (1983) Pollution problems in Norwegian fish farming. ICES C.M.1983/F:26. International Council for the Exploration of the Sea, Copenhagen, 11pp.

Brett, J.R. and Zala, C.A. (1975) Daily pattern of nitrogen excretion and oxygen consumption of sockeye salmon (*Oncorhynchus nerka*) under controlled conditions. *Journal of the Fisheries Research Board of Canada* 32, 2479–2486.

Brooks, K.M. (1991) Environmental sampling at Sea Farm Washington Inc., netpen facility II in Port Angeles Harbor, WA during 1991. Produced for the Washington Department of Natural Resources, Olympia, Washington, 16pp.

Brooks, K.M. (1992) Environmental sampling at the Sea Farm Washington Inc., netpen facility II in Port Angeles Harbor, WA during 1992. Produced for the Washington Department of Natural Resources, Olympia, Washington, 7pp.

Brooks, K.M. (1993a) Environmental sampling at Sea Farm Washington Inc., netpen facility II in Port Angeles Harbor, WA during 1993. Produced for the Washington Department of Natural Resources, Olympia, Washington, 18pp.

Brooks, K.M. (1993b) Environmental sampling at Paradise Bay Salmon Farm located in Port Townsend Bay, WA January 1993 following abandonment of the site. Produced for the Washington State Department of Natural Resources, Olympia, Washington.

Brooks, K.M. (1994a) Environmental sampling at Sea Farm Washington, Inc., netpen facility II in Port Angeles Harbor, Washington during 1994. Produced for the Washington Department of Natural Resources, Olympia, Washington, 18pp.

Brooks, K.M. (1994b) Environmental sampling at Global Aqua USA Inc., saltwater II salmon farm located in Rich Passage, WA 1994. Prepared for Global Aqua USA Inc., 600 Ericksen Avenue N.E., Suite 370, Bainbridge Island, Washington, 20pp.

Brooks, K.M. (1995a) Environmental sampling at Sea Farm Washington Inc., netpen facility II in Port Angeles Harbor, WA during 1995. Produced for the Washington Department of Natural Resources, Olympia, Washington, 20pp.

Brooks, K.M. (1995b) Environmental sampling at Global Aqua USA Inc., saltwater II salmon farm located in Rich Passage, WA 1994. Prepared for Global Aqua USA Inc., 600 Ericksen Avenue, N.E., Suite 370, Bainbridge Island, Washington, 20pp.

Brooks, K.M. (2000a) Salmon farm benthic and shellfish effects study 1996–1997. Aquatic Environmental Sciences, 644 Old Eaglemount Road, Port Townsend, Washington, 117pp.

Brooks, K.M. (2000b) Sediment concentrations of zinc near salmon farms in British Columbia, Canada during the period June through August 2000. British Columbia Salmon Farmers Association, 1200 West Pender Street, Vancouver, British Columbia, Canada, 12pp.

Brooks, K.M. (2000c) Database report to the Ministry of Environment describing sediment physicochemical response to salmon farming in British Columbia, 1996 through April 2000. British Columbia Salmon Farmers Association, 1200 West Pender Street, Vancouver, British Columbia, Canada, 41pp.

Brooks, K.M. (2000d) Determination of copper loss rates from Flexgard XI™ treated nets in marine environments and evaluation of the resulting environmental risks. Report to the Ministry of Environment for the British Columbia Salmon Farmers Association, 1200 West Pender Street, Vancouver, British Columbia, Canada, 24pp.

Brooks, K.M. (2000e) Sediment concentrations of sulfides and total volatile solids near salmon farms in British Columbia, Canada during the period June through August 2000, and recommendations for additional sampling. Report to the Ministry of Environment prepared for the British Columbia Salmon Farmers Association, 1200 West Pender Street, Vancouver, British Columbia, Canada, 16pp.

Brooks, K.M. (2000f) Results of the June 2000 interim salmon farm monitoring at Stolt Sea Farm, Inc. salmon aquaculture tenures located in British Columbia. Submitted to the Ministry of Environment for Stolt Sea Farm, Inc., 1261 Redwood Street, Campbell River, British Columbia, Canada.

Brooks, K.M. (2000g) Literature review and model evaluation describing the environmental effects and carrying capacity associated with the intensive culture of mussels (*Mytilus edulis galloprovincialis*). Technical appendix to an Environmental Impact Statement produced for Taylor Resources, Southeast 130 Lynch Road, Shelton, Washington, USA.

Brown, J.R., Gowen, R.J. and McLusky, D.S. (1987) The effect of salmon farming on the benthos of a Scottish sea loch. *Journal of Experimental Marine Biology and Ecology* 109, 39–51.

Burridge, L.E. and Haya, K. (1993) The lethality of ivermectin, a potential agent for treatment of salmonids against sea lice, to the shrimp *Crangon septemspinosa*. *Aquaculture* 117, 9–14.

Chow, K.W. and Schell, W.R. (1978) The minerals. In: *Fish Feed Technology*. A series of lectures presented at the FAO/UNDP training course in fish feed technology held at the College of Fisheries, University of Washington, Seattle, Washington, 9 October–15 December 1978. FAO Publication ADCP/REP/80/11.

Collier, L.M. and Pinn, E.H. (1998) An assessment of the acute impact of the sea lice treatment ivermectin on a benthic community. *Journal of Experimental Marine Biology and Ecology* 230, 131–147.

Costello, M.J. (1993) Review of methods to control sea lice (Caligidae: Crustacea) infestations on salmon (*Salmo salar*) farms. In: Boxshall, G.A. and Defaye, D. (eds) *Pathogens of Wild and Farmed Fish: Sea Lice*. Ellis Horwood, Chichester, UK.

Crema, R., Prevedelli, D., Valentini, A. and Castelli, A. (2000) Recovery of the macrozoobenthic community of the Comacchio lagoon system (northern Adriatic Sea). *Ophelia* 52, 143–152.

Crisp, D.J. (1964) The effects of the severe ice winter of 1962–63 on marine life in Britain. *Journal of Animal Ecology* 33, 165–210.

Cross, S.F. (1990) *Benthic Impacts of Salmon Farming in British Columbia. Summary Report (Volume I)*. Prepared for the Ministry of Environment, Water Management Branch, 765 Broughton Street, Victoria, British Columbia, Canada, 78pp.

Cross, S.F. (1993) *Oceanographic Characteristics of Net-Cage Culture Sites Considered Optimal for Minimizing Environmental Impacts in Coastal British Columbia*. Prepared for the Ministry of Agriculture, Fisheries and Food, Courtenay, British Columbia, Canada. Prepared by Aquametrix Research Ltd, Sidney, British Columbia, Canada, 86pp.

Davies, I.M., McHenery, J.G. and Rae, G.H. (1997) Environmental risk from dissolved ivermectin to marine organisms. *Aquaculture* 158, 263–275.

Davies, I.M., Gillibrand, P.A., McHenery, J.G. and Rae, G.H. (1998) Environmental risk of ivermectin to sediment dwelling organisms. *Aquaculture* 163, 29–46.

DFO (Department of Fisheries and Oceans, Canada) (1996) Monitoring of sea lice treatment chemicals in southwestern New Brunswick. DFO Science High Priority Project. Project Code 9019. Project leaders: W. Watson-Wright and B. Chang. DFO Street, Andrews Biological Station, New Brunswick, Canada.

Di Toro, D.M., Mahony, J.D., Hansen, D.J., Scott, K.J., Carlson, A.R. and Ankley, G.T. (1992) Acid volatile sulfide predicts the acute toxicity of cadmium and nickel in sediments. *Environmental Science and Technology* 26, 96–101.

Eagle, R.A. (1975) Natural fluctuations in a soft bottom community. *Journal of the Marine Biology Association of the United Kingdom* 55, 865–878.

EAO (Environmental Assessment Office, Canada) (1997) *British Columbia Salmon Aquaculture Review*. Environmental Assessment Office, Government of British Columbia, 836 Yates Street, Victoria, British Columbia, Canada.

Earll, R.C., James, G., Lumb, C. and Pagett, R. (1984) *A Report on the Effects of Fish Farming on the Marine Ecology of the Western Isles*. Report to the Nature Conservancy Council. Contract MF3/11/9. Marine Biological Consultants Ltd.

Einen, O., Holmefjord, I., Asgard, T. and Talbot, C. (1995) Auditing nutrient discharges from fish farms: theoretical and practical considerations. *Aquaculture Research* 26, 701–713.

Ellis, D. (1996) *Net Loss: The Salmon Netcage Industry in British Columbia*. A report to the David Suzuki Foundation, Suite 219, 2211 West Fourth Avenue, Vancouver, British Columbia, Canada, 146pp.

Enell, M. and Ackefors, H. (1992) Development of Nordic salmonid production in aquaculture and nutrient discharges into adjacent sea areas. *Aquaculture Europe* 16, 6–11.

Enell, M. and Lof, J. (1983) Environmental impact of aquaculture: sedimentation and nutrient loadings from fish cage culture farming. *Vatten* 39, 346–375.

EPA (US Environmental Protection Agency) (1986) *Quality Criteria for Water – 1986*. EPA 440/5-86-001. US Environmental Protection Agency, Office of Water, Regulations and Standards, Washington, DC.

EPA (US Environmental Protection Agency) (1994) Briefing report to the EPA science advisory board on the EqP approach to predicting metal bio-availability in sediment and the derivation of sediment quality criteria for metals. EPA 822/D-94/002. US Environmental Protection Agency, Washington, DC.

EPA (US Environmental Protection Agency) (1995) *Ambient Water Quality Criteria – Saltwater Copper Addendum*. US Environmental Protection Agency, Office of Water, Office of Science and Technology, Washington, DC.

ERT, Ltd (1997) *Ivermectin Field Trials: Impact on Benthic Assemblages, Incorporating Additional Data*. Report to the Scottish Salmon Growers Association. ERT Ltd, Edinburgh, ERT 97/029.

ERT, Ltd (1998) *Ivermectin Field Trials: Impact on Benthic Assemblages*. Report to the Scottish Salmon Growers Association. ERT Ltd, Edinburgh, ERT 97/223.

Ervik, A., Johannessen, P. and Aure, J. (1985) *Environmental Effects of Marine Norwegian Fish Farms*. ICES C.M. 1985 F:37. International Council for the Exploration of the Sea, Copenhagen.

Ervik, A., Hansen, P.K., Aure, J., Stigebrandt, A., Johannessen, P. and Jahnsen, T. (1997) Regulating the local environmental impact of intensive marine fish farming. I. The concept of the MOM system. *Aquaculture* 158, 85–94.

Findlay, R.H. and Watling, L. (1994) Toward a process level model to predict the effects of salmon netpen aquaculture on the benthos. In: Hargrave, B.T. (ed.) *Modeling Benthic Impacts of Organic Enrichment from Marine Aquaculture*, pp. 47–77. *Canadian Technical Report of Fisheries and Aquatic Sciences, 1949*, 125pp.

Folke, C., Kautsky, N. and Troell, M. (1994) The costs of eutrophication from salmon farming: implications for policy. *Journal of Environmental Management* 40, 173–182.

Fox, W.P. (1990) Modelling of particulate deposition under salmon netpens. In: *Final Programmatic Environmental Impact Statement: Fish Culture in Floating Netpens (Technical Appendices)*. Washington State Department of Fisheries, 115 General Administration Building, Olympia, WA 98504, 15pp.

GESAMP (Joint Group of Expert on Scientific Aspects of Marine Environmental Protection) (1996) Monitoring the ecological effects of coastal aquaculture wastes.

GESAMP Reports and Studies No. 57. Food and Agriculture Organization of the United Nations, Rome, 38pp.

Gormican, S.J. (1989) Water circulation, dissolved oxygen, and ammonia concentrations in fish net-cages. MSc. thesis, University of British Columbia, Vancouver, British Columbia, Canada.

Gowen, R.J. and Bradbury, N.B. (1987) The ecological impact of salmonid farming in coastal waters: a review. *Oceanography and Marine Biology Annual Review* 25, 563–575.

Gowen, R.J., Brown, J., Bradbury, N. and McLusky, D.S. (1988) *Investigation into Benthic Enrichment, Hypernutrification and Eutrophication Associated with Mariculture in Scottish Coastal Waters (1984–1988)*. Report by the Department of Biological Sciences, University of Stirling, Scotland.

Gowen, R.J., Weston, D.P. and Ervik, A. (1991) Aquaculture and the benthic environment: a review. In: Cowey, C.B. and Cho, C.Y. (eds) *Nutritional Strategies and Aquacultural Waste*. Fish Nutrition Research Laboratory, Department of Nutritional Sciences, University of Guelph, Ontario, Canada, pp. 187–205.

Gowen, R.J., Smyth, D. and Silvert, W. (1994) Modelling the spatial distribution and loading of organic fish farm waste to the seabed. In: Hargrave, B.T. (ed.) *Modeling Benthic Impacts of Organic Enrichment From Marine Aquaculture. Canadian Technical Report in Fisheries and Aquatic Science 1949*, pp. 19–39.

Goyette, D. and Brooks, K.M. (1999) *Creosote Evaluation. Phase II, Sooke Basin Study: Baseline to 535 Days Post-construction, 1995–1996*. Commercial Chemicals Division, Environment Canada, Pacific and Yukon Region, 568pp.

Grant, A. and Briggs, A.D. (1998a) Use of ivermectin in marine fish farms: some concerns. *Marine Pollution Bulletin* 36, 566–568.

Grant, A. and Briggs, A.D. (1998b) Toxicity of ivermectin to estuarine and marine invertebrates. *Marine Pollution Bulletin* 36, 540–541.

Hansen, P.K., Pittman, K. and Ervik, A. (1990) Effects of organic waste from marine fish farms on the sea bottom beneath the cages. ICES C.M. 1990/F:34, International Council for the Exploration of the Sea, Copenhagen, 9pp.

Hargrave, B.T. (1994) A benthic enrichment index. In: Hargrave, B.T. (ed.) *Modeling Benthic Impacts of Organic Enrichment from Marine Aquaculture. Canadian Technical Report in Fisheries and Aquatic Science 1949*, pp. 79–91.

Hargrave, B.T., Phillips, G.A., Doucette, L.I., White, M.J., Milligan, T.G., Wildish, D.J. and Cranston, R.E. (1995) Biogeochemical observations to assess benthic impacts of organic enrichment from marine aquaculture in the Western Isles region of the Bay of Fundy, 1994. *Canadian Technical Report in Fisheries and Aquatic Science 2062*, 159pp.

Hargrave, B.T., Phillips, G.A., Doucette, L.I., White, M.J., Milligan, T.G., Wildish, D.J. and Cranston, R.E. (1997) Assessing benthic impacts of organic enrichment from marine aquaculture. *Water, Air and Soil Pollution* 99, 641–650.

Henderson, A.R. and Ross, D.J. (1995) Use of macrobenthic infaunal communities in the monitoring and control of the impact of marine cage fish farming. *Aquaculture Research* 26, 659–678.

Johannessen, P.J., Botnen, H.B. and Tvedten, O.F. (1994) Macrobenthos: before, during and after a fish farm. *Aquaculture and Fisheries Management* 25, 55–66.

Johnsen, F. and Wandsvik, A. (1991) The impact of high energy diets on pollution control in the fish farming industry. In: Cowey, C.B. and Cho, C.Y. (eds) *Nutritional Strategies and Aquaculture Waste. Proceedings of the 1st International Symposium on*

Nutritional Strategies in Management of Aquaculture Waste. University of Guelph, Ontario, Canada, pp. 51–62.

Johnsen, R.I., Grahl-Nelson, O. and Lunestad, B.T. (1993) Environmental distribution of organic waste from a marine fish farm. *Aquaculture* 118, 229–244.

Johnson, S.C. and Margolis, L. (1993) Efficacy of ivermectin for control of the salmon louse *Lepeophtheirus salmonis* on Atlantic salmon. *Diseases of Aquatic Organisms* 17, 101–105.

Kadowaki, S., Kasedo, T., Nakazono, T., Yamashita, Y. and Hirata, H. (1980) The relation between sediment flux and fish feeding in coastal culture farms. *Memoirs of the Faculty of Fisheries Kagoshima University (Kagoshimadai Suisangakubu Kiyo)* 29, 217–224.

Karakassis, I., Hatziyanni, E., Tsapakis, M. and Plaiti, W. (1999) Benthic recovery following cessation of fish farming: a series of successes and catastrophes. *Marine Ecology Progress Series* 184, 205–218.

Kontali (1996) Introduction of new vaccines in the production of salmon – analysis of the consequences. Kontali analyses. Industriv. 18, 6500 Kr..sund Norway, 25pp.

Levings, C.D. (1997) Waste discharge. In: *British Columbia Salmon Aquaculture Review.* Environmental Assessment Office, Vancouver, British Columbia, Canada, pp. WD1–47.

Levings, C.D., Ervik, A., Johannessen, P. and Aure, J. (1994) Ecological criteria used to help site fish farms in fjords. *Estuaries* 18(1A), 81–90.

Lewis, A.G. and Metaxas, A. (1991) Concentrations of total dissolved copper in and near a copper-treated salmon netpen. *Aquaculture* 99, 269–276.

Long, E.R., MacDonald, D.D., Smith, S.L. and Calder, F.D. (1995) Incidence of adverse biological effects within ranges of chemical concentrations in marine and estuarine sediments. *Environmental Management* 19, 81–97.

Lu, L. and Wu, R.S.S. (1998) Recolonization and succession of marine macrobenthos in organic-enriched sediment deposited from fish farms. *Environmental Pollution* 101, 241–251.

Lunz, J.D. and Kendall, D.R. (1982) Benthic resources assessment technique: a method for quantifying the effects of benthic community changes on fish resources. In: *Oceans '82 Conference Record: Industry, Government, Education – Partners in Progress*, 20–22 September 1982. Washington, DC, pp. 1021–1027.

MacDonald, D.D. (1994) *Approach to the Assessment of Sediment Quality in Florida Coastal Waters.* Florida Department of Environmental Protection, Tallahassee, Florida.

Mahnken, C.V.W. (1993) Benthic faunal recovery and succession after removal of a marine fish farm. PhD dissertation, University of Washington, Seattle, Washington, 290pp.

Mayer, I. and McLean, E. (1995) Bioengineering and biotechnological strategies for reduced waste aquaculture. *Water Science and Technology* 31, 85–102.

Mazzola, A., Mirto, S. and Danovaro, R. (1999) Initial fish-farm impact on meiofaunal assemblages in coastal sediments of the western Mediterranean. *Marine Pollution Bulletin* 38, 1126–1133.

Meijer, L.E. and Avnimelech, Y. (1999) On the use of micro-electrodes in fish pond sediments. *Aquaculture Engineering* 21(2), 71–83.

Merican, Z.O. and Phillips, M.J. (1985) Solid waste production from rainbow trout (*Salmo gairdneri* Richardson) cage culture. *Aquaculture and Fisheries Management* 1, 55–69.

Mills, E.L. (1969) The community concept in marine zoology, with comments on continua and instability in some communities: a review. *Journal of the Fisheries Research Board of Canada* 26, 1415–1428.

Morrisey, D.J., Gibbs, M.M., Pickmere, S.E. and Cole, R.G. (2000) Predicting impacts and recovery of marine-farm sites in Stewart Island, New Zealand, from the Findlay–Watling model. *Aquaculture* 185(3–4), 257–271.

NSSP (National Shellfish Sanitation Program) (1997) *Manual of Operations, I. Sanitation of Shellfish Growing Areas*. US Department of Health and Human Services, Public Health Service, Food and Drug Administration, Washington, DC.

Parametrix (1990) Final programmatic environmental impact statement fish culture in floating netpens. Prepared by Parametrix Inc., for Washington State Department of Fisheries, 115 General Administration Building, Olympia, Washington, 161pp.

Parsons, T.R., Rokeby, B.E., Lalli, C.M. and Levings, C.D. (1990) Experiments on the effect of salmon farm wastes on plankton ecology. *Bulletin of the Plankton Society of Japan* 37, 49–57.

Pearson, T.H. (1986) Disposal of sewage in dispersive and non-dispersive areas: contrasting case histories in British coastal waters. In: Kullenberg, G. (ed.) *The Role of the Oceans as a Waste Disposal Option*. D. Reidel Publishing Company, Dordrecht, The Netherlands, pp. 577–595.

Pearson, T.H. and Rosenberg, R. (1978) Macrobenthic succession in relation to organic enrichment and pollution of the marine environment. *Oceanography and Marine Biology Annual Review* 16, 229–311.

Pease, B.G. (1977) The effect of organic enrichment from a salmon mariculture facility on the water quality and benthic community of Henderson Inlet, Washington. PhD disseration, University of Washington, Seattle, Washington, 145pp.

Persson, G. (1988) *Relationship Between Feed, Productivity and Pollution in the Farming of Large Rainbow Trout (Salmo gairdneri)*. National Swedish Environmental Protection Board Report No. 3534.

Peterson, L.K., D'Auria, J.M., McKeown, B.A., Moore, K. and Shum, M. (1991) Copper levels in the muscle and liver tissue of farmed chinook salmon, *Oncorhynchus tshawytscha*. *Aquaculture* 99, 105–115.

Poole, N.J., Wildish, D.J. and Kristmanson, D.D. (1978) The effects of the pulp and paper industry on the aquatic environment. *Critical Reviews in Environmental Control* 8, 153–195.

Pridmore, R.D. and Rutherford, J.C. (1992) Modeling phytoplankton abundance in a small-enclosed bay used for salmon farming. *Aquaculture and Fisheries Management* 23, 525–542.

PSEP (Puget Sound Estuary Protocols) (1986) Recommended protocols for measuring selected environmental variables in Puget Sound. Puget Sound Water Quality Authority, PO Box 40900, Olympia, Washington.

Rensel, J.E. (1988) Environmental sampling at the American Aqua foods netpen site near Lone Tree Point in north Skagit Bay. Prepared by Rensel Associates, Seattle, Washington USA, Pacific Aqua Foods, Vancouver, British Columbia, Canada and Washington Department of Natural Resources, Olympia, Washington, 7pp.

Rensel, J.E. (1989) Phytoplankton and nutrient studies near salmon netpens at Squaxin Island, WA. In: *Technical Appendices to the Final Programmatic Environmental Impact Statement, Fish Culture in Floating Netpens*. Prepared for the Washington Department of Fisheries, Olympia, Washington, 33pp.

Ritz, D., Lewis, M.E. and Shen, M. (1989) Response to organic enrichment of infaunal macrobenthic communities under salmonid sea cages. *Marine Biology (New York)* 103, 211–214.

Roberts, R.J. (1978) *Fish Pathology*. Baillière Tindall, University Press, Aberdeen, UK, 318pp.

Rosenthal, H.D., Weston, D., Gowen, R. and Black, E. (1988) *Environmental Impact of Mariculture. Cooperative Research Report*. ICES:154. International Council for the Exploration of the Sea, Copenhagen.

Rosenthal, H., Scarratt, D.J. and McInerney-Northcott, M. (1995) Aquaculture and the environment. In: Boghen, A.D. (ed.) *Cold-water Aquaculture in Atlantic Canada*, 2nd edn. CIRRD, Moncton, New Brunswick, Canada, pp. 451–500.

Roth, M., Richards, R.H. and Sommerville, C. (1993) Current practices in the chemotherapeutic control of sea lice infestations in aquaculture: a review. *Journal of Fish Diseases* 16, 1–26.

Samuelsen, O.B., Ervik, A. and Solheim, E. (1988) A qualitative and quantitative analysis of the sediment gas and diethylether extract of the sediment from salmon farms. *Aquaculture* 74, 277–285.

SEPA (Scottish Environment Protection Agency) (1997) Cage fish farms: sea lice treatment chemicals risk assessment of azamethiphos. SEPA Policy No. 17.

SEPA (Scottish Environment Protection Agency) (1998a) Ivermectin: a review of the laboratory and field data available to SEPA.

SEPA (Scottish Environment Protection Agency) (1998b) The use of cypermethrin in marine cage fish farming risk assessment, EQS, and recommendations. SEPA Policy No. 30.

SEPA (Scottish Environment Protection Agency) (1999a) Emamectin benzoate use in marine fish farming. SEPA, Fish Farm Advisory Group, SEPA Report 66/99.

SEPA (Scottish Environment Protection Agency) (1999b) Calicide (teflubenzuron): authorization for use as an in-feed sea lice treatment in marine cage salmon farms. Risk assessment, EQS, and recommendations. SEPA Policy No. 29.

SEPA. (Scottish Environment Protection Agency) (2000) Regulation and monitoring of marine cage fish farming in Scotland: a manual of procedures. Internet document http://www.sepa.org.uk/publications/fishfarmmanual.htm

Silvert, W. (1994a) Modeling benthic deposition and impacts of organic matter loading. In: Hargrave, B.T. (ed.) *Modeling Benthic Impacts of Organic Enrichment from Marine Aquaculture. Canadian Technical Report of Fisheries and Aquatic Sciences 1949*, pp. 1–30.

Silvert, W. (1994b) Simulation models of finfish farms. *Journal of Applied Ichthyology* 10, 349–352.

Silvert, W. and Sowles, J.W. (1996) Modeling environmental impacts of marine finfish aquaculture. *Journal of Applied Ichthyology* 12, 75–81.

Skalski, J.R. and McKenzie, D.H. (1982) A design for aquatic monitoring programs. *Journal of Environmental Management* 14, 237–251.

Smith, P.R., Moloney, M., McElligott, A., Clarke, S., Palmer, R., O'Kelly, J. and O'Brien, F. (1993) The efficiency of oral ivermectin in the control of sea lice infestations of farmed Atlantic salmon. In: Boxshall, G.A. and Defaye, D. (eds) *Pathogens of Wild and Farmed Fish – Sea Lice*. Shorewood Publishing, New York.

Sowles, J.W., Churchill, L. and Silvert, W. (1994) The effect of benthic carbon loading on the degradation of bottom conditions under farm sites. In: Hargrave, B.T. (ed.) *Modeling Benthic Impacts of Organic Enrichment From Marine*

Aquaculture. Canadian Technical Report of Fisheries and Aquatic Sciences 1949, pp. 31–79.

Stanley, S.O., Leetley, J., Miller, D. and Pearson, T.H. (1980) Chemical changes in the sediments of Loch Eil arising from the input of cellulose fiber. In: Albaiges, J. (ed.) *Analytical Techniques in Environmental Chemistry*. Pergamon Press, Oxford, UK, pp. 409–418.

Striplin Environmental Associates, Inc. (1996) *Development of Reference Value Ranges for Benthic Infauna Assessment Endpoints in Puget Sound*. Final Report prepared for the Washington State Department of Ecology, Sediment Management Unit, Olympia, Washington, USA, 45pp.

Sutherland, T.F., Martin, A.J. and Levings, C.D. (2000) *The Characterization of Suspended Particulate Matter Surrounding a Salmonid Netpen in the Broughton Archipelago, British Columbia*. Department of Fisheries and Oceans, West Vancouver Laboratory, West Vancouver, British Columbia, Canada.

Taylor, F.J.R. (1993) Current problems with harmful phytoplankton blooms in British Columbia waters. In: Smayda, T.J. and Shimizu, Y. (eds) *Toxic Phytoplankton Blooms in the Sea*. Elsevier Science Publishers, Amsterdam, pp. 699–703.

Taylor, F.J.R. and Horner, R. (1994) Red tides and other problems with harmful algal blooms in Pacific Northwest coastal waters. In: Wilson, R.C.H., Beamish, R.J., Aitkens, F. and Bell, J. (eds) *Review of the Marine Environment and Biota of Strait of Georgia, Puget Sound, and Juan de Fuca Strait. Canadian Technical Report of Fisheries and Aquatic Sciences 1948*, pp. 175–186.

Taylor, L.A., Chapman, P.M., Miller, R.A. and Pym, R.V. (1998) The effects of untreated municipal sewage discharge to the marine environment off Victoria, British Columbia, Canada. Water quality international 1998. IAWQ 19th Biennial International Conference, 21–26 June 1998, Vancouver, Canada.

Thain, J.E., Davies, I.M., Rae, G.H. and Allen, Y.T. (1997) Acute toxicity of ivermectin to the lugworm *Arenicola marina*. *Aquaculture* 159, 47–52.

Tsutsumi, H., Kikuchi, T., Tanaka, M., Higashi, T., Imasaka, K. and Miyazaki, M. (1991) Benthic faunal succession in a cove organically polluted by fish farming. *Marine Pollution Bulletin* 23, 233–238.

WAC (Washington Administrative Code) 173-204. *Sediment Management Standards*. State of Washington Administrative Code, WAC 173-204 (1991), Olympia, Washington, 61pp.

Wang, F. and Chapman, P.M. (1999) Biological implications of sulfide in sediment: a review focusing on sediment toxicity. *Environmental Toxicology and Chemistry* 18, 2526–2532.

Warrer-Hansen, I. (1982) Evaluation of matter discharged from trout farming in Denmark. In: Alabaster, J.S. (ed.) *Workshop of Fish-farm Effluents*. Silkeborg, Denmark, 26–28 May 1981. EIFAC Tech. Pap. 41. Food and Agriculture Organization of the United Nations, European Inland Fisheries Advisory Committee, Rome, pp. 57–63.

Weston, D. (1986) *The Environmental Effects of Floating Mariculture in Puget Sound*. Report prepared for the Washington State Department of Fisheries and the Washington State Department of Ecology, Olympia, Washington, 148pp.

Weston, D.P. (1990) Quantitative examination of macrobenthic community changes along an organic enrichment gradient. *Marine Ecology Progress Series* 61, 233–244.

Weston, D.P. and Gowen, R.J. (1988) *Assessment and Prediction of the Effects of Salmon Netpen Culture on the Benthic Environment. Final Programmatic Environmental*

Impact Statement, Fish Culture in Floating Netpens. Prepared for the Washington State Department of Fisheries, Olympia, Washington, 62pp.

Wildish, D.J., Akagi, H.M., Hamilton, N. and Hargrave, B.T. (1999) A recommended method for monitoring sediments to detect organic enrichment from mariculture in the Bay of Fundy. *Canadian Technical Report of Fisheries and Aquatic Sciences, No. 2286*, 34pp.

Winsby, M., Sander, B., Archibald, D., Daykin, M., Nix, P., Taylor, F.J.R. and Munday, D. (1996) *The Environmental Effects of Salmon Netcage Culture in British Columbia.* Prepared for the Ministry of Environment, Lands and Parks, Environmental Protection Department, Industrial Waste/Hazardous Contaminants Branch, Victoria, British Columbia, Canada, 214pp.

Issues Associated with Non-indigenous Species in Marine Aquaculture

11

Robert R. Stickney

Director, Texas Sea Grant College Program, 2700 Earl Rudder Highway South, Suite 1800, College Station, TX 77845, USA

Abstract

Non-indigenous species have long been employed by aquaculturists. In some instances these species had been introduced and become accepted members of the local ecosystem long before they were reared by aquaculturists. Tilapia in Asia are a good example. In other cases, problems associated with the culture of local species led culturists to find related non-indigenous species as alternatives. Examples are the introduction of Latin American shrimp species into the United States and Atlantic salmon (*Salmo salar*) to the west coast of North America.

The use of non-indigenous species is more widely accepted in association with inland aquaculture than marine aquaculture. A number of diseases, primarily viral in nature, have occurred on shrimp farms around the world. Countries that have used non-indigenous shrimp species have not been immune to the problem, so diseases have been imported along with the shrimp. There is a major concern that these diseases will spread to native populations.

The lack of assurance that animals cultured in the marine environment can be contained is a major factor. Escape as a result of culture system failure, vandalism, accidents and carelessness cannot be completely avoided. Concerns that escapees will become established and possibly displace native species have prompted the marine aquaculture community to reconsider the use of non-indigenous species in general and to concentrate on local species.

Introduction

Non-indigenous (sometimes called non-native) species are herein defined as organisms that have become established in a location that is outside of

(eds R.R. Stickney and J.P. McVey)

their native range. The term does not distinguish between species that are introduced from the waters of one nation to those of another (often termed exotic introductions) and those that become established in a location out of their native range but within the same nation (translocation). Narrowing the scope of the discussion further, this chapter deals only with marine, estuarine, diadromous or euryhaline species that are associated in some way with mariculture activities. Non-indigenous species may be the target species of the mariculturist or may be a competitor, predator, parasite or pathogen associated with the target species.

The association of non-indigenous species with aquaculture may be intentional (as in the rearing of Atlantic salmon in netpens in the Pacific Northwest of the United States and Canada as well as in Chile) or accidental (such as appears to have been the case with at least some disease organisms that were introduced in association with the dumping of ballast water).

There are probably less than 200 species of animals currently being reared by aquaculturists around the world for human food. Of these, far fewer than half are reared in brackish or marine waters, and only a fraction of the latter group are non-indigenous. Further, some of the non-indigenous species had become established in the regions where they are being cultured long before mariculture industries had been established. An excellent example are a group of fishes with the common name tilapia, many of which are highly euryhaline. *Oreochromis mossambicus*, for example, have been reported from salinities of 60 and 120 ppt (Potts *et al.*, 1967; Assem and Hanke, 1979). Tilapia are currently being reared in both freshwater and saltwater culture systems in several nations, primarily in the tropics. *O. mossambicus* were spread throughout the Far East after World War II, and may have originated with as few as five escapees from an aquarium (Balarin and Hatton, 1979). Other species were introduced subsequent to the establishment of *O. mossambicus*. Most residents of such nations as Indonesia, Malaysia, the Philippines and Thailand do not realize that tilapia are non-indigenous.

Tilapia are very popular in many regions where they have been introduced. The same cannot always be said of non-indigenous species, though most of the species that have caused problems have been associated with activities other than aquaculture (take, for example, common carp, *Cyprinus carpio*, that were introduced to the United States over a century ago).

Ballast water releases account for large numbers of non-indigenous species introductions. Perhaps the most widely publicized introduction in terms of environmental problems was that of the freshwater zebra mussel, *Dreissena* sp., which has caused significant problems in the Great Lakes region and has spread well beyond that part of North America.

Ballast water is a primary source of non-indigenous species as confirmed by a number of studies. One of those studies was conducted by Subba Rao *et al.* (1994), who analysed ballast water from 86 foreign vessels that visited the upper St Lawrence River and Great Lakes along the St Lawrence in 1990 and 1991. A total of 102 phytoplankton taxa were recognized, several of which had not previously been found in St Lawrence waters. Some 69 diatom and 30 dinoflagellate species were among the 102. Some 21 bloom-forming, red tide and/or toxigenic algal species were present and the authors felt these posed a potential threat to mariculture on the east coast of Canada.

In a later study, the Canadian Department of Fisheries and Oceans (DFO, 1999) reported that in 1995 over 700 ships from 30 countries visited the ports on the St Lawrence River estuary and the Gulf of St Lawrence. More than 66% of the species found in the sediments at those locations and in the ballast water of the ships were non-indigenous.

Ballast water is a global issue. It is also a difficult one to deal with for a number of reasons which are beyond the scope of this chapter. Aquaculture, and particularly marine aquaculture, provides a much more fertile opportunity for the focus of attacks by those opposed to the introduction of non-indigenous species.

Some Examples of Introductions

The intentional movement of aquatic species to locations outside of their native ranges for aquaculture or stocking purposes has a long history, much of which has been documented in the United States (Stickney, 1996). Many of the species that were introduced to new parts of that nation for culture were used as broodstock, with the progeny released in stocking programmes that supported commercial and recreational fisheries. Further, the majority of these introductions were freshwater species.

Atlantic salmon (from both anadromous and landlocked Atlantic coast populations) were introduced to the Pacific Northwest in the latter part of the 19th century, and Pacific salmon were introduced to the east coast. In neither case were reproducing populations established. Pacific salmon were successfully introduced to the Great Lakes in the 1980s, though populations are sustained largely through hatchery programmes in that region, augmented to some extent by natural reproduction.

Atlantic salmon were moved to the west coast in modern times when the National Marine Fisheries Service (NMFS) in Washington State began working with broodstock of this species and several others beginning in 1971 (Mighell, 1981). The Atlantic salmon work was in conjunction with recovery attempts in association with the Endangered Species Act. Commercial production in Washington began with Pacific salmon but by the late 1980s only Atlantic salmon were being reared in Puget Sound netpens. The British Columbia, Canada, netpen industry is also dominated by Atlantic salmon production.

The NMFS studies also involved other non-indigenous species. Included were masu (cherry salmon; *Oncorhynchus masou*) and hybrids of chinook × masu and pink × masu salmon. None of these species and hybrids has been used in aquaculture in the United States.

Escape from netpens has been major news both in the United States and in Canada. Fears that escapees will reproduce successfully and displace native Pacific salmon have been expressed (Gross, 1998), though there seems to be little recognition that attempts to establish breeding populations of Atlantic salmon in regions where they were not native took place over several decades and failed in all instances (Stickney, 1996).

Significant numbers of Atlantic salmon have escaped from netpens in both British Columbia, Canada, and the state of Washington. The first reported catches of Atlantic salmon from British Columbia waters were in 1987 and undoubtedly represented escapes from culture facilities, though the first reported escapes in that province (about 2000 fish) occurred a year later (McKinnell *et al.*, 1997). Washington had an established industry by 1987 and could easily have been the source of the fish captured during that year. Between 1988 and 1997, the number of escapees reported neared 100,000. As of 1997 there had been no reports of successful reproduction of Atlantic salmon in the North Pacific (McKinnell *et al.*, 1997), though alarms expressing fear that Atlantic salmon will successfully reproduce and overwhelm native species of salmon continue to be sounded.

Fears of consequences associated with Atlantic salmon escapees from netpens interacting with native salmon are not restricted to the Pacific Northwest of North America. Interactions between wild and cultured Atlantic salmon on the east coast of North America and in Europe have also been the focus of concern. In Northern Ireland, the annual average percentage of cultured salmon in the capture fishery has ranged from 0.26 to 4.04% (Crozier, 1998). A proportion of the Atlantic salmon farmed in Scotland have also escaped and later been recovered in capture fisheries (Webb and Youngson, 1992). Some of these losses were catastrophic as a result of storms or equipment failure (Black, 1996). Methods of identifying farmed salmon escapees include an abdominal marker related to intra-abdominal vaccination (Lund *et al.*, 1995) and scale morphology (Friedland *et al.*, 1994).

Escapees from culture have been shown to survive, home to their river of origin, and spawn with other escaped returnees as well as wild fish (Crozier, 1993; Webb *et al.*, 1993; Clifford *et al.*, 1997). However, Fleming *et al.* (1996) reported that studies with fifth-generation farmed Atlantic salmon demonstrated that reproductive success was reduced considerably in comparison with wild fish and concluded that domesticated fish were reproductively inferior. Regardless, the use of sterile salmon in culture has been advocated in some circles. Cotter *et al.* (2000) conducted research to evaluate the use of triploids in Ireland.

With the percentage of salmon under culture on the east coast of Canada exceeding 90% of the total population, and with the developmental and environmental biology of the wild and cultured fishes being different, Gross (1998) suggested that cultured Atlantic salmon be treated as an exotic (non-indigenous) species, and that it might be called *Salmo domesticus*.

Striped bass (*Morone saxatilis*), from stocks originating on the east coast of the United States, were established in California in the 19th century. Like tilapia in Asia, striped bass have now been established for so long in California that they are considered to be a native species.

Attempts were made several years ago to evaluate the potential for culture of American lobsters (*Homarus americanus*) in California, though no industry was ever established for that non-indigenous species, nor have wild populations become established. The oyster industry on the west coast of the United States is based primarily on a non-indigenous species, the Pacific oyster (*Crassostrea gigas*). The same is true of British Columbia, Canada (Quayle, 1988). Pacific oysters have also been introduced to the South Pacific, the United Kingdom and France (Mann, 1979).

Because of declines in east coast American oyster (*C. virginica*) stocks due to disease and environmental factors, there is a considerable amount of interest in introducing Pacific oysters to parts of the east coast. Pacific oysters are considered to be adaptable to such regions as Chesapeake Bay and are resistant to some of the diseases that have plagued American oyster populations (Gottlieb and Schweighofer, 1996). Experimental trials have been conducted with primarily triploid Pacific oysters as well as Suminoe oysters, *C. ariakensis* (Calvo and Luckenbach, 1998; Calvo *et al.*, 1998, 1999). Allen and Guo (1997) found that the methods used to produce triploid oysters were not totally effective. It would be necessary to evaluate each oyster produced to certify that all animals stocked were triploids if regulations called for stocking only triploid animals.

During the 1970s, there was considerable interest in the United States in the culture of the freshwater shrimp *Macrobrachium rosenbergii*, which is native to Southeast Asia. Larval development of that species requires saline water, so there is a direct tie to marine aquaculture. Because of marketing and other problems, the small industry that developed collapsed. However, marine shrimp culture has been developed to some extent in the United States. While the industry is very small compared with those in China, Ecuador, Thailand and several other nations, it is a multimillion dollar business. Commercial hatchery and/or growout facilities exist in Arizona, Hawaii, Florida, South Carolina and Texas, plus the territory of Puerto Rico. Virtually all of the shrimp being produced are non-indigenous species from Latin America or Asia. The dominant species is *Litopenaeus vannamei*, though there is also some production of *L. stylirostris* and *Penaeus monodon* (Granvil Treece, personal communication).

A great deal of research was conducted in the 1960s and 1970s on native shrimp species, particularly by investigators at the National Marine Fisheries Service laboratory in Galveston, Texas (Stickney, 1996). Research activity was curtailed when the industry turned to non-indigenous species, though a modest amount of work with native species did continue (e.g. Sandifer *et al.*, 1992).

In many instances, such as in association with tilapia, introductions occurred and species became established long before they were incorporated into aquaculture systems in the areas where the introductions occurred. The original introductions were sometimes accidental but often they were intentional. Many nations now have restrictions on the introduction of non-indigenous aquatic species. Others have no restrictions, and in some the determination has been made that the aquaculture industry should be based largely or in part on non-indigenous species.

Manila clams (*Tapes philippinarum*), which are currently cultured in Washington, were probably introduced to that state incidentally in shipments of Pacific oysters. Those shipments began prior to World War II and continued beyond 1970 (Westley, 1975). Similarly, the varnish clam (*Nuttallia abscurata*) was unintentionally introduced to British Columbia during the 1980s and 1990s and became widely distributed (Heath, 1998). Licensing of the species for culture was under consideration by the Ministry of Fisheries in 1998. The non-indigenous bay scallop (*Argopecten irradians irradians*) was evaluated as an aquaculture candidate in Georgia during the 1980s (Heffernan *et al.*, 1988).

In Sri Lanka various species were introduced to augment the availability of fish in rural communities. There, tilapia are grown in fresh and brackish waters, as well as in rice fields (de Zylva, 1999). In Brunei, attempts were made in 1979 and 1980 to introduce the green mussel (*Perna viridis*). Problems associated with a lack of natural spatfall and the presence of toxic algal blooms were encountered (Lindley and Currie, 1982). In other parts of the Middle East, nations such as Egypt, Saudi Arabia and Israel (Colorni, 1989) have been developing shrimp culture based on non-indigenous species. Egypt has had successful rearing trials with three penaeid shrimp species. Freshwater shrimp are usually grown in polyculture with tilapia (Wassef, 2000).

Non-indigenous mullet (*Mugil cephalus*) and oysters (*Crassostrea rhizophorae*) were established as brackish and marine culture species in Colombia some three decades ago (FAO, 1977). More recently, freshwater and marine shrimp culture has been developed in that country (Angarita Zerda, 1989). Chile has become a major producer of Atlantic and Pacific salmon. There have also been attempts to introduce the scallop, *Pecten maximus* (Illanes *et al.*, 1999), to Chile.

Marine aquaculture options were studied in the Madeira Archipelago of Portugal and it was concluded that the best option was to culture non-indigenous fishes in offshore cages (Andrade, 1996). Many other nations, without doing such studies, have come to the same conclusion.

It is likely that until recent years, there was little or no regulation in most nations with regard to the introduction of aquatic species. There are exceptions. The federal Lacey Act of 1900 and subsequent amendments serve as the basis for existing regulations on the introduction of species to the United States (Clugston, 1986). In 1977, an executive order was signed that instructed federal agencies to restrict introductions of non-indigenous species into federally owned or controlled lands and waters to the extent legally possible. In 1999, another executive order was issued that established the National Invasive Species Council and directed it to provide national leadership on invasive species, including ensuring that efforts at the federal level are coordinated and effective. With respect to intentional introductions, the Council, according to its draft management plan, will undertake development of a comprehensive screening system for evaluating intentionally introduced non-native species. Screening will be conducted to establish that the risk of intentionally introducing a particular species into the United States is acceptable. Miglarese and DeVoe (1992) reviewed state policies and regulations with regard to the introduction and transfer of non-indigenous species in the United States.

A code of practice was developed in Ireland and governs introductions and transfers by aquaculturists in that nation. The code was touted for its usefulness in reducing the risks associated with introductions (Minchin and Sheehan, 1995), though the major threat appears to come from ballast water discharges. The Global Aquaculture Alliance (Boyd, 1999) has prepared a code of conduct for responsible shrimp aquaculture, which includes the guiding principle which states that those engaged in shrimp farming, 'Shall take all reasonable steps to ascertain that permissible introductions of exotic species are done in a responsible and acceptable manner and in accordance with appropriate regulations'.

The Case Against Introductions

The introduction of non-indigenous species has been objected to on a number of grounds. Most common among the objections involves the introduction of diseases with the non-indigenous species. The concern is that these diseases might be transmitted to native species. Another major issue revolves around the potential establishment of escapees. Interbreeding of non-indigenous animals with indigenous congeners and the resulting population genetic alterations has also been raised as an issue (Naylor *et al.*, 2000). Another concern associated with the conscious introduction of non-indigenous species involves the possibility that, in addition to the possible introduction of diseases, there could be introduction of predators, competitors or pests (Minchin, 1996).

Disease issues

In terms of disease transmission, one of the loudest alarms has been sounded in conjunction with the use of non-indigenous shrimp species in aquaculture. A number of examples concerning where non-indigenous shrimp are being cultured have previously been discussed (above). Viral diseases emerged as a devastating problem for the shrimp industries during the 1990s (Lightner, 1999). Viral diseases in native species in Asia and Latin America have seriously affected production in such nations as China, India, Indonesia, Korea, Taiwan, Thailand, Ecuador and others. Even Asia is not without problems related to the rearing of non-indigenous species. Taura syndrome virus (TSV), for example, was so severe in Taiwan in the non-indigenous Pacific white shrimp (*Litopenaeus vannamei*) during 1998 and 1999 that 90% of the shrimp ponds were abandoned a month or month and a half after being stocked (Tu *et al.*, 1999).

Viruses such as IHHN (infectious hypodermal and haematopoietic necrosis), WSSV (white spot syndrome virus), YHV (yellow head virus) and TSV have occurred in non-indigenous shrimp species reared in the United States. When challenged with WSSV and YHV, native penaeid shrimp species were found to be susceptible (Lightner *et al.*, 1998), though YHV affected primarily juveniles, with postlarvae being less sensitive. Overstreet *et al.* (1997) found that native white shrimp (*Litopenaeus setiferus*) in the Gulf of Mexico and along the southeast coast of the United States are susceptible to mortality from TVS, while brown shrimp (*L. aztecus*) and pink shrimp (*L. duorarum*) are not.

IHHN, HPV (hepatopancreatic parvo-like virus) and MBV (monodon-type baculovirus) were probably introduced to Israel with the introduction of non-indigenous shrimp. These viruses have also occurred in native shrimp in the waters off that nation (Colorni, 1989).

Some of the viruses that have been identified in cultured shrimp can also infect other crustaceans, often without causing signs of disease. Examples include infection of various decapods with WSSV in Taiwan (Wang *et al.*, 1998) and Thailand (Supamattaya *et al.*, 1998).

In situations where non-indigenous culture species are being introduced, the probability of also introducing exotic diseases can be reduced through a quarantine process as suggested for Hawaii by Davidson and Brick (1988). Quarantine has been implemented in some instances; for example in Australia (Garland, 1988) and Indonesia (Arthur and Shariff, 1991), though a general lack of such procedures in Southeast Asia has led to the introduction of various diseases associated with the introduction of non-indigenous species.

For those regions where non-indigenous species continue to be cultured, further introductions of exotic diseases might be curbed by developing broodstock and avoiding further importation. This approach and quarantine have been employed by the Texas shrimp farming industry, but even then, virus outbreaks have occurred. Since Texas shrimp processors handle wild, domestically cultured non-indigenous, and imported shrimp, one theory is that the viruses that made their way into shrimp ponds, possibly from seagull faeces (Garza *et al.*, 1997), came from processing plant wastes.

Transmission of mollusc diseases to native populations from introduced species has been reported. Manila clams were introduced into culture in England and Wales because of their rapid growth, high demand by consumers and high market value. An additional advantage was that the species was judged unable to establish reproducing populations (Spencer *et al.*, 1991). However, a *Vibrio* that was the probable cause of 'brown ring disease' in native clams appears to have been introduced with the non-indigenous Manila clam (Edwards, 1998). The state of Alaska has a permit system for the importation of non-indigenous species, with the Pacific oyster being the only shellfish that can be obtained with a permit. Further, no imports can come from Korea, the Gulf of Mexico or the Atlantic coast of North America (Meyers, 1989).

The potential introduction of non-indigenous diseases in association with the farming of salmon was raised as an issue early in the last decade (Folsom and Sanborn, 1993; Kent, 1994). This potential appears to have been realized a few years later when the bacterium *Piscirickettsia salmonis* was blamed for high mortalities of the non-indigenous coho salmon (*Oncorhynchus kisutch*) in Chile (Getchell, 1998).

The disease issues associated with salmon culture are not restricted to the introduction of exotic diseases. In addition to that concern, Kent (1994) expressed the view that diseases in cultured fish can also impact wild populations by increasing the prevalence of diseases associated with endemic pathogens.

Establishment

In a paper cautioning the introduction of *Crassostrea gigas* to New England, Andrews (1980) reflected the general concern about non-indigenous species introductions. He indicated that such an introduction could have 'consequences to native biota and coastal ecosystems'. Beveridge *et al.* (1997) commented on the direct and indirect impacts of mariculture on biodiversity. Among the impacts they recognized was 'the translocation of exotic plants and animals'. As indicated by Lodge *et al.* (1998), predicting which introduced species might cause ecological change is a problem for natural resource managers. They suggest that managers should focus on preventing introductions in freshwater systems because eradication is often not possible. The same statement is equally or even more valid in marine systems.

There is little doubt that cultured species will escape from aquaculture facilities. Biosecurity is an excellent concept, and it may even be possible to achieve true biosecurity in a closed recirculating water system that is located within a closed building and equipped with the proper control devices to keep everything from gametes to adult animals from being released in the effluent stream, intermittent or minuscule as it may be. Outdoor facilities on land are always susceptible to escape, both in the effluent and through the activities of careless predators or poachers who let the animals they capture escape.

When a facility is established in open water in cages or netpens, the chances of escape increase dramatically. A major storm can destroy an entire facility, potentially releasing thousands, or even millions of culture animals into the wild. Smaller losses can occur through failures of a single netpen or cage, often helped along by predators that rend the netting in their attempts to capture the culture animals. Carelessness and accidents during fish handling can also account for losses, though the numbers of animals involved are typically small.

While there is no known foolproof way to totally eliminate escapes from marine culture facilities, the question of whether escapees will establish self-sustaining populations is a question that cannot be answered in any convincing way as the outcome will vary from one species to another and one environment to another.

Addressing the Issues

Disease

The statement by Elston (1989) that the risk posed from exotic diseases can be reduced by controlling the importation of non-indigenous culture species can be extended to other objections as well. However, as noted above, the policy in some countries is to seek out non-indigenous species for use in mariculture, particularly when appropriate native species either do not occur or there is insufficient information on their culture requirements to support establishment of a commercial industry. In cases where introductions of non-indigenous species are made, there is a clear need to conduct risk assessments and develop risk management strategies (Kern and Rosenfield, 1988).

In the United States, laws associated with non-indigenous species introductions vary considerably, though in general, those laws have become more restrictive in recent years. Many states require permits for importation of non-indigenous species.

Avoidance and prevention of the spread of diseases can be achieved to some extent by employing specific pathogen-free (or high-health) stocks and establishing biosecure culture facilities, as has been discussed in relation to shrimp culture, for example, by Hopkins and Sandifer (1993), Wyban (1993), Wyban *et al.* (1993), Pruder (1994), Lotz (1997) and Clifford (1999). These techniques work particularly well when the entire life cycle of the culture species is conducted in an environment that is totally under the control of the culturist. An example would be an indoor flow-through or recirculating water system supplied with well water.

There are instances where a non-indigenous species became established in nature, apparently without bringing new diseases with them. An example is the presence of non-indigenous blue mussels (*Mytilus edulis* and *M. galloprovincialis*) in British Columbia, Canada, waters. When further introductions were proposed for culture purposes, the proposal was made to determine if non-indigenous mussels already present could be used as broodstock (Yanick and Heath, 1998).

Establishment

As indicated above, there is no doubt that animals can be expected to escape from marine aquaculture facilities, particularly those located outdoors, such as in ponds and netpens. Thus, we are back to the question: Will the escapees become established in the natural environment? Whether established or not, there are concerns that escapees from marine aquaculture systems will have negative impacts on native flora or fauna. That possibility was examined by the Department of Ecology in the state of Washington (Anonymous, 1999). It was concluded that there was no unacceptable risk to native Pacific salmon from the activities of escaped Atlantic salmon in the following areas: colonization, competition, disease transmission, hybridization and predation. However, one cannot extrapolate that conclusion to other situations.

The argument can be made in some inland situations that introduction of certain non-indigenous species can lead to the extirpation of one or more native species through competition or predation. If the extirpated species is one with a highly confined distribution, the threat of extinction is a very real one. In the marine environment, it seems more difficult to argue that an introduced species could lead to extinction of a native species. The argument that an introduced species could eliminate some genetically distinct population is one that can be made, however.

The most responsible approach to marine aquaculture is to limit the species reared to those that occur naturally in the environment where the aquaculture activity is to be conducted. Some nations apply that principle at present, though as shown above, many countries have concluded that their marine aquaculture industry must, of necessity, be based on the rearing of non-indigenous species.

Even when the decision is made to rear only native species, there is opposition to the practice in some circles on the basis that escapees will reproduce with wild congeners and that the genetics of the wild population will be detrimentally altered. The level of concern escalates when there is even a hint of using genetically modified organisms (GMOs). The subject of GMOs is considered in another chapter of this book and is a topic that is likely to become even more controversial in the future.

Ultimately, neither extreme is likely to win the day. Taking the position that there should be no introductions of non-indigenous species into marine aquaculture is unreasonable because, in many instances, introductions have already been made. Opening the door sufficiently wide to allow all introductions without any control is irresponsible. When a new introduction is being considered, a risk assessment should be conducted. Societal need should be balanced against potential environmental impact.

Some will argue that the natural environment should be maintained unaltered at all cost. Again, that horse is not only out of the stable, it has left the ranch. Human activities have altered and will continue to alter our environment, including that of the coasts. That does not mean that human activities should not be conducted without a considerable amount of planning and with the intent of causing as little environmental impact as possible.

Human population continues to grow, and with it the demand for seafood will also grow. That is irrefutable. It is also a fact that humans are exploiting the world ocean up to, and in many cases beyond, the capability of the ocean to sustain the fishing pressure that is placed upon it. Increased seafood supplies can only come from aquaculture. Decrying the practice of marine aquaculture is not going to have much impact on those who demand more seafood, and particularly those who *depend* upon a continuous and growing supply of seafood to meet their routine food needs. In the final analysis, those whose primary goal is environmental protection and maintenance of the status quo with respect to human impacts on the coastal environment and those who are interested in enhancing the food supply through marine aquaculture need to work together. Collaboration and cooperation are the logical approaches, but the fact is that humans do not always do what is logical.

References

Allen, S.K., Jr. and Guo, X. (1997) Can we have our oyster and eat it too? A case for aquaculture parks using non-native species. *Journal of Shellfish Research* 16, 257.

Andrade, C.A.P. (1996) A fish farm pilot-project I Madeira Archipelago, northeastern Atlantic. 1. The offshore option. In: Polk, M. (ed.) *Proceedings of an International Conference on Open Ocean Aquaculture*, Portland, Maine, 9–10 May. New Hampshire/Maine Sea Grant College Program Rpt. UNHMP-CP-SG-96-9, pp. 371–376.

Andrews, J.D. (1980) A review of introductions of exotic oysters and biological planning for new importations. *Marine Fisheries Review* 42, 1–11.

Angarita Zerda, E. (1989) Las acciones de PROEXPO en acuicultura, y su relacion con la introduction de especies. In: *Memorias Taller Sobre Introduccioin de Expecies Hidrobiologicas a la Acicultura*. Bogota, Colombia, August, pp. 75–78.

Anonymous (1999) Are farmed salmon pollutants? *Pacific Fishing* 20(3), 27.

Arthur, R. and Shariff, M. (1991) Towards international fish disease control in Southeast Asia. *Infofish International* 3, 45–48.

Assem, H. and Hanke, W. (1979) Concentrations of carbohydrates during osmotic adjustment of the euryhaline teleost *Tilapia mossambica*. *Comparative Biochemistry and Physiology* 64A, 5–16.

Balarin, J.B. and Hatton, J.P. (1979) *Tilapia: a Guide to their Biology and Culture in Africa*. University of Stirling, UK, 174pp.

Beveridge, M.C.M., Ross, L.G. and Stewart, J.A. (1997) The development of mariculture and its implications for biodiversity. In: Ormond, R.F.G., Gage, J.D. and Angel, M.V. (eds) *Marine Biodiversity: Patterns and Processes*. Cambridge University Press, New York, pp. 372–393.

Black, K.D. (1996) Escaped farmed salmon in western Scotland. In: *Aquaculture and Sea Lochs*. Scottish Association for Marine Science, Oban, UK, pp. 89–93.

Boyd, C.E. (1999) *Codes of Practice for Responsible Shrimp Farming*. Global Aquaculture Alliance, St Louis, Missouri, 42pp.

Calvo, G.W. and Luckenbach, M.W. (1998) Non-native oysters survive and grow in Virginia: evaluating the performance of *Crassostrea gigas* against *Crassostrea virginica*, in relation to salinity, in Chesapeake Bay and Atlantic coastal waters. *Journal of Shellfish Research* 17, 320–321.

Calvo, G.W., Luckenbach, M.W. and Burreson, E.M. (1998) Non-native oyster studies in Virginia. *Journal of Shellfish Research* 17, 349.

Calvo, G.W., Luckenbach, M.W. and Burreson, E.M. (1999) Evaluating the performance of non-native oyster species in Virginia. *Journal of Shellfish Research* 18, 303.

Clifford, H.C. III (1999) A review of diagnostic, biosecurity and management measures for the exclusion of white spot virus disease from shrimp culture systems in the Americas. In: Cabrera, T., Jory, D. and Silva, M. (eds) *Aquaculture '99*, Vol. 1. Congreso sur American de Acuicultura, Puerto la Cruz, Venezuela, pp. 134–171.

Clifford, S.L., McGinnity, P. and Ferguson, A. (1997) Genetic changes in an Atlantic salmon population resulting from escaped juvenile farm salmon. *Journal of Fish Biology* 52, 118–127.

Clugston, J.P. (1986) Strategies for reducing risks from introductions of aquatic organisms: the federal perspective. *Fisheries* 11(2), 26, 28, 29.

Colorni, A. (1989) Penaeid pathology in Israel: problems and research. In: *Actes Colloq.* IFREMER, No. 9, Brest, France, pp. 89–96.

Cotter, D., O'Donovan, V., O'Maoileidigth, N., Rogan, G., Roche, N. and Wilkins, N.P. (2000) An evaluation of the use of triploid Atlantic salmon (*Salmo salar* L.) in minimising the impact of escaped farmed salmon on wild populaitons. *Aquaculture* 186, 61–75.

Crozier, W.W. (1993) Evidence of genetic interaction between escaped farmed salmon and wild Atlantic salmon (*Salmo salar* L.) in a northern Irish river. *Aquaculture* 113, 19–29.

Crozier, W.W. (1998) Incidence of escaped farmed salmon, *Salmo salar* L., in commercial salmon catches and fresh water in Northern Ireland. *Fishery Management and Ecology* 5, 23–29.

Davidson, J.R. and Brick, R.W. (1988) Dispersal of exotic aquatic animals for aquacultural purposes with special emphasis on the Hawaii experience. *Journal of Shellfish Research* 7, 553–554.

DFO (1999) In: *The St Lawrence Marine Environment. Knowledge and Action: 1993–1998.* Department of Fisheries and Oceans Marine Science Research Center, Mont-Joli, Canada, pp. 55–70.

Edwards, E. (1998) Disease hits North European clams. *Fish Farming International* 25(3), 16.

Elston, R. (1989) Aquaculture health management and disease control in the Pacific Northwest. *Journal of Shellfish Research* 8, 321.

FAO (1977) Informe sobre la situacion de la acuicultura en Colombia. In: *Proceedings of the Symposium on Aquaculture in Latin America*, Vol. 3. Food and Agriculture Organization of the United Nations, Rome, pp. 37–48.

Fleming, I.A., Jonsson, B., Gross, M.R. and Lamberg, A. (1996) An experimental study of the reproductive behaviour and success of farmed and wild Atlantic salmon (*Salmo salar*). *Journal of Applied Ecology* 33, 893–905.

Folsom, W.B. and Sanborn, E.A. (1993) The difficulties of salmon culture. *Aquaculture Magazine* 19(1), 51–63.

Friedland, K.D., Esteves, C., Hansen, L.P. and Lund, R.A. (1994) Discrimination of Norwegian farmed, ranched and wild-origin Atlantic salmon, *Salmo salar* L., by image processing. *Fishery Management and Ecology* 1, 117–128.

Garland, C.D. (1988) A microbiological perspective of Australian mariculture hatcheries and nurseries. In: Evans, L.H. and O'Sullivan, D. (eds) *Proceedings, First Australian Shellfish Aquaculture Conference*, Perth, Australia, pp. 147–160.

Garza, J.R., Hasson, K.W., Poulos, B.T., Redman, R.M., White, B.L. and Lightner, D.V. (1997) Demonstration of infectious Taura syndrome virus in the feces of seagulls collected during an epizootic in Texas. *Journal of Aquatic Animal Health* 9, 156–159.

Getchell, R. (1998) New salmon pathogen: sleuthing uncovers *Piscirickettsia salmonis*. *Fish Farming News* 6, 6.

Gottlieb, S.J. and Schweighofer, M.E. (1996) Oysters and the Chesapeake Bay ecosystem: a case for exotic species introduction to improve environmental quality? *Estuaries* 19, 639–650.

Gross, M.R. (1998) One species with two biologies: Atlantic salmon (*Salmo salar*) in the wild and in aquaculture. *Canadian Journal of Fisheries and Aquatic Sciences* 55 (Suppl. 1), 131–144.

Heath, W.A. (1998) The varnish clam (*Nuttallia obscurata*): another commercial species. *Journal of Shellfish Research* 17, 1281.

Heffernan, P.B., Walker, R.L. and Gillespie, D.M. (1988) Biological feasibility of growing the northern bay scallop, *Argopecten irradians irradians* (Lamarck, 1819), in coastal waters of Georgia. *Journal of Shellfish Research* 7, 83–88.

Hopkins, J.S. and Sandifer, P.A. (1993) Development of environmentally sensitive shrimp aquaculture. In: Carrillo, M., Ddahle, L., Morales, J., Sorgeloos, P., Svennevig, N. and Wyban, J. (eds) *Special Publication of the European Aquaculture Society 19*. Ostend, Belgium, p. 95.

Illanes, J., Uribe, E. and Solar, C. (1999) First attempt of introduction of *Pecten maximus* (L.) to Chile. 12th International Pectinid Workshop, Bergen, Norway, 5–11 May.

Kent, M.L. (1994) The impact of diseases of pen-reared salmonids on coastal marine environments. In: Ervik, A., Hansen, P.K. and Wennevik, V. (eds) *Proceedings of the Canada–Norway Workshop on Environmental Impacts of Aquaculture. Fisken Havet Saernummer* 13, 85–89.

Kern, F.G. and Rosenfield, A. (1988) Shellfish health and protection. *Journal of Shellfish Research* 7, 557–558.

Lightner, D.V. (1999) The penaeid viruses TSV, IHHN, WSSV, and YHV: current status in the Americas, available diagnostic methods, and management strategies. *Journal of Applied Aquaculture* 9, 27–52.

Lightner, D.V., Hasson, K.W., White, B.L. and Redman, R.M. (1998) Experimental infection of Western Hemisphere penaeid shrimp with Asian white spot syndrome and Asian yellow head virus. *Journal of Aquatic Animal Health* 10, 271–281.

Lindley, R.H. and Currie, D.J. (1982) Studies on the growth and aquaculture potential of green mussel *Perna viridis* in Brunei waters. *Monographs of the Brunei Museum Journal* 5, 125–133.

Lodge, D.M., Stein, R.A., Brown, K.M., Covich, A.P., Broenmark, C., Garvey, J.E. and Klowiewski, S.P. (1998) Predicting impact of freshwater exotic species on native biodiversity: challenges in spatial scaling. *Australian Journal of Ecology* 23, 52–67.

Lotz, J.M. (1997) Viruses, biosecurity and specific pathogen-free stocks in shrimp aquaculture. *World Journal of Microbiology and Biotechnology* 13, 405–413.

Lund, R.A., Midtlying, P.J. and Hansen, L.P. (1995) Identifiscring av romt oppdrettslaks ved effeckter av vaksinering. *Nina Fragrapp* 12, 1–14.

Mann, R. (ed.) (1979) *Exotic Species in Mariculture*. MIT Press, Cambridge, Massachusetts, 363pp.

McKinnell, S., Thomson, A.J., Black, E.A., Wing, B.L., Guthrie, C.M. III, Koerner, J.F. and Helle, J.H. (1997) Atlantic salmon in the North Pacific. *Aquaculture Research* 28, 145–157.

Meyers, T.R. (1989) Certification policy for importation of *Crassostrea gigas* spat into the state of Alaska. *Journal of Shellfish Research* 8, 323.

Mighell, J.L. (1981) Observation on non-native broodstock reared in netpens in Puget Sound, Washington. In: Nosho, T. (ed.) *Proceedings: Salmonid Broodstock Maturation*. WSG-WO-80-1. Washington Sea Grant, Seattle, Washington, pp. 27–30.

Miglarese, J. and DeVoe, M.R. (1992) In: *Aquaculture '92: Growing Toward the 21st century*. World Aquaculture Society, Baton Rouge, Louisiana, p. 161.

Minchin, D. (1996) Management of the introduction and transfer of marine molluscs. *Aquatic Conservation: Marine and Freshwater Ecosystems* 6, 229–244.

Minchin, D. and Sheehan, J. (1995) The significance of ballast water in the introduction of exotic marine organisms to Cork Harbour, Ireland. ICES Council Meeting Papers, Council Meeting, International Council for the Exploration of the Sea, Copenhagen, Denmark, 21–29 September, 15pp.

Naylor, R.L., Goldburg, R.J., Primavera, J.H., Kautsky, N., Beveridge, M.C.M., Clay, J., Folke, C., Lubchenco, J., Mooney, H. and Torell, M. (2000) Effect of aquaculture on world fish supplies. *Nature* 405, 1017–1024.

Overstreet, R.M., Lightner, D.V., Hasson, K.W., McIlwain, S. and Lotz, J.M. (1997) Susceptibility to Taura syndrome virus of some penaeid shrimp species native to the Gulf of Mexico and the southeastern United States. *Journal of Invertebrate Pathology* 69, 165–176.

Potts, W.I.M., Foster, M.A., Rady, P.P. and Howell, G.P. (1967) Sodium and water balance in the cichlid teleost *Tilapia mossambica*. *Journal of Experimental Biology* 47, 461–470.

Pruder, G.D. (1994) High health shrimp stocks: an advance, an opportunity – but not a panacea. *World Aquaculture* 25, 26–28.

Quayle, D.B. (1988) Pacific oyster culture in British Columbia. *Canadian Bulletin of Fisheries and Aquatic Science* 218, 247.

Sandifer, P.A., Browdy, C.L., Baird, M., Hopkins, J.S. and Stokes, A.D. (1992) Intensive pond culture of the white shrimp, *Penaeus setiferus*, in South Carolina: advantages and limitations. In: *Aquaculture '92: Growing toward the 21st century*. World Aquaculture Society, Baton Rouge, Louisiana, p. 200.

Spencer, B.E., Edwards, D.B. and Millican, P.F. (1991) Cultivation of Manila clams. Laboratory Leaflet, Director of Fishery Research (Great Britain), 65, 1–30.

Stickney, R.R. (1996) *Aquaculture in the United States, a Historical Survey*. John Wiley & Sons, New York, 372pp.

Subba Rao, D.V., Sprules, W.G., Locke, A. and Carlton, J.T. (1994) Exotic phytoplankton species from ship's ballast waters: risk of potential spread to mariculture sites on Canada's east coast. *Canadian Data Report in Fisheries and Aquatic Science* 937, 55pp.

Supamattaya, K., Hoffmann, R.V., Boonyaratpalin, S. and Kanchanaphum, P. (1998) Experimental transmission of white spot syndrome virus (WSSV) from black tiger shrimp *Penaeus monodon* to the sand crab *Portunus pelagicus*, mud crab *Scylla serrata* and krill *Acetes* sp. *Diseases of Aquatic Organisms* 32, 79–85.

Tu, C., Huang, H.-T., Chuang, S.-H., Hsu, J.-P., Kuo, S.-T., Li, N.-J., Hsu, T.-L., Li, M.-C. and Lin, S.-Y. (1999) Taura syndrome in Pacific white shrimp *Penaeus vannamei* cultured in Taiwan. *Diseases of Aquatic Organisms* 38, 169–171.

Wang, Y.C., Lo, C.F., Chang, P.S. and Kou, G.H. (1998) Experimental infection of white spot baculovirus in some cultured and wild decapods in Taiwan. *Aquaculture* 164, 22–231.

Wassef, E.A. (2000) Status of aquaculture in Egypt. *World Aquaculture* 31, 29–34.

Webb, J.H. and Youngson, A.F. (1992) Reared Atlantic salmon, *Salmo salar* L., in the catches of a salmon fishery on the western coast of Scotland. *Aquaculture and Fishery Management* 23, 393–397.

Webb, J.H., McLaren, I.S., Donaghy, M.J. and Youngson, A.F. (1993) Spawning of farmed Atlantic salmon, *Salmo salar* L., in the second year after escape. *Aquaculture and Fishery Management* 24, 557–561.

Westley, R.E. (1975) Past, present, and future trends in cultural techniques and oyster production in the state of Washington. *Proceedings of the World Mariculture Society* 6, 213–219.

Wyban, J.A. (1993) High health shrimp production. In: Carrillo, M., Ddahle, L., Morales, J., Sorgeloos, P., Svennevig, N. and Wyban, J. (eds) *Special Publication of the European Aquaculture Society* 19. Ostend, Belgium, p. 181.

Wyban, J.A., Swingle, J.S., Sweeney, J.N. and Pruder, G.D. (1993) Specific pathogen free *Penaeus vannamei*. *World Aquaculture* 24, 39–45.

Yanick, J. and Heath, D. (1998) Growth and survival of introduced and native blue mussels (*Mytilus* spp.) in Georgia Strait, BC: prospects for mussel aquaculture. *Journal of Shellfish Research* 17, 14–16.

Zylva, E.R. de (1999) The introduction of exotic fish in Sri Lanka with special reference to tilapia. *Naga* 22, 4–8.

Genetic Changes in Marine Aquaculture Species and the Potential for Impacts on Natural Populations

12

William K. Hershberger

Director, USDA/ARS/National Center for Cool and Cold Water Aquaculture, 11876 Leetown Road, Kearneysville, WV 25430, USA

Abstract

Although many aspects of aquaculture have changed very dramatically over the last 50 years, one feature that has not changed very much is the use of natural genetic resources as the primary supply of 'raw material' for use in production facilities. Consequently, some serious attention should be paid to the factors that affect natural aquatic genetic resources and their conservation. Three major factors are identified that can have a major impact on the natural genetic resources in the marine environment: habitat alteration, changes in the commercial fishery, and use of aquaculture to restore or enhance natural populations. The genetic effects of each of these factors are discussed. With this diversity in potential effectors, there is some 'genetic risk' associated with introducing an additional component to the natural system, e.g. intensive aquaculture systems. Minimization of genetic risk can be accomplished by maintaining genetic variability within natural populations and a model is discussed to evaluate the characteristics of an aquaculture species that can be utilized to minimize loss. Only through understanding how major genetic alterations are introduced, what their impacts are, and how these are addressed can a marine aquaculture operation be managed in a viable and sustainable manner.

Introduction

There is little question that aquaculture has changed dramatically over the last 50 years. Aquaculture is a practice that started out as an extensive, low-input method of maintaining animals for food, decoration or recreation. Over the past century it has developed into an intensive, high-input industry that is

(eds R.R. Stickney and J.P. McVey)

being increasingly employed for the direct production of food. Initially, relatively few species were utilized in aquaculture and they were mostly freshwater species. Currently, estimates of the number of species grown in aquaculture of one type or another range between 400 and 500, and culture systems occupy most usable types of aquatic environments, including recent expansion into the open ocean portion of the marine environment. On a worldwide basis, production has increased about 5% annually, from about 3.7 Mkg in 1990 to about 18 Mkg in 1998.

One aspect of aquaculture that has not changed very much is the use of natural aquatic resources as sources of genetic material. Currently, most of the industry relies on the genetic resources provided by the natural environment. While fewer aquaculture operations are relying directly on a supply of organisms for stocking from naturally produced populations, there has not been much effort aimed at developing truly domesticated and genetically selected broodstocks specifically for aquaculture production. Consequently, production stocks are generally initiated from natural populations, and when problems arise from unfavourable production responses in aquacultured populations they are often addressed by 'mining' available natural animal (or plant) resources. This will probably remain the primary source of genetic material for the foreseeable future since considerable time and financial investment is needed to develop the domesticated stocks to supply and meet the needs of the industry. Consequently, it is mandatory that some serious attention be paid to the natural aquatic genetic resources and the factors that impact them.

Sources and Magnitude of Genetic Impact

The first order of business in evaluation of the factors that impact natural aquatic genetic resources is to identify those that hold the potential to introduce major changes. In a rather 'broad brush' approach, I feel there are three types of changes that have a major role in causing the genetics of natural populations to be altered.

Environmental changes

The genetic constitution of an organism is the information it uses for survival and development. Development proceeds in response to environmental opportunity and, thus, each species has been shaped by natural selection in accordance with the demands of its natural habitat. Consequently, the genetic diversity within and among individuals provides the resilience to survive natural environmental change. However, the scale and the rate of change produced by man-made disturbances are often beyond the genetic capacity of a species to respond without some modifications. As a result, changes in habitats or degradation of environments can lead to significant genetic alteration, even to extinction of a natural resource.

The physical, chemical and biotic features of the aquatic environment of the world are being altered at an alarming rate (Reisner, 1986). Some of the more highly visible alterations (e.g. dams and harbours, dredging or filling of waterways) have had documented major effects on the genetic resources of aquatic species. For example, as a result of the construction of hydroelectric dams on the Columbia River in the Pacific Northwest region of the United States, it is estimated that as many as 200 populations of Pacific salmon (*Oncorhynchus* spp.), uniquely adapted to specific freshwater environments, were lost (NPPC, 1987). These and other environmental changes have led to more than 100 stocks of anadromous salmonid fishes in the states of California, Oregon, Idaho and Washington being characterized as at 'high risk of extinction' (Nehlsen *et al.*, 1991). Other less apparent changes can also have major effects. While it is not the purpose of this chapter to delve into the problems associated with the degradation of aquatic environments, it is imperative to emphasize that the abuse of these environments needs to be halted if aquatic genetic resources are to be of any value or to have any future.

Capture fisheries

The second factor that has had an effect on the genetic composition of natural fish populations is the changes that have taken place in the commercial capture fisheries. In the past 20 years the market demand for fish products has increased rather dramatically; for example, per capita consumption in the United States increased 25% between 1980 and 1990 (NRC, 1992). Per capita consumption has recently levelled off somewhat, but increases in population numbers have caused a steady increase in demand and it is projected that this will escalate even more (New, 1991). The escalating demand, accompanied by increased costs and diminishing resources, resulted in the evolution of the capture fisheries from small, subsistence-level operations to sophisticated, highly mechanized fish capture systems.

The increased harvesting capacity and improved efficiency of the capture fisheries have led to the rapid depletion of some fish resources and, in some cases, their complete elimination. The 3000 or so species that are harvested by the world's capture fisheries (86% of which are marine species) are at a modest risk of losing genetic variation as a consequence of exploitation (Smith *et al.*, 1991). However, they are at extreme risk of being overexploited. Overexploitation reduces the number of reproducing individuals, which subjects the population to increased potential for inbreeding and reduced genetic diversity, raising the spectre of further loss. Furthermore, the recurring pattern of resource depletion has led to increased management regulation that mandated increasingly selective fishing practices. These practices have had a demonstrable effect on salmonid genetic resources (Ricker, 1981) although the impact on marine species is less defined. The extent of genetic change caused by capture fisheries can only be estimated in a few instances since genetic data are available for only a relatively small sample of the species that are harvested. Thus, a much more extensive database is needed to evaluate these impacts and a better understanding of, particularly, marine species is required to allow definition of the steps needed to conserve natural genetic resources.

Aquaculture

Since the latter half of the 19th century, aquaculture has been utilized in efforts to restore or enhance production from the natural environment that has been lost owing to pollution, environmental modifications, or overfishing (Eschmeyer, 1955). The history of the success of this use of aquaculture has been extremely variable for a variety of reasons (Radonski and Martin, 1986). However, with improvements in technology and better scientific information, this approach has been successful in a number of commercial capture and sport fisheries (Liao, 1988; Sandifer, 1988). Furthermore, it has been suggested that this approach to increasing fish production in the world will grow in the future (New, 1991). Such an increase will present problems for the natural genetic resources if changes are not made in some operational procedures.

Inherent in attempting to raise aquatic organisms under artificial conditions and introducing them into the natural environment are several potential genetic consequences that can have a long-term impact on natural genetic resources. These problems, while most extensively studied in salmonids, are not unique to a particular group of species (Harlan, 1981). Several reviews have been published that cover the specifics of the genetic impacts that may be realized in association with fish (Allendorf and Ryman, 1987; Hindar *et al.*, 1991; Waples, 1991). Waples (1991) summarized the concerns in three issues about the genetic interaction between cultured and natural populations. The first is a direct genetic effect that can be introduced by hybridization of the two groups or by the introgression of genes into natural populations. Next are the indirect genetic effects that may be induced by altered selection regimes or reductions in population sizes brought about by competition, predation, disease, or other factors. Finally, genetic changes in enhanced populations may intensify the consequences of mating with natural fish. It has been adequately documented that these issues are important and need to be factored into management decisions.

Intensive captive aquaculture development in the marine environment also poses potential risks for genetic resource management. First, as mentioned previously, the loss of natural populations presents serious problems for maintaining genetic diversity and providing a source for new genetic material for future utilization. Second, the likelihood of escape of domestic stocks into the natural habitat poses some as yet unquantified risk to the genetic integrity of wild stocks (Hansen *et al.*, 1991). Domestication of aquacultural stocks, including selection for improvement of production-related traits, can reduce their capacity for survival and reproduction in the natural environment (Doyle, 1983). Thus, the issues are similar to those defined for restoration and enhancement aquaculture except there are differences with respect to the major causative factors. With restoration and enhancement the major effector is the number of genetically modified fish in the natural environment and the frequency of their interaction with the natural aquatic resources. Alternatively, while production aquaculture may not mean that large numbers escape into the natural environment, the magnitude of the genetic differences between the cultured and natural populations is probably much greater. Consequently, the potential for genetic change needs to be evaluated for both sources of impact.

Genetic Risk

With the variety and strength of the potential impacts on the natural genetic resources mentioned above, it is not too difficult to deduce that the opportunity for introducing genetic change is very real. The potential for introducing genetic change in a natural living system comprises the 'genetic risk' of conducting the activity being considered.

The term 'genetic risk' is a rather general one that can be defined using a variety of parameters that measure genetic alteration from the status quo. However, when we want to consider the sustainability of natural resources a measurement is needed that will evaluate long-term success. A strong case can be made for this value being comparable to the genetic diversity contained within these aquatic resources. The genetic structure of a species or population is the repository of biological variability and this variability is what provides a species with the resilience to survive natural environmental change. As stated by Conrad (1983), 'Variability of biological matter is the *sine qua non* for the ability of organisms to cope with the uncertainty of the environment'. Hence, there is a need to conserve as much of the existing genetic variability as possible, especially since there are no realistic alternatives to this goal (Ryman, 1991).

How can genetic risk be evaluated in terms of genetic variability, and what are the component parts of the system that have a major influence on this parameter? A model has been developed for evaluating the need to conserve genetic variability in aquatic resources that is particularly informative (Thorpe *et al.*, 1995). As shown in Fig. 12.1, this model incorporates four major components for the evaluation of genetic risk: nature of the resource, components of diversity, genetic processes, and human interventions. Briefly, what this model provides is a framework on which to base assessment of relative genetic risks and to highlight areas where steps could be taken to minimize the loss of genetic variability.

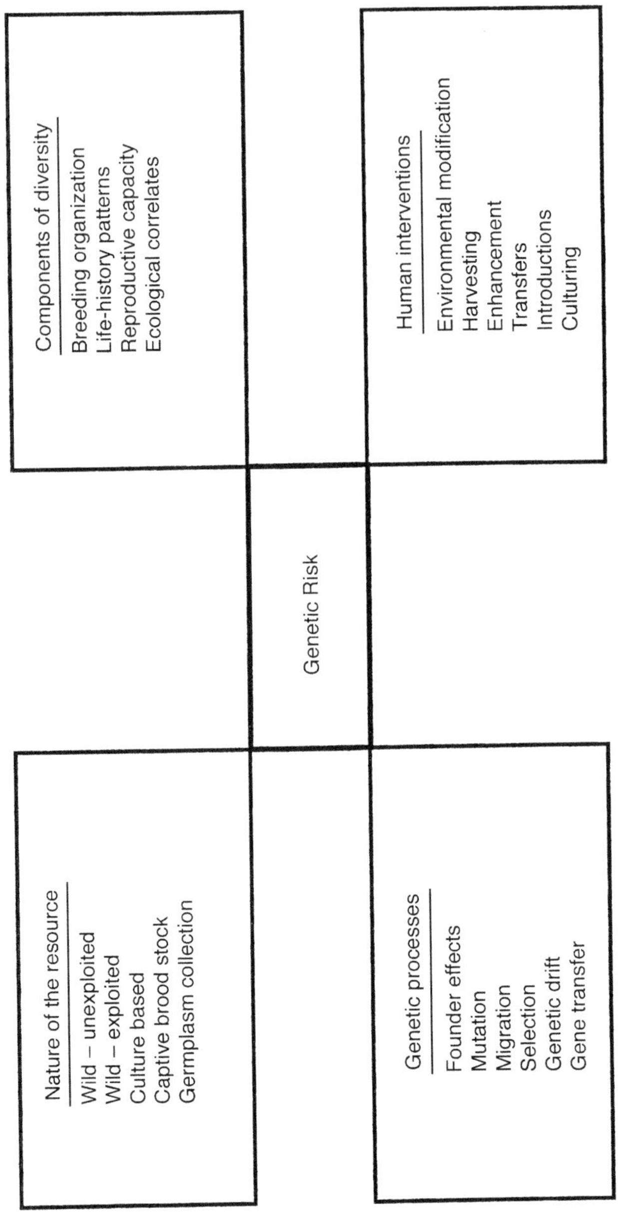

Fig. 12.1. Model illustrating the four components that determine genetic risk in aquatic species (after Thorpe *et al.*, 1995).

An example of the use of this model to define areas where attention is needed to minimize the genetic risk to natural populations is shown in Table 12.1. It contains the author's evaluation of the important characteristic for each of the four risk factors in the model for two major aquaculture species. This illustration is based on experience in the Pacific Northwest and was chosen to illustrate a point; it utilizes, by design, two species with very different characteristics. Looking first at the nature of the two resources, natural Pacific salmon genetic resources are comprised of, primarily, wild and exploited populations, while those of the cupped oysters are based almost solely on cultured populations. Next, the components of diversity are very different for these two species: one has a complex breeding structure that plays a role in gene diversity, while the other has little microstructure but diversity is defined primarily by the ecological diversity of its habitat. Major human interventions are (arguably) harvesting and culture for the salmon, while environmental modifications are those crucial for the cupped oysters. The defining genetic process could be identified for Pacific salmon and cupped oysters as, respectively, migration and founder effects. These conclusions suggest that the emphasis for maintaining genetic variability should be placed on the level of exploitation with the salmon and on the methods of artificial reproduction for the oysters. Evaluations such as these provide the opportunity to reach some conclusions about the level of genetic risk that specific activities may have on natural resources, and developing quantification techniques may permit more objective criteria on which to base management decisions.

Role of Aquaculture

Aquaculture occupies a rather unique position with respect to the need for genetic conservation and the risks inherent in ignoring its role in affecting genetic variability. As indicated earlier, most species under cultivation have been sampled only recently from the natural populations (within the last 30 years) and the only source of new genetic material is from the wild. However, most of these species are also harvested in the capture fisheries and this raises

Table 12.1. Evaluation of two aquaculture species for the four components affecting genetic risk.

Organism	Nature of the resource	Components of diversity	Human interventions	Genetic processes
Pacific salmon	Wild and exploited	Breeding organization	Harvesting and culture	Migration
Cupped oysters	Cultured	Ecological correlates	Modification of environments	Founder effects

some political and social concerns about the genetic impact of escaped and released fish on the genetic integrity (fitness) of the wild stocks (Waples *et al.*, 1990; Hansen *et al.*, 1991). Add to this mixture the use of aquaculture to compensate for losses in natural production or to enhance the natural productivity of capture fisheries and there is a rather interesting dilemma. On the one hand, aquaculture provides opportunities to reduce exploitation and thus conserve wild stocks, while, on the other hand, the natural populations are possibly exposed to some level of risk.

The marine aquaculture industry has a number of opportunities to address this risk in developing approaches to the formation, handling and maintenance of their captive brood stocks. These are best considered on the basis of the timeframe in which the activities can be applied.

Immediate activities

The first step that should be taken is to utilize a model such as mentioned in this chapter to evaluate and, perhaps, quantify the genetic risks to be considered in developing broodstocks with new species and in maintaining the genetic 'robustness' of both the captive and natural populations. The impacts of factors such as the breeding organization of the candidate species, the proximity of populations of the same or other species, or the level of previous commercial harvesting can be taken into consideration. Methods of addressing those with the maximum potential for introducing genetic change can be devised to assist in keeping viable captive and natural populations.

Next, the methods of initiating and maintaining broodstocks should be evaluated and modified, if necessary. Often stocks are derived from narrow samples of the ancestral population either because of limited sampling or owing to the low reproductive success in the captive fish. In addition, the processes of domestication and stock maintenance can result in high levels of genetic drift and inbreeding. Furthermore, owing to the very high reproductive capacity of many marine species, the cost of maintaining adult fish, and the fear of spreading infectious diseases, production populations are generally reproduced from limited numbers of adults. All of these factors can lead to severe decreases in genetic variability, can maximize the genetic differences with natural populations, and can possibly lead to untenable difficulties in the future. Captive populations should be initiated with a reasonable sampling of genetic variation and breeding approaches should be used to ensure adequate variation for the future.

Finally, record keeping is essential to the development of a reliable captive population that will retain its genetic viability. With the combination of very high reproductive output per female and the likelihood of non-random distribution of performance traits among families, it is essential to develop a reliable and thorough set of records to eliminate major losses of genetic variation within a captive population. At the very least, fish destined for reproducing the population should be marked and utilized in a manner that will minimize mating with close relatives.

Longer-term activities

For an aquaculture production system to be successful on a long-term basis it will be necessary to develop domesticated strains that are genetically adapted to the rigours of growing under production conditions. Not only will this provide benefits to the efficiency of production, but the animals can be made sufficiently distinct to allow their unequivocal identification among natural populations. With the current lack of differentiation among natural and captive populations, the ability to evaluate their interaction and even the extent of mixing is lacking. Further development of molecular technology to make genetic analysis even more precise will provide the tools to satisfactorily address this need.

Closely associated with the formation of domesticated strains should be development of techniques to control reproduction and eliminate the potential for genetically intermixing with natural populations. To date the best approach is the production of triploid organisms, but the technology for producing such animals is as yet relatively undeveloped for a lot of marine species. In addition, the performance of sterile triploids under production conditions varies dramatically among species. More predictable methods of introducing sterility in production animals without impacting the biological functioning of the organism are needed.

To protect the investment in development of domesticated strains of animals and to limit the need to 'mine' the natural genetic resources for future variation, techniques by which the germplasm of these strains could be 'banked' would be a definite asset to the industry. There has been rapid progress in developing techniques for the cryopreservation and storage of sperm from aquatic animals (Tiersch and Mazik, 2000). However, a great deal more research and development will be needed for the species that will comprise the marine aquaculture industry of the future.

Conclusions

With the relative infancy of the marine aquaculture industry there is a unique opportunity to develop a very robust food production system and at the same time maintain viable systems of natural living aquatic resources. The approaches to achieve this will have to be multifaceted and include the concerns of the environmental, commercial harvest, sport fishing and aquaculture communities. None of these will remain viable or sustainable without genetic resources that are adequately variable. Thus, careful attention to initial genetic evaluation of the genetic raw material used to initiate a programme, to the management of the genetic building blocks chosen for use, and to the impacts various activities may have on the external genetic resources will yield beneficial results for all parties with a vested interest in the production from the aquatic environment.

References

Allendorf, F.W. and Ryman, N. (1987) Genetic management of hatchery stocks. In: *Genetic Management of Hatchery Stocks. Population Genetics and Fishery Management.* University of Washington Press, Seattle, Washington, pp. 111–159.

Conrad, M. (1983) *Adaptability: the Significance of Variability from Molecule to Ecosystem.* Plenum Press, New York.

Doyle, R.W. (1983) An approach to the quantitative analysis of domestication selection in aquaculture. *Aquaculture* 33, 167–185.

Eschmeyer, R.W. (1955) Fish conservation fundamentals. In: Fish conservations highlights of the 1954 issue of *Sport Fishing Institute Bulletin*, No. 38 (Jan.). Sport Fishing Institute, Washington, DC, pp. 79–109.

Hansen, L.P., Hastein, T., Naevdal, G., Saunders R.L. and Thorpe, J.E. (eds) (1991) Interactions between cultured and wild Atlantic salmon. *Aquaculture* 98, 1–324.

Harlan, J.R. (1981) Who's in charge here? *Canadian Journal of Fisheries and Aquatic Sciences* 39, 1459.

Hindar, K., Ryman, N. and Utter, F.M. (1991) Genetic effects of cultured fish on natural fish populations. *Canadian Journal of Fisheries and Aquatic Sciences* 48, 945–957.

Liao, I-C. (1988) East meets west: an eastern perspective of aquaculture. *Journal of the World Aquaculture Society* 19, 62–73.

Nehlsen, W., Williams, J.E. and Lichatowich, J.A. (1991) Pacific salmon at the crossroads: stocks at risk from California, Oregon, Idaho, and Washington. *Fisheries* 16, 4–21.

New, M.B. (1991) Turn of the millennium aquaculture. Navigating troubled waters or riding the crest of the wave? *World Aquaculture* 22, 28–49.

NPPC (1987) Columbia River basin fish and wildlife program. Northwest Power Planning Council, Portland, Oregon.

NRC (National Research Council, US) (1992) *Marine Aquaculture: Opportunities for Growth.* National Academy Press, Washington, DC.

Radonski, G.C. and Martin, R.G. (1986) Fish culture is a tool, not a panacea. In: *Fish Culture in Fisheries Management.* American Fisheries Society, Bethesda, Maryland, pp. 7–13.

Reisner, M.P. (1986) *Cadillac Desert.* Viking Press, New York, 582pp.

Ricker, W.E. (1981) Changes in average size and average age of Pacific salmon. *Canadian Journal of Fisheries and Aquatic Sciences* 38, 1636–1656.

Ryman, N. (1991) Conservation genetics and considerations in fisheries management. *Journal of Fish Biology*, 39 (Suppl. A), 211–224.

Sandifer, P.A. (1988) Aquaculture in the west, a perspective. *Journal of the World Aquaculture Society* 19, 73–84.

Smith, P.J., Francis, R.I.C.C. and McVeagh, M. (1991) Loss of genetic diversity due to fishing pressure. *Fisheries Research* 10, 309–316.

Tiersch, T.R. and Mazik, P.M. (eds) (2000) Cryopreservation in aquatic species. *Advances in World Aquaculture* 7, 1–439.

Thorpe, J.E., Gall, G.A.E., Lannan, J.E. and Nash, C.E. (eds) (1995) *Conservation of Fish and Shellfish Resources: Managing Diversity*. Academic Press, New York, 206pp.

Waples, R.S. (1991) Conservation genetics of Pacific salmon. II. Effective population size and the rate of loss of genetic variability. *Journal of Heredity* 81, 267–276.

Waples, R.S., Winans, G.A., Utter, F.M. and Mahnken, C. (1990) Genetic approaches to the management of Pacific salmon. *Fisheries* 15, 19–25.

What Role Does Genetics Play in Responsible Aquaculture?

13

L. James Lester

Environmental Institute of Houston, University of Houston – Clear Lake, 2700 Bay Area Boulevard, Houston, TX 77058, USA

Abstract

Aquaculturists and their organizations can be assessed in comparison to other people and organizations on moral, ethical and stewardship criteria. This chapter considers the criticisms directed towards aquaculture on ethical and stewardship grounds. Ethically irresponsible behaviour would lead to harm to humans in terms of health, economic or social well-being. Irresponsible stewardship would lead to harm to future generations of humans or to the sustainability of natural resources. Charges of irresponsible actions are assessed here only as they relate to the role of genetics in management of the production system.

Genetics plays two roles in aquaculture: monitoring and modification. Use of genetic monitoring is generally associated with ethical operations exhibiting good stewardship. The genetic modifications of aquacultured animals are compared to those of maize in the context of responsibility. Genetic technologies for modifications include selective breeding, inbreeding, hybridization, chromosomal manipulation, gynogenesis, control of sex determination, cloning and production of transgenics. Aquaculturists have genetic responsibility for domestic populations and for wild stocks that are linked genetically to cultured animals. A large difference exists between genetic modification accomplished as research and that which is applied in commercial operations.

In general, commercial aquaculture has not exhibited the same drive to commodification of genetic resources as the crop breeding industry. Selection of outbreeding populations is the principal technology employed. Traits selected for improvement are chosen for improvement of yield, not for mechanization or marketing advantages. Few production systems use more complex genetic technologies. Genetic modifications appear to improve the ethical responsibility of aquaculture operations. The analysis of responsibility for stewardship is less clear. The long-term impact of genetic management of aquaculture stocks may be loss of genetic variation. Genetics permits economic valuation of traits, but no method exists for incorporation of ethical or stewardship values. Aquaculturists should avoid genetic modification for irresponsible reasons and adopt precautionary management of genetic resources.

Introduction

Aquaculture is the fastest-growing component of agriculture and, unlike many high-yield systems, occurs primarily in the developing world. Aquaculture has the potential to provide food security to many of the poorest regions. It utilizes many different species and production systems. Each production system makes different use of technology and exists in a different socioeconomic context. Despite justifiable criticism of some poorly planned and managed aquaculture operations, most systems have the potential for minimal or positive environmental impacts. Codes of practice and regulations are needed to ensure the realization of aquaculture as a positive force in achieving sustainable development. Aquaculture policies should promote management practices that are environmentally and socially responsible (FAO, 2000).

Aquaculture and the genetic modification of aquacultured organisms have a long history. As a species, goldfish, *Carassius auratus*, rivals the dog, *Canis familiaris*, in the degree to which genetic modification has been achieved. Common carp (*Cyprinus carpio*) is the best example of a domesticated species used in aquaculture for food (Kirpichnikov, 1981). Common carp has many geographic strains and modern varieties that have been selected for appearance and performance in different environments. Since common carp were first cultured, humans have added hundreds of species to the list of aquacultured organisms, but few have achieved the level of genetic modification that is associated with full domestication as in sheep, cattle and chickens.

Animal domestication was well advanced before the promulgation of Mendelian genetics and Darwinian concepts of selection. Husbandmen knew they could modify their stock, but they did not understand the principles. Now we are in possession of genetic knowledge and technology that permits us to implement genetic modifications at a rate and of a type that challenges the imagination. This potential to modify can be used irresponsibly and result in effects that are detrimental to people or other species. It is appropriate for the scientists and managers responsible for the direction of domestication of aquacultured organisms to reflect on the implications of their actions and their responsibilities on several levels.

Charges of Irresponsible Aquacultural Practices

Aquaculture has been criticized for damaging ocean and coastal resources. Coastal pond and open-water pen facilities have been accused of using methods that waste fishery resources and contaminate water. Also, coastal production systems have been charged with decreasing fishery resources because they were sited in mangrove ecosystems. Marine and coastal aquaculture has been criticized for introducing exotic species and diseases that threaten native populations and ecosystems. Charges have been made that aquaculture undermines human nutrition and health because it uses fishery resources on feeds for cultured animals that could go to feed poor humans (Naylor *et al.*, 2000). Coastal aquaculture has been accused of social displacement and generating economic inequities (Stonich *et al.*, 1997). Management of the organism's genetics may seem irrelevant to these critiques, but it is central. Genetics is at the heart of the developing systems and all of the components of aquaculture production systems are interrelated.

Forms of Responsibility

Making a judgement on what constitutes responsible behaviour is a matter of social or individual standards. Sometimes these are codified in laws, policies or rules; sometimes there is a cultural consensus; sometimes an individual decides for him- or herself. Responsibility must be defined in relation to someone or something. I define it at three levels. Moral codes are a matter of how an individual views responsibility to him- or herself or a deity. Codes of ethics describe how an individual may responsibly relate to society or other humans in general. There is also a stewardship responsibility to future generations and to the health of the biosphere. The stewardship concept has developed in recent years under the term sustainable development. If our actions are responsible in a stewardship context, they will not impair the access of future generations to necessary resources or lead to the extinction of other species.

There are moral codes that judge the killing of animals to be immoral and therefore irresponsible behaviour. Obviously persons with such a moral code would find all animal aquaculture objectionable. Professional associations are not in the business of espousing moral codes. Each aquaculturist has a moral code and should accommodate the moral positions of others to whatever extent is practical.

Ethical codes express the views of groups of people. Ethically irresponsible behaviour would harm human health or diminish social or economic well-being. Many professions, including legal and medical, operate according to ethical codes. Attempts have been made to codify ethical behaviour for fishers and aquaculturists. The FAO produced the *Code of Conduct for Responsible Fisheries* (FAO, 1995), which embodies a set of ethics for fishers and aquaculturists. There have been other efforts to codify ethical behaviour for aquaculturists (Donovan, 1997; Pawaputanon, 1997), and there were several presentations at the Aquaculture 2001 meeting on codes of conduct and best management practices that incorporate ethical standards.

Stewardship will be embodied in similar codes. However, sustainability science is in its infancy (Kates, 2000) and the knowledge required to define unsustainable practices is limited. One difficulty is that the quality of stewardship cannot be judged immediately. Determining what is responsible stewardship currently requires a time lag that allows us to see the response of interacting systems. We can make use of historical results from similar practices, but environmental change and complexity will make this a very challenging judgement.

Good stewardship is expressed by managing the cultured organism and production systems so that they do not use resources unsustainably, do not damage ecosystems, and maintain their productive capacity in perpetuity. The impact of consumption systems, like an aquaculture farm, on the biosphere is called its ecological footprint. Good stewardship would involve the benefit of the culture system exceeding the cost of the ecological footprint measured, not in economic but in ecological units. Guidelines have been proposed for sustainable aquaculture despite the uncertainty of evaluation methods (Anonymous, 1998).

Criteria for Assessing Ethical and Stewardship Practices

In order to determine whether a practice should be considered good or bad, there must be a value system. In discussing the values associated with genetic practices in aquaculture, a few simple criteria will be used. First, genetic changes should be evaluated according to their impact on human health. Do the genetic practices change the nutritional value of the cultured organism or affect the disease status of the consumer? Second, genetic practices should be assessed for the impact they make on decreasing or increasing social equity, e.g. changes in economic or property status of different groups. Third, genetic modifications can result in alterations of aquacultural practices that change the efficiency of resource utilization. The effect of genetic modifications on the ecological footprint of the production system is an important consideration. Fourth, genetic practices should be evaluated on their contribution to increasing or decreasing environmental quality. Pollution can have both stewardship and ethical implications. Fifth, the degree to which a genetic practice maintains or decreases genetic diversity should be evaluated. Diversity is the raw material for adaptations and change and should be valued as a resource. Loss of genetic diversity can reduce the ability of a production system to operate in a sustainable fashion.

Objectives for Responsible Use of Genetics in Aquaculture

The content of the *Code of Conduct for Responsible Fisheries* (FAO, 1995) can be paraphrased to make it explicit for genetic management of cultured organisms. The objectives that are derived by this process can be stated as follows:

- Establish principles for responsible practice of genetics in aquaculture activities, taking into account all their relevant biological, technological, economic, social, environmental and commercial aspects.
- Establish principles and criteria for the elaboration and implementation of policies for responsible conservation of aquaculture genetic resources and for responsible aquaculture stock management and development.
- Promote the contribution of aquaculture genetics to food security and food quality, giving priority to the nutritional needs of local communities.
- Promote protection of genetic diversity and the environments and coastal areas that support it.
- Promote genetic research on aquaculture species and culture practices, as well as on potentially impacted ecosystems and relevant environmental factors.
- Provide standards of conduct for all persons involved in the management of genetic resources in the aquaculture sector.

Recent deliberations have expanded the concepts of ethics and stewardship contained in the original code. Genetics is now seen as an integral component of the development of sustainable aquaculture. Improvement of stocks can provide benefits in the form of better use of resources and environmental protection. High priority is recommended for the application of genetics to aquaculture. This application includes the usual increase in productivity, but also the application of genetics to conservation of aquatic biodiversity and taking a safe, precautionary approach because there are implications for human and environmental health (FAO, 2000).

In the *Bangkok Principles* (FAO, 2000), there is a section on improving environmental sustainability. The policies suggested for sustainable aquaculture would ensure environmental sustainability. The principles emphasize environmentally sound technologies and resource-efficient farming systems. Improvements in environmental sustainability can be achieved through implementation of environmental, economic and social sustainability assessment criteria for aquaculture development; improved management practices; efficient use of water, land, fry, and feed inputs; ensuring aquaculture developments are within local and regional carrying capacities; promotion of good practices for environmental management of aquaculture; and promotion of aquaculture as a means of improving environmental quality.

Historical Lessons on Ethics and Stewardship from Crop Genetics

Mendel and Darwin published their observations in the latter half of the 19th century and provided a theoretical framework for applying genetics to domestication and improvement of food and fibre production. Prior to their work, domestication of plants and animals occurred through the accumulated actions of many breeders who did not understand the effects of their choices. After 1900, plant and animal breeders were able to design genetic improvement programmes based on scientific knowledge of the organism's genotype. As the 20th century progressed more and more genetic techniques became available for genetic modification of plant and animal resources.

The results have been so dramatic that the dissemination of high-yielding strains of wheat, rice and maize in the latter half of the 20th century has been called the Green Revolution. The knowledge that went into this exceptional example of genetic modification provides an opportunity to examine the role of genetics in responsible food production. Maize (*Zea mays*) is the focus of consideration here because its genetic system is based on outcrossing like the animals reared in aquaculture, not inbreeding like wheat and rice.

Until the 20th century, maize breeding was incorporated into the farming process. Most farmers produced their own seed. The crop existed in many local races that differed as a result of simple selection of parent plants by local farmers. The first efforts to apply Mendelian genetics and select superior strains were disappointing. The solution proposed by a leading geneticist was the use of hybrids of inbred lines (Shull, 1909). This controversial solution was adopted for a variety of economic and political reasons. There is evidence from their publications that maize geneticists chose the approach of hybridizing inbred lines because it produced a proprietary product for capitalization of farming. Hybrid maize led to the commodification of seed and the requirement that farmers buy seed every year. Some geneticists believe that the same yield improvements could have been obtained by recurrent selection of outcrossing populations. Using the latter method would have avoided the need to buy seed every season. The economics of production inputs would have been different.

Once maize seed production was in the hands of corporate breeders, the goals of genetic modification changed. Development of phytoengineering has led geneticists to place emphasis on selecting characteristics that enhance mechanical harvesting and cultivation. Nutritional value was not a breeding objective of the early breeders. Much effort was put into designing maize for mechanical harvest. The traits selected encouraged a shift from hand labour to machine labour and encouraged the shift from small, family farms to large, corporate farms (Kloppenburg, 1988).

As the application of plant genetics shifted from public to private organizations, the competition for market share led to more secrecy and the growth of conglomerates with considerable control over what was planted. Efforts by public organizations to certify the performance of strains were diminished. Concern over the direction of plant breeding led the United States Food and Drug Administration (FDA) to propose that newly developed plant varieties come under the generally recognized as safe (GRAS) classification in order to guard against a decline in nutritional value. The plant science community and seed companies, however, forced the FDA to rescind the regulations (Hanson, 1974). Genetic modification programmes are able to focus on cosmetic characteristics, e.g. coloration, and ignore effects on nutritional value.

Successful marketing of hybrid seed by large suppliers resulted in a shift from local races to the monoculture of genetically identical plants over large areas. The homogeneity of the maize crop introduces a genetic vulnerability that is not present when crops are genetically diverse. Diseases and pests able to damage the common genotype spread in epidemics. Geneticists are constantly striving to introduce disease resistance genes to the inbred lines used to generate the uniform hybrids. Increased risk of epidemics justifies increased use of chemical pesticides, many of which have been shown to degrade environmental quality.

Lack of adaptation to local environments justifies additional external inputs in the form of water, fertilizers and soil treatments. Pesticides have been implicated in harm to humans and wildlife. Fertilizers contribute to eutrophication of water and the death of aquatic organisms from lack of oxygen. Expansion of mechanized maize production contributed to the justification for dam construction to provide irrigation water. Reservoirs have their own set of environmental and social impacts, e.g. loss of habitat and displacement of people. Some assessments of dam and irrigation projects consider the impacts on humans unethical and the ecological effects poor stewardship (Kloppenburg, 1988).

The success of genetic modification of maize in developed countries led to an effort to disseminate the technology to developing countries in the 1960s and 1970s. Just as in the United States, the technology transfer led to the disuse and extinction of races of maize that were locally adapted. Seed production was no longer an option for the farmer and more commercial inputs were required for the promised high yields. The Green Revolution raised production dramatically in places where infrastructure and capitalization favoured large-scale mechanized farming. In some other places, it was an economic and social failure leading to the same displacement of small farmers seen in the United States (McNeill, 2000).

Now the biotechnology revolution is contributing to genetic modification of maize. Environmental objections to pesticides have sparked the development of transgenic maize with an internal insecticide in its tissues. This has created a controversy about the poisoning of non-target insects. Another biotech approach is the insertion of genes for herbicide resistance. This permits the application of higher doses of herbicides to remove unwanted plants from a farm field. Both types of traits enhance the business of large corporations with patents on genetic research products and have implications for environmental stewardship.

After studying the role of genetics in the development of commercial maize production, there can be no doubt that the decisions on how to genetically modify an organism have significant effects on the ethics of the business and the sustainability of the industry. Maize agriculture might still be a domain for small farmers using more labour, fewer pesticides and less fertilizer if genetic improvement had been accomplished by recurrent selection of open-pollinated populations at public agriculture experiment stations. Maintaining maize genetics in the public domain might have maintained the profitability of small farms. The social diaspora from the farm to the city might have included fewer farming families.

Inbreeding and biotechnology have resulted in large increases in yield per hectare that have contributed to greater food security in parts of the world. The importance of food security and food exports in the competition called the Cold War is underappreciated. Increased yields per unit area made it possible to maintain per capita world grain production around 300 kg for the last 30 years despite population growth (Brown *et al.*, 2000). This has kept malnourishment relatively constant. Per capita meat and fish production have increased over the last 30 years based to a large degree on higher yields of grain (Brown *et al.*, 2000). Nevertheless, industrial maize production is currently dependent on synthetic chemicals and petroleum in a way that cannot be sustainable and will require substitute technologies, including biotechnology advances, in the future.

How should the results of hybrid maize development and the Green Revolution be judged from the perspective of ethical and stewardship impacts? Was it genetically responsible to use techniques that made all of the improvements into proprietary property? Was it responsible to emphasize selection for mechanization over selection for environmental adaptation and nutritional value? One's personal perspective on technology and economics will guide their assessment of the responsibility exhibited by decision-makers during the development process. In large part, the answer depends on how the cost–benefit analysis is constructed.

Could the Green Revolution have been more responsible from a genetic perspective? Yes: it could have recognized and avoided some of the social and environmental impacts of the genetic modifications and production systems being disseminated. Is this an appropriate lesson for aquaculture? Yes: we are in danger of applying the same model of industrial agriculture to our development of genetically improved strains.

Role of Genetics in Aquaculture

There are two ways of using genetics in the practice of aquaculture: monitoring and modification. Monitoring of genetic characteristics is typically tied to irresponsible behaviour by its omission rather than its commission. Genetic monitoring is used to assess diversity and similarity of wild and captive stocks and to identify species, hybrids, stocks, families and individuals. Management of gene pools (populations) cannot be properly performed without information from genetic monitoring. Management of genetic modification utilizes knowledge of the genotypes and sometimes pedigrees of broodstock to effectively design and assess improvement programmes. From the perspectives of ethical and stewardship behaviour, monitoring is usually positive.

Genetic modification accompanies domestication and human-mediated change in cultured organisms. The processes employed by geneticists and breeders can be summarized as human intervention in evolution. Modification can be accomplished in many ways. The classic approach is selective breeding, which has been practised for centuries on common carp and decades on Atlantic salmon (*Salmo salar*). Other techniques that alter the genetics of individuals or gene pools are inbreeding, hybridization, polyploidization, other chromosome modifications, alteration of sex determination, gynogenesis or androgenesis, and transgenics or recombinant DNA.

Using genetic technology and practices on a population, we can alter

- The physical appearance of an organism.
- The physiological tolerances of an organism.
- The frequency of alleles or genotypes.
- The reproductive success of particular genotypes.
- The diversity of genotypes and gene pools.
- The number of genes or chromosomes in a genotype.
- The way genes are packaged into chromosomes.
- The sex ratio.
- The similarity of combining nuclei.
- The presence of genes or DNA sequences from another species.

Certainly there are more critiques of genetic modification of cultured animals than of monitoring. There is the moral critique based on changing creation and playing God, which is outside the realm of this discussion. There are ethical considerations related to the characteristics modified and the ownership of the results. Historically, animals were modified over many generations for performance under regional environmental conditions. Modern animal husbandry has encouraged the widespread culture of genotypes in production systems that require such major environmental inputs as antibiotics, vaccines, pesticides and special diets. This combination of genetic and environmental technology has been criticized from a stewardship perspective. Proprietary strains are often synthesized by combining genes from many sources that were in the public domain. This conversion of common to private property and subsequent modification for profit of the breeder can be criticized from an ethical perspective.

Genetic Technologies and Implications for Responsibility

Genetic monitoring

In the 1970s, genetic monitoring of genotypes primarily employed protein electrophoresis. Today monitoring is based on biochemical techniques that focus on differences in DNA sequences. These include various assays of nucleic acids for sequence or base number differences, e.g. restriction fragment length polymorphism (RFLP), randomly amplified polymorphic DNA (RAPD), microsatellite number and DNA sequencing. The methods differ in their probability of detecting genetic diversity. The more diversity the technique assays, the greater potential it has to differentiate groups and individuals. Monitoring diversity is an important check on the genetic status of a group for sustainable use. DNA attributes can be correlated to commercial traits and used in marker assisted selection.

Some genetic monitoring techniques can be used to screen for DNA segments of commercial interest. Under current law, it is possible to patent genetic sequences discovered in a wild or cultured organism and privatize the information/sequence.

Selective breeding

Dog breeds are a good example of the power of selective breeding. This ancient method has the power to modify the phenotype in major ways. Selective breeding is simply the addition of human choice to natural selection. Selection can be focused on comparison of individuals, families, or a combination and can involve one, few or many traits. Powerful analytical and statistical technologies have been incorporated into modern selection programmes.

One conundrum with selective breeding is that it reduces genetic diversity by eliminating genes that contribute to low-value phenotypes. Under most selection schemes, the population maintains significant genetic variation for response to environmental change, but very intensive culling and marker assisted selection can rapidly reduce genetic variation in parts of the genome. One must evaluate the benefits and costs of improving performance while reducing future options.

Inbred lines

Inbreeding favours the creation of homozygous over heterozygous genotypes. This is accomplished by breeding close relatives. In traditional breeding schemes, lines are repeatedly inbred to eliminate genetic variation within the lines. In a large population, inbreeding may not result in loss of genetic variation, but in a small population genetic diversity is likely to decline with inbreeding. Once inbred lines are achieved, they are usually crossed to test for heterosis of commercial traits. Heterosis exists when the hybrid performs better than the two parents.

If uniformity is important for the efficiency of the production system, it can be a justifiable approach. Marketing hybrids of inbred lines may create significant heterosis or it may simply be a method to enforce proprietary ownership of genetic resources.

Gynogenesis and androgenesis

There are methods to produce offspring with genes from only the mother or only the father. The methods are equivalent to self-fertilization. They represent extreme forms of inbreeding. Again this technology does not eliminate genetic variation if a genetically diverse population is maintained for production of homozygotes.

Hybridization

Hybridization refers to crossing individuals from two distinct genetic populations. There are two forms of hybridization potentially employed in aquaculture: hybridization between strains and interspecific hybridization.

Usually interstrain hybrids are chosen for culture because they show heterosis. Dependence on hybrids for mass culture necessitates the development and maintenance of parental lines. Breeders are managing two strains instead of one.

Hybridization can be followed by selection of the combined gene pool or the system can be developed for culture of the first-generation hybrids only. In the former case, a grower could manage his own stock. In the second case, the system is designed for a seed producer to control the production and sale of seedstock.

Interspecific hybridization has been used in plant and animal breeding, but is rare. Several examples are found in commercial aquaculture, e.g. tilapia hybrids (*Oreochromis* spp.), and striped bass (*Morone saxatilis*) × white bass (*Morone chrysops*). If the hybrid is fertile, then the initial cross may be a way of adding desirable genetic variation. A simple selection process can follow. If the hybrid is partially or fully sterile, it will be produced in each generation. Hybridization of tilapia is considered below as a technique to alter sex determination.

Chromosome manipulation

Polyploidy

Many cultured plants have polyploid origins, but no common domestic animals are polyploid. Research has shown that simple manipulations of fertilized eggs of aquatic organisms can cause fusion of the egg nucleus and polar bodies or inhibit cell division and result in polyploidy (Allen *et al.*, 1986). There are two common polyploidies: triploid with one extra set of chromosomes and tetraploid with two extra sets. Triploids have such unbalanced chromosome complements that gamete production is nearly impossible. They are almost completely sterile in most species. Triploidization as a sterilization technique cannot be guaranteed 100% effective because there is some probability that meiosis will yield a viable gamete. Tetraploids have balanced chromosome sets and can produce functional gametes in most cases. They are often used in crosses with normal diploids to produce triploids without physical manipulation of the gametes or zygotes.

Aneuploidy

It is also possible to obtain chromosomal mutations in an organism that differ from the normal karyotype by only one or a few chromosomes (Guo *et al.*, 1998). Fixing this type of trait in a population would act to isolate that population reproductively from other populations of the same species. It is a difficult trait to establish, but there is evidence that it can occur during domestication.

Genetic manipulation of sex ratio

Aquaculturists have employed the shifting of phenotypic sex with several fish species to exploit sexually dimorphic growth patterns, e.g. monosex tilapia culture. Hormonal shifting is a non-genetic technology with genetic implications because it changes the potential for crossing certain individuals. In aquaculture it has been possible through hormonal manipulation of sex to produce parental types that, when crossed, produce highly biased sex ratios or only one sex of offspring (e.g. Mair *et al.*, 1993).

Cloning

Progress has been made in the clonal propagation of plants and animals, i.e. the production of new individuals from existing individuals without fertilization. Tissue culture propagation of plants is a common commercial practice. Production of cloned animals from somatic cells has shown less commercial promise, but this technique may be more appropriate for aquatic animals. Cloning presumes that a 'best' genotype can be identified and does not consider different environmental conditions. This approach can limit the genetic variation of the stock more than any other. If used to produce strains for monoculture, it maximizes genetic vulnerability to diseases and pests. The culture system could require additional inputs for risk reduction.

Transgenics/recombinant DNA

Genetic engineering usually involves the addition of gene sequences for messages not currently produced by the organism. This is accomplished by constructing a sequence of DNA consisting of a foreign gene for a desirable trait, a regulatory sequence (promoter), a reporter gene with an easy to assay product and a vector sequence. In research, the vector is often a virus, but there is reluctance to use viral sequences for potential food organisms. The construct is inserted into gametes or zygotes, and individuals that show incorporation and stable expression are selected for phenotypic qualities. These individuals are then bred to determine whether the transgenic trait will be passed to offspring. The resultant organisms are known as transgenics, the products of transfer of a gene from a source other than their own gene pool.

Production of a true breeding line of transgenics effectively isolates them from the genetic diversity contained in the remainder of the species. Crossing a transgenic with a normal dilutes the impact of the transgene in the offspring. There is limited value in the non-transgenics for commercial breeders. Thus, cultured populations containing valuable genetic diversity may be discarded.

Of more concern is the effect of transgenic production on the non-transgenic population of the subject species. If there are viable wild populations of the same species, there is the potential for escapees to immigrate into these gene pools.

Evaluation of Current Aquaculture Genetic Practices Based on Ethics and Stewardship

Management of wild stocks

Most marine aquaculture operations involve the exploitation of wild stocks either as the source of broodstock or animals for stocking culture systems. Rarely does the aquaculture industry actively manage the wild stocks on which they depend.

Genetic monitoring can improve the identification of these stocks. Management decisions should opt to maintain the genetic diversity and reproductive potential of these stocks. Most wild stocks are the common property of a nation or community. It is good ethical practice to limit damage to shared resources. Stewardship for future generations also requires maintenance of this resource.

Some wild stocks on which aquaculture depends have been poorly managed and aquaculturists have been responsible for some of the impact. For example, movement of tilapia among lakes and rivers in Africa has resulted in hybridization that has reduced the number of distinct gene pools available in this family (Pullin, 1988; Pullin and Capili, 1988). The Asian shrimp aquaculture industry has placed very high fishing pressure on stocks of *Penaeus monodon* owing to dependence on gravid females for seedstock.

Tilapia, carp, trout, oysters and other cultured species have also been introduced by culturists to non-native habitats and contributed to the loss of genetic diversity of native fish and shellfish. Some of the depleted species had been the subjects of fisheries or culture. Feral populations of Mozambique tilapia (*Oreochromis mossambicus*) have become established in many subtropical regions and reached abundance levels that suggest negative impacts on indigenous species. *The Atlas of North American Freshwater Fishes* lists ten species of cichlids that were introduced into US waters from fish farms or by government agencies. Six of these species are noted as abundant, dominant or replacing native species in their new habitats (Lee *et al*., 1980). Rainbow, cut throat, brown and brook trout are often mixed by intentional stockings. Some of these species hybridize with each other and with other indigenous trout species. In some cases, introgression is sufficient to merge the gene pools.

Culture of penaeid shrimp in Asia and Latin America has created fishing pressures on the wild stock that may modify or decrease the genetic diversity. Those marine species that are still sourced from the wild may suffer high selection pressures from the capture methods. Exchange of disease organisms between cultured and wild populations has probably impacted the gene pools of the wild stocks.

For many species that are still collected from the wild for stocking or used as broodstock, the growers or breeders are striving to reduce their dependence on wild stocks and moving towards captive reproduction and domestication, e.g. Atlantic salmon (*Salmo salar*) and Pacific white shrimp (*Litopenaeus vannamei*). Dependence on hatcheries has increased. Efforts are being made to create genetically improved stocks.

There is no indication that the aquaculture industry has assumed a stewardship role for the wild populations on which it depends. The responsibility lapses shown by aquaculturists in the management of wild stocks are a small part of the continuing legacy of fisheries management that encouraged the movement of gene pools and genetic alteration of fish stocks. For example, salmon were introduced in the wild in Chilean waters long before they were grown there in netpens. Hatchery stocks were intentionally added to wild stocks for fishing enhancement before there was concern about genetically modified escapees from aquaculture facilities. These practices are becoming unacceptable for fisheries management and should be discouraged for stocks used in aquaculture. Genetic monitoring will assist fisheries managers in assessing whether introgression of genes from cultured to wild populations has occurred and is serious. Domestication of cultured stocks should make survival in the wild less likely. This issue is addressed more fully in Chapter 12.

I found no evidence of unethical actions taken by aquaculturists for genetic monitoring or modification and impacting wild populations. If demands for a foundation population for genetic improvement resulted in harvest of seedstock or broodstock creating a food shortage for local humans or denying fishers access to the resource, it would be unethical. But genetic programmes can be started with modest populations. Utilization of wild populations for genetic programmes has not been demonstrated to impact the human health or social equity of a region. Impact of genetically modified organisms on wild conspecifics is covered in Chapter 12.

Management of domestic stocks

Aquaculture managers have control of the genetic resources of many of the common species produced by them. Carp species are domesticated and exist as many local races. Atlantic salmon stocks have also been domesticated. Channel catfish (*Ictalurus punctatus*) exist in selected strains and fry production is managed by farm operators. Rainbow trout (*Oncorhynchus mykiss*) have genetically modified strains. Spat production for oyster aquaculture is controlled by breeders in many places. There are multiple genetic improvement projects starting for both freshwater and marine shrimp. The aquaculture industry is embarked on a massive effort to genetically improve the stocks used in many production systems.

There are 12 aquaculture species that are among the top ten on the FAO lists for either yield or crop value or both. They are listed in Table 13.1. Of the top ten species for production quantity, five are domesticated carp that have been subject to a long period of genetic modification. These stocks are widely viewed as beneficial to human nutrition and socioeconomic status (Goldburg and Triplett, 1997). However, introductions of domesticated carp to areas where they are not indigenous have adverse stewardship effects.

The Pacific oyster (*Crassostrea gigas*) has recently been subjected to intensive genetic modification that is beginning to manifest in commercial strains. Oysters are seldom criticized for ethical or stewardship considerations. Oyster culture operations using exotic species are subject to some criticism for contributing to a loss of biodiversity. Introduced oysters occupy benthic habitat that was historically available to native species of oysters in some parts of the Pacific Northwest of the United States (Couch and Hassler, 1989) and Australia. However, the sustainability of food production is not threatened by

Table 13.1. Global production of aquaculture species in 1998. Top species by production weight.

Common name	Scientific name	Production (million tonnes)
Kelp	*Laminaria japonica*	4.17
Pacific oyster	*Crassostrea gigas*	3.44
Silver carp	*Hypophthalmichthys molitrix*	3.31
Grass carp	*Ctenopharyngodon idella*	2.89
Common carp	*Cyprinus carpio*	2.47
Bighead carp	*Aristichthys nobilis*	1.58
Japanese carpet shell	*Ruditapes philippinarum*	1.43
Crucian carp	*Carassius carassius*	1.04
Yesso scallop	*Pecten yessoensis*	0.86
Nile tilapia	*Oreochromis niloticus*	0.79
Atlantic salmon	*Salmo salar*	0.69
Giant tiger prawn	*Penaeus monodon*	0.58

the manner in which the oyster industry operates. In some cases, oyster and other types of shellfish culture can contribute to reduction of eutrophication and improved water quality by increasing the density of suspension feeders in an area.

The life cycle of an extremely valuable aquaculture product, the giant tiger shrimp (*Penaeus monodon*), has only recently been closed and genetic improvement programmes initiated. Most giant tiger shrimp production is carried out in tropical regions and the product is exported to developed economies. This particular production system provides a significant contribution to some Asian economies, but has been criticized for environmental and social impacts.

Cultured Atlantic salmon are produced almost entirely from breeding programmes managed by geneticists. Production of Atlantic salmon in coastal netpen production systems has been criticized for environmental impacts, both water pollution and biological pollution of wild salmon stocks. The system has also been criticized for shifting fishing yield from human consumption to fish feed.

Application of Genetic Technology

As previously mentioned, the uses of genetic technologies for monitoring normally have positive ethical and stewardship implications. Recently, microsatellite marker technology has been employed for minimizing inbreeding in rainbow trout and obtaining information on sire parentage in a marine shrimp (*Marsupenaeus japonicus*). Quite a few species are the subjects of genome mapping efforts that will enhance monitoring and modification capabilities. Aquaculture industries using mapping include salmon, tilapia, channel catfish and shrimp (Davis and Hetzel, 2000). This type of activity can enhance sustainability and does not impact human health or social equity.

Aquaculture research has demonstrated the possibility of using for modification all of the genetic techniques mentioned above. The numbers of genetic methodologies actually applied by the commercial industry are few. Credit is due to the industry for the fact that genetic improvement has not moved in directions leading to ethical conflicts or poorer stewardship.

Only 1–2% of fish and shellfish aquaculture is based on genetically improved stocks (Gjedrem, 2000). Even the largest aquaculture crops have received very little effort aimed at genetic improvement. Few commercially cultivated macroalgae have been the focus of genetic improvement. Nori (*Porphyra yezoensis*) is an exception. It has been genetically modified by selection for taste, colour, texture and adaptation to local growing environments (Minocha, 1999).

In traditional culture of domesticated species, e.g. carp, small private and some large public breeding operations employ selective breeding to accomplish genetic modification. Mass selection or a combination of among and within family selection is being applied to most of the newly domesticated species. Nowhere is a private breeder attempting to follow the pattern of proprietary inbred hybrids controlled by large industrial breeders. Thus, there has been no pressure towards monoculture of identical genotypes over large areas.

Goals of selective breeding programmes are usually growth, food conversion efficiency (FCE) and disease resistance. Breeding goals for salmon and trout are FCE, growth, disease resistance, flesh quality (i.e. fat content, flesh colour, fat distribution and tenderness) and age at sexual maturation (Gjedrem, 2000). Channel catfish have been the subject of genetic improvement programmes for more than 25 years. Technologies used in improvement include mass and family selection, strain crossing and interspecific hybrids (Ligeon *et al.*, 2000). Examining a variety of fish selection programmes for size at harvest shows a consistent response to selection above 10% per generation (Gjedrem, 2000). Selection for resistance to pathogens has been successful, e.g. oyster resistance to diseases caused by *Haplosporidium nelsoni* and *Perkinsus marina* (Ragone Calvo *et al.*, 1997). This is a record of success without proprietary approaches based on limited access to stocks that are simple to manage.

Another interesting feature of genetic improvement programmes in aquaculture is the focus on production traits rather than those related to marketing or labour reduction. A survey of commercial and research experts in Australian aquaculture produced a list of biological traits for breeding objectives. Twelve of the traits recommended by the group are production related and only two could be considered cosmetic: flesh colour and shell shape. Size at harvest was the first choice (Lymbery, 2000). There has been no equivalent to genetic modification for easier mechanization of harvest. I am aware of no evidence that aquaculturists have adopted selection goals that will negatively impact human health, directly contribute to social inequity or decrease the sustainability of the culture system.

In the Australian survey reported by Lymbery (2000), strategies for genetic improvement were ranked. The list places family selection and mass selection at the top; followed by cross-breeding, selection among stocks, and marker assisted selection; then transgenics, chromosome manipulation and sex ratio control are listed (Lymbery, 2000). Methods that increase the proprietary nature of genetically improved strains are not elevated in the list.

A recent assessment of the US catfish industry recommended a marker assisted selection programme as the top research priority for catfish (Ligeon *et al.*, 2000). This approach combines DNA monitoring techniques and standard selection methods. The approach is also known as quantitative trait locus (QTL) selection. Some QTL programmes that have been reported include selection for temperature tolerance in rainbow trout, growth in Kuruma prawn, and infectious haematopoietic necrosis (IHN) resistance in rainbow trout (Davis and Hetzel, 2000). Efforts are underway to provide QTL capabilities for breeding disease-resistant penaeid shrimp (Alcivar-Warren, 2000).

Ethical and stewardship issues arise regarding the choice of traits to be modified and the degree of modification. If a breeder selects cosmetic appearance at the expense of nutritional value, the process would be unethical according to our definition. If the breeder selects for an organism that requires input of more resources to obtain the same output it would violate the principles of stewardship. Selective breeding can also produce an animal that can only be cultured in specialized environments and, therefore, remove it from among the options available to poor farmers. This could contribute to social inequity. The goals of aquaculture selective breeding programmes are typically, but not always, greater and more efficient production in a variety of environments.

There are examples of breeding programmes that have focused on coloration to the exclusion of production traits. The red coloration of *Oreochromis mossambicus* and hybrid *Oreochromis* has been fixed in production lines at the expense of production traits, as shown when growth comparisons are conducted. By the mid-1980s, several researchers had documented negative attributes of red tilapia for food culture, and these studies were reported at the Second International Symposium on Tilapia in Aquaculture (Pullin *et al.*, 1988). Many red mutants were lethal or seriously deleterious in the homozygous form; therefore fertility of reds was less than in fish with normal coloration (El Gamal *et al.*, 1988). A negative correlation was reported between red coloration and incresed growth rate (Behrends *et al.*, 1988). A red strain was reported to have lower viability juveniles compared to normal coloration (El Gamal *et al.*, 1988). The inbreeding necessary to fix the red trait in a population was reported to decrease growth performance (Galman *et al.*, 1988). Nevertheless, development of red strains continued due to marketing potential.

External coloration and flesh coloration are components of selection programmes for other species, e.g. gilthead seabream (*Sparus aurata*) and Atlantic salmon. These genetic goals may be the few examples of unethical or unsustainable choices made for genetic modification of aquaculture production systems.

Inbred lines

Inbreeding is viewed as detrimental and a process to be avoided in most aquaculture production systems. This view is based on observations of inbreeding depression for a variety of organisms. Laminarialean kelp has a mainly outcrossing breeding system and exhibits high inbreeding depression (Kloareg *et al.*, 1999). Inbreeding depression of salmon and trout has been estimated at 3–6% per 10% increase in inbreeding coefficient for different traits (Gjedrem, 2000). Many geneticists directing aquaculture programmes recommend that mating of close relatives be avoided. In addition, any plans to follow the model of hybrid maize would require a great investment in a breeding facility unlike any aquaculture production facility in the world.

The question for ethical consideration is whether the selected line was inbred for the benefit of the breeder, the grower or the consumer. The question for stewardship assessment is whether genetic groups outside selected inbred lines are discarded and diversity is lost. Random-mating populations can be maintained for future use, but often are discarded once successful inbred lines are established. I could find no evidence of aquaculture breeders following the model of inbred hybrids. However, there are applications of gynogenesis, which is equivalent to self-fertilization.

Gynogenesis

There have been many research studies of gynogenesis in fish and molluscan species. In some cases, gynogenesis is a component of more complex genetic modifications. The ethical and stewardship questions are the same as for inbreeding. In both cases, the application of these genetic technologies makes production of stock by small farmers much less likely and changes the social and economic relationships.

Hybridization

Aquaculture has made use of hybridization for some very practical reasons. Interspecific hybrids of tilapia species have been used to reduce pond overcrowding from early maturation and reproduction. Certain interspecific crosses of tilapia (such as *Oreochromis aureus* × *Oreochromis niloticus*) yield nearly monosex progeny, but the hybrids must be produced each generation from selected parents to maintain this effect. This production system is currently in competition with systems that use hormonal sex reversal or genetically modified broodstock that can produce sex-biased progeny in intraspecific crosses. There is no indication that production of hybrid interspecific fry will be employed by large industrial breeders to the detriment of small farmers who have other options.

An interspecific cross between striped and white bass has been the basis for a culture system in the United States. However, there is now commercial effort underway to domesticate striped bass under selective breeding without hybridization (Woods *et al.*, 1999; Woods, 2000).

An interspecific cross of channel × blue catfish (*I. punctatus* × *I. furcatus*) appears to be heterotic and shows improvement for several culture traits: faster growth, better feed conversion, disease resistance and tolerance of low DO (Ligeon *et al.*, 2000).

For most culture systems, hybridization was considered on its production merits and rejected. Studies in Norway have shown no significant heterosis between strains of Atlantic salmon (Gjedrem, 2000). Intraspecific crosses of gilthead seabream yielded no significant heterosis and mass selection was employed for dramatic improvements in weight gain (Knibb *et al.*, 1998). The US catfish industry is based mainly on channel catfish stocks despite many studies of interspecific hybrids (Smitherman and Dunham, 1995). Interspecific crosses of penaeid shrimp have been successful, but have demonstrated no major advantage for culture (Lawrence *et al.*, 1984; Bray *et al.*, 1990). The global shrimp and oyster culture industries are based on stocks of single species that are not intentionally inbred or produced by hybridization of strains. It appears that the use of hybridization for ensuring intellectual property rights has been avoided in the aquaculture industry.

The same issues apply to interstrain and interspecific hybrids regarding ethics and stewardship. Does hybridization increase or decrease human welfare? Intraspecific hybrids have been used to combine important commercial traits for better performance and disease resistance into one cultured stock. Interspecific hybrids have been used to achieve important production goals. When used for commercial purposes in aquaculture, hybridization has not been detrimental to human nutrition.

Does the hybrid production system change the social or economic status of growers? This approach discriminates against small farmers because it adds significant complexity to the management of the stock. The relationship between the breeder and grower and between both and the consumer can involve ethical conflicts. If the hybridization is a unique event, the genetic process is less likely to have a major effect on the production system.

Does hybridization affect genetic variation or the viability of wild populations? The stewardship of genetic resources does not appear to have suffered from the limited examples of commercial applications of hybridization.

Chromosomal manipulation

Chromosomal manipulation has been employed in fish and molluscs to create sterile culture organisms, to avoid loss of product quality during reproduction, and to increase size. Polyploids have been used commercially in culture systems of Pacific and Sydney rock oysters (Allen *et al.*, 1986; Nell and Maguire, 1998; Supan, 2000). Use of triploids can make oyster meat quality and production less seasonal and may enhance economic opportunity for labourers. It appears that triploids produced from crosses of tetraploid and diploid lines may become a widespread commercial practice for oysters (Guo and Allen, 1994; Guo *et al.*, 1996). Triploidy is being investigated for application in non-molluscan production systems, e.g. penaeid shrimp (Fu-Hua *et al.*, 1999).

There are ethical issues associated with marketing triploid seedstock. This method of seed production tends to take spat production away from the farmer. Also, triploid oysters are not completely sterile. Oysters and grass carp (*Ctenopharyngodon idella*) have demonstrated the probability of gamete production from triploids. Pacific oysters produced as triploids show a high frequency of heteroploid mosaics which are capable of producing a small proportion of fertile gametes (Allen and Guo, 1996). Thus, marketing those animals as completely sterile and incapable of reproduction is misleading. There is usually some potential for creation of a feral population.

Sex manipulation

Sex manipulation is possible in many fish species, and is a major component of tilapia culture. Genetic manipulations of the sex-determining system have resulted in crosses that can reliably produce all-male improved offspring. This system is usually beyond the abilities of small farmers. It has both ethical and sustainability issues associated with it. The system is designed for production of brood fish by large genetic operations, which modifies the economic relationship of breeders and producers. Escapees of super male fish could impact the demographics of wild populations. Sex ratio manipulation does not necessarily impact genetic diversity in the domestic population.

Cloning

Vegetative propagation of algae should be practical, but Kloareg *et al.* (1999) reported that laminarialean kelp doubled the DNA content of its chromosomes when cloned. This could be problematic for efficient growth. Cloning of kelp could have ethical and stewardship implications if the genetic system is significantly altered.

Cloning of aquatic animals is less of a technical challenge than cloning mammals. For example, Galbreath *et al.* (1997) demonstrated the ability to propagate clones of hybrids between Atlantic salmon and brown trout (*Salmo trutta*) gynogenetically. Apparently diploid eggs were fertilized with inactivated sperm resulting in offspring genetically identical to their mothers. Longwell (1997) has suggested a method based on tetraploidy of precursors to oocytes, gynogenetic induction of spawned eggs and sex reversal of clonal individuals to yield a broodstock capable of producing genetically identical individuals. Such a system does not appear to be of commercial interest for food production at this time. There may be situations in the aquarium trade in which it could be financially justified.

Clearly, distribution of a commercial clone would result in very high genetic vulnerability to genotype-sensitive disease organisms. This approach could contribute to less sustainable production systems if genetic diversity is discarded in conjunction with clonal propagation.

Transgenics

More than 20 species of aquatic animals have been the subject of successful transgenic experiments (Goldburg, 2000). Insertion of growth-promoting genes has been the most common goal. Atlantic salmon with growth-promoting sequences from other fish have been created and tested. Proposals have been made for use of such fish in commercial production systems. This genetic manipulation has ethical implications similar to transgenic maize with the addition of potential human health impacts.

It is interesting that the technology for production of transgenic macroalgae has not been established (Minocha, 1999). Typically plants are more amenable to genetic manipulation than animals. One report concludes that transgenic kelp is not an attractive alternative because they are too large for tank culture and transfer to the marine environment is ethically unacceptable (Kloareg *et al.*, 1999).

Extensive research on aquaculture organisms has shown that transgenics can be produced and can be created to express commercially useful traits. Maclean and Rahman (1994) listed 11 species of fish subjected to transgenic induction experiments and five species that demonstrated transmission to an F_1 generation. Goldburg (2000) reported that 21 species of fish have been genetically engineered. The types of fish genes that are candidates for insertion into new species include growth hormone genes, antifreeze proteins, metallothionein and gonadotrophin (Maclean and Rahman, 1994).

To date, transgenic stocks have not been adopted by commercial aquaculturists. Research priorities reported from a survey of the Australian aquaculture community placed establishment of genetic markers at the top and transgenics at the bottom (Lymbery, 2000). The economic backing for widespread application of transgenic technology in aquaculture appears to be missing. It is not certain that approval of these organisms for human consumption would lead to widespread adoption by commercial culturists. The demand for approval appears to be coming from the research and development companies interested in creating the stock rather than from the industry.

Ethical considerations arise from questions about human health, economic inequity, potential loss of genetic variation and the consequences of feral reproduction. Proper guidelines and precautionary testing can ensure a low risk of human health effects from transgenic food organisms. The choice of DNA sequences for the inserted construct should be guided by responsible practices.

Dunham (1999) stated that genetically improved fish generated by recombinant DNA technology probably do not pose any greater risk to the environment than fish that are genetically improved through traditional selective breeding. However, concern over the ethical use and release of transgenics has been high for some time. The American Fisheries Society has established a policy regarding transgenic fish (Kapuscinski and Hallerman, 1990). In other countries the process of applying this technology in aquaculture seems to be moving faster. For example, the risk of transgenic tilapia to human health and feral stocks of the same species has been evaluated by Cuban researchers (Guillen *et al.*, 1999). They concluded that the widespread use of transgenic tilapia in Cuba posed neither human health nor environmental risks.

Summary

A wide range of genetic manipulations has been performed on aquacultured species. Research suggests that fish and shellfish are more amenable to more genetic manipulations than birds and mammals. The majority of genetic manipulations have been selective breeding. Selection goals have not been detrimental to nutritional quality or farm labour requirements. Aquaculture breeders have occasionally experienced pressures from marketing concerns to emphasize cosmetic traits, e.g. red coloration, rather than production traits. Trading production efficiency for market appeal may reduce sustainability of an aquaculture industry.

Chromosomal and sex-determination manipulations have been introduced to some types of commercial culture, e.g. oysters and tilapia. These do not appear to have ethical issues, but they may have stewardship issues.

Hybridization and biotechnology are tools that may be more important for the commodification of the organism than for the improvement of properties. Aquaculture breeding has not incorporated inbreeding, hybridization and biotechnology to the extent that research has demonstrated potential applications in those areas. Thus, in most aquaculture production systems, the grower can also be a breeder.

Aquaculture exhibits variety in production systems. So far the industry has operated with genetic diversity and there are no miracle strains sweeping the locally adapted strains out of existence. There are a few proponents of high-input, high-yield varieties. Many growers are using wild stock and locally developed strains. Change from this status should be done carefully with a perspective on the ethics of social and environmental impacts and sustainability.

Aquaculture production systems in developed countries are already experiencing a shift towards industrial production. The catfish industry is experiencing consolidation, which translates to fewer, larger firms, increased land space used, and increased yield from intensive production methods. The industry is moving towards production and marketing contracts from large agribusiness firms (Ligeon *et al.*, 2000). Despite the potential socioeconomic changes associated with industrial production, there is still a belief in this form of aquaculture. Gjedrem (2000) recommended commercialization of the breeding programme after the concept is proven and the foundation stock is started. Lymbery (2000) concluded that government should start any genetic improvement programme in Australia and transfer it to industry. This trend from public ownership of genetic resources to private ownership contains some inherent ethical issues.

In the future, aquaculture genetics will become involved in the patenting of life forms and genes. When geneticists obtain patent protection for breeding or biotech methodologies that prevent others from developing useful genetic types, there are justifiable ethical challenges. Also, in the future, aquaculture geneticists will face the question of ownership of genetic resources obtained in other countries. Crop seed producers have been accused of genetic piracy for transferring germplasm from countries of origin to commercial breeders in other countries and not sharing the economic rewards.

Interaction of Genetics and Industrial Agriculture

> What is profitable is not always coterminus with what is socially optimal.
>
> Kloppenburg (1988)

The fundamental problem with using genetics as an ethical steward is that it has been developed as a tool of industrial agriculture. For example, Davis and Hetzel (2000) stated: 'The ultimate aim of a genetic improvement programme for an enterprise or industry is to improve profit. . . . the breeding objective can be defined as a weighted combination of traits that have economic value. . . . the economic value of a trait can be determined by specifying the production and marketing system'. Genetic engineering is viewed as a field with ethical implications, but there is nothing taught in quantitative genetics about the ethics of the results.

It is clear that the most common form of genetic modification in commercial aquaculture is selective breeding. Everyone who learns how to calculate a selection index knows that you must weight the traits by their economic value. If the improvement of economic value has a negative relationship with the nutritional value or health value of the trait, there is no correction term in the equation. If the improvement of economic value for the breeder results in an economic loss for the grower, the grower suffers because there is no control over the genetics. A negative correlation between improving the economic value of a trait and improving the environmental quality on and around the farm cannot be factored into standard quantitative genetics. Certainly externalities that cannot be converted into economic terms cannot be included in the equations. If the profit motive results in unethical or unsustainable decisions, the science of genetics is not designed to counter the results.

Science and technology are grounded in the exploitation of resources for human use. Scientists contribute to the conversion of organisms to resources. Conservation is a new and quasi-science that is not integrated into the world view of most scientists (Rolston, 1990). Thus, the idea of designing a genetic improvement programme to create a sustainable culture system rather than one that maximizes profit is rarely considered. Aquaculture could be the exception because it is so early in its development, but it will have to create its own model.

In my opinion, the application of genetics to aquaculture organisms has been quite responsible to date. My principal criticism is that there has been too little effort to use genetic improvement for the benefit of current and future producers and consumers. I would like to see future genetic applications chart a new course towards sustainable systems that contribute to health, social equity and resource stewardship.

The shift to sustainable development is primarily an ethical shift.

Kothari (1990)

References

Alcivar-Warren, A. (2001) SHRIMPMAP: a genetic approach to understand immune response and disease resistance in shrimp. In: *Book of Abstracts. Aquaculture: a Fantasy Come True*. Disney's Coronado Springs Resort, 21–25 January 2001. World Aquaculture Society, Baton Rouge, Louisiana, p. 11.

Allen, S.K. and Guo, X. (1996) Triploids for biological containment: the risk of heteroploid mosaics. In: Levin, M., Grimm, C. and Angle, J.S. (eds) *Biotechnology Risk Assessment: USEPA/USDA/Environment Canada/Agriculture and Agri-Food Canada. Proceedings of the Biotechnology Risk Assessment Symposium*. University of Maryland Biotechnology Institute, pp. 336–356.

Allen, S.K. Jr, Downing, S.L. and Beattie, J.H. (1986) Chemically and pressure-induced triploidy in the Pacific oyster *Crassostrea gigas*. *Aquaculture* 57, 359–360.

Anonymous (1998) The Holmenkollen guidelines for sustainable aquaculture. In: *Proceedings of the Second International Symposium on Sustainable Aquaculture*. AA Balkema, Rotterdam, Netherlands. Also: http://www.ntnu.no/ntva/rapport/aqua/recom.htm

Asian Development Bank and Network of Aquaculture Centres in Asia-Pacific (1996) *Aquaculture Sustainability Action Plan*. Regional study and workshop on aquaculture sustainability and the environment (RETA 5534). Manila, Asian Development Bank, and Bangkok, Network of Aquaculture Centres in Asia-Pacific, 21pp.

Bartley, D. (2000) Biodiversity and genetics. In: *FAO Fishery Resources Division Review of the State of World Aquaculture*. FAO Fisheries Circular no.886, FAO, Rome, 163 pp.

Behrends, L.L., Kingsley, J.B. and Price, A.H. III (1988) Bidirectional-backcross selection for body weight in a red tilapia. In: Pullin, R.S.V., Bhukaswan, T., Tonguthai, K. and McLean, J.L. (eds) *The Second International Symposium on Tilapia in Aquaculture*. ICLARM Conference Proceedings 15. Department of Fisheries, Bangkok, Thailand and International Center for Living Aquatic Resources Management, Manila, Philippines, pp. 125–133.

Bray, W.A., Lawrence, A.L., Lester, L.J. and Smith, L.L. (1990) Hybridization of *Penaeus setiferus* (Linnaeus, 1767) and *Penaeus schmitti* Burkenroad, 1936 (Decapoda). *Journal of Crustacean Biology* 10, 278–283.

Brown, L.R., Renner, M. and Halwell, B. (2000) *Vital Signs 2000: the Environmental Trends That Are Shaping Our Future*. WW Norton and Co., New York, 192pp.

Couch, D. and Hassler, T.J. (1989) Species profiles: life histories and environmental requirements of coastal fishes and invertebrates (Pacific Northwest): Olympia oyster. Biological Report 82(11.124), TR EL-82-4. United States Fish and Wildlife Service, National Wetlands Research Center, Slidell, Louisiana, 98pp.

Davis, G.P. and Hetzel, D.J.S. (2000) Integrating molecular genetic technology with traditional approaches for genetic improvement in aquaculture species. *Aquaculture Research* 31, 3–10

Donovan, D.J. (1997) Environmental code of practice for Australian prawn farmers. July 1997. 32 pp. Cited in *The State of World Fisheries and Aquaculture 1998*. FAO website (http://www.fao.org/docrep/w9900e/w9900e00.htm).

Dunham, R.A. (1999) Utilization of transgenic fish in developing countries: potential benefit and risks. *Journal of the World Aquaculture Society* 30, 1–11.

El Gamal, A.A., Smitherman, R.O. and Behrends, L.L. (1988) Viability of red and normal-colored *Oreochromis aureus* and *O. niloticus* hybrids. In: Pullin, R.S.V., Bhukaswan, T., Tonguthai, K. and McLean, J.L. (eds) *The Second International Symposium on Tilapia in Aquaculture*. ICLARM Conference Proceedings 15. Department of Fisheries, Bangkok, Thailand, and International Center for Living Aquatic Resources Management, Manila, Philippines, pp. 153–157.

FAO (Food and Agriculture Organization) (1995) *Code of Conduct for Responsible Fisheries*. FAO, Rome, 41pp.

FAO (Food and Agriculture Organization) (1997) *Aquaculture Development*. FAO Technical Guidelines for Responsible Fisheries No. 5, FAO, Rome, 40pp.

Fu-hua, L., Ling-hua, Z., Jian-hai, X., Xu-dong, L. and Jin-zhao, Z. (1999) Triploidy induction with heat shocks to *Penaeus chinensis* and their effects on gonad development. *Chinese Journal of Oceanology and Limnology* 17, 57–61.

Galbreath, P.F., Adams, K.J., Wheeler, P.A. and Thorgaard, G.H. (1997) *Clonal Atlantic Salmon × Brown Trout Hybrids Produced by Gynogenesis*. Washington Sea Grant Publication, WASHU-R-97-016, Seattle, Washington, 9pp.

Galman, O.R., Moreau, J. and Avtalion, R.R. (1988) Breeding characteristics and growth performance of Philippine red tilapia. In: Pullin, R.S.V., Bhukaswan, T., Tonguthai, K. and McLean, J.L. (eds) *The Second International Symposium on Tilapia in Aquaculture*. ICLARM Conference Proceedings 15. Department of Fisheries, Bangkok, Thailand and International Center for Living Aquatic Resources Management, Manila, Philippines, pp. 169–175.

Gjedrem, T. (2000) Genetic improvement of cold-water fish species. *Aquaculture Research* 31, 25–33.

Goldberg, R. and Clay, J. (1998) Draft guidelines for sustainable shrimp aquaculture. http://www.aquanet.com/aqua/topics/aq6a_98.htm

Goldburg, R. (2000) Something fishy. Accessed 10/6/2000 at http://www.environmentaldefense.org/pubs/Reports/Aquaculture/transgenic.html

Goldburg, R. and Triplett, T. (1997) *Murky Waters: Environmental Effects of Aquaculture in the United States*. Environmental Defense Fund, Washington, DC, 195pp.

Guillen, I., Berlanga, J., Valenzuela, C.M., Morales, A., Toledo, J., Estrada, M.P., Puentes, P., Hayes, O. and de la Fuente, J. (1999) Safety evaluation of transgenic tilapia with accelerated growth. *Marine Biotechnology* 1, 2–14.

Guo, X. and Allen, S.K. Jr (1994) Viable tetraploids in the Pacific oyster (*Crassostrea gigas* Thunberg) produced by inhibiting polar body 1 in eggs from triploids. *Molecular Marine Biology and Biotechnology* 3, 42–50.

Guo, X., Debrosse, G.A. and Allen, S.K. Jr (1996) All-triploid Pacific oysters (*Crassostrea gigas* Thunberg) produced by mating tetraploids and diploids. *Aquaculture* 142, 149–161.

Guo, X., Yang, H., Landau, B.J., Debrosse, G.A. and Allen, S.K. Jr (1998) The creation of aneuploid families in the Pacific oyster, *Crassostrea gigas* Thunberg. *Journal of Shellfish Research* 17, 328.

Hanson, C.H. (1974) The effect of FDA regulations (GRAS) on plant breeding and processing. Crop Science Society of America Special Publication 5. Crop Science Society of America, Madison, Wisconsin.

Kapuscinski, A.R. and Hallerman, E.M. (1990) American Fisheries Society position statement on transgenic fishes. *Fisheries* 15, 2–4.

Kates, R. (2000) Our common journey: a transition toward sustainability. Paper presented at National Conference on Science, Policy and the Environment. National Council for Science and the Environment, Washington, DC, 9pp.

Kirpichnikov, V.S. (1981) *Genetic Bases of Fish Selection.* Springer-Verlag, Berlin, 410pp.

Kloareg, B., Ar Gall, E., Asensi, A., Billot, C., Crepineau, F., Moulin, P., Boven, C. and Valero, M. (1999) Molecular and cellular approaches of reproduction biology and genetic improvement in laminarialean kelps. *World Aquaculture* 30(1), 23–25.

Kloppenburg, J.R. Jr (1988) *First the Seed: the Political Economy of Plant Biotechnology, 1492–2000.* Cambridge University Press, Cambridge, 349pp.

Knibb, W., Gorshkova, G. and Gorshkova, S. (1998) Selection and crossbreeding in Mediterranean cultured marine fish. In: *Proceedings of the Seminar of the CIHEAM Network on Technology of Aquaculture in the Mediterranean,* CIHEAM, Zaragoza, Spain, pp. 47–60.

Kothari, R. (1990) Environment, technology and ethics. In: Engel, J.R. and Engel, J.G. (eds) *Ethics of Environment and Development.* University of Arizona Press, Tucson, Arizona, pp. 27–35.

Lawrence, A.L., Bray, W.A., Wilkenfeld, J.S. and Lester, L.J. (1984) Successful interspecific cross of two species of marine shrimp, *Penaeus stylirostris* × *Penaeus setiferus*. *Proceedings of the World Mariculture Society* 1984, 39.

Lee, D.S., Gilbert, C.R., Hocutt, C.H., Jenkins, R.E., McAllister, D.E. and Stauffer, J.R. Jr (1980) *Atlas of North American Freshwater Fishes.* North Carolina State Museum of Natural History, Raleigh, North Carolina, 854pp.

Ligeon, C., Jolly, C., Crews, J., Morley, Z. and Dunham, R. (2000) The US catfish industry in the new century. *Aquaculture Magazine* 26(5), 39–43.

Longwell, A.C. (1997) Toward cloning aquacultured fish. *Reviews in Fisheries Science* 5, 341–365.

Lymbery, A J. (2000) Workshop summary: genetic improvement in the Australian aquaculture industry. *Aquaculture Research* 31, 145–149.

Maclean, N. and Rahman, A. (1994) Transgenic fish. In: Maclean, N. (ed.) *Animals with Novel Genes.* Cambridge University Press, New York, pp. 63–105.

Mair, G.C., Capili, J.B., Beardmore, J.A. and Skibinski, D.O.F. (1993) The YY male technology for production of monosex male tilapia, *Oreochromis niloticus* L. In: Penman, D.J., Roongratri, N. and McAndrew, B.J. (eds) *Proceedings of the International Workshop on Genetics in Aquaculture and Fisheries Management,* University of Stirling, 31 August–4 September 1992 ASEAN–EEC Aquaculture Development and Coordination Programme, Bangkok, Thailand, pp. 93–95.

McNeill, J.R. (2000) *Something New Under the Sun: an Environmental History of the Twentieth-century World.* WW Norton, New York, 421pp.

Minocha, S.C. (1999) Genetic engineering of marine macroalgae: current status and future perspectives. *World Aquaculture* 30(1), 29–30, 57.

National Marine Fisheries Service (1997) *Implementation Plan for the Code of Conduct for Responsible Fisheries.* United States Department of Commerce, NOAA, NMFS, Silver Spring, Maryland, 20pp.

Naylor, R.L., Goldburg, R.J., Primavera, J.H., Kautsky, N., Beveridge, M.C.M., Clay, J., Folke, C., Lubchenco, J., Mooney, H. and Troell, M. (2000) Effect of aquaculture on world fish supplies. *Nature* 405, 1017–1023.

Nell, J.A. and Maguire, G.B. (1998) *Commercialisation of Triploid Sydney Rock and Pacific Oysters, Part 1: Sydney Rock Oysters.* Final Report to Fisheries Research and Development Corporation (FRDC), New South Wales Fisheries Final Report Series, Report 10, NSW Fisheries, Port Stephens Research Centre, Taylors Beach, NSW, Australia, 121pp.

Network of Aquaculture Centres in Asia-Pacific and Food and Agriculture Organization (2000) *Aquaculture Development Beyond 2000: The Bangkok Declaration and Strategy.* Conference on Aquaculture in the Third Millennium, 20–25 February 2000, Bangkok, Thailand. NACA, Bangkok and FAO, Rome, 27pp.

Pawaputanon, O. (1997) Manual for harmonization of good shrimp farm practice. ASEAN Fisheries Network Project. ASEAN Secretariat, Jakarta, 45pp.

Pullin, R.S.V. (ed.) (1988) *Tilapia Genetic Resources for Aquaculture.* ICLARM Conference Proceedings 16. International Center for Living Aquatic Resource Management, Manila, Philippines, 108pp.

Pullin, R.S V. and Capili, J.B. (1988) Genetic improvement of tilapias: problems and prospects. In: Pullin, R.S.V., Bhukaswan, T., Tonguthai, T. and Maclean, J.L. (eds) *Second International Symposium on Tilapia in Aquaculture.* ICLARM Conference Proceedings 15. Department of Fisheries, Bangkok, Thailand, and International Center for Living Aquatic Resource Management, Manila, Philippines, pp. 256–266.

Pullin, R.S.V., Bhukaswan, T., Tonguthai, K. and Maclean, J.L. (eds) (1988) *Second International Symposium on Aquaculture.* ICLARM Conference Proceedings 15, Department of Fisheries, Bangkok, Thailand and International Center for Living Aquatic Resources Management, Manila, Philippines, 623pp.

Ragone Calvo, L.M., Harmon, V. and Burreson, E.M. (1997) Selection of oysters for resistance to two protozoan parasites. *Journal of Shellfish Research* 16, 327–328.

Rolston, H. III (1990) Science-based versus traditional ethics. In: Engel, J.R. and Engel, J.G. (eds) *Ethics of Environment and Development.* University of Arizona Press, Tucson, Arizona, pp. 63–72.

Shehadeh, Z.H. and Pedini, M. (2000) Issues and challenges. In. *FAO Fisheries Department Review of the State of World Aquaculture.* FAO Fisheries Circular no. 886, FAO, Rome, 163pp.

Shull, G.H. (1909) A pure-line method in corn breeding. In: *Annual Report of the American Breeders Association,* Vol. 5. American Breeders Association, Washington, DC, pp. 51–59.

Smitherman, R.O. and Dunham, R.A. (1995) Relationships among cultured and naturally occurring populations of freshwater catfish in the United States. In: Collie, M.R. and McVey, J.P. (eds) *Interactions between Cultured Species and Naturally Occurring Species in the Environment: Proceedings of the 22nd US–Japan Meeting on Aquaculture,* Homer, Alaska, 21–22 August 1993. Fairbanks, Alaska. University of Alaska Sea Grant College Program, [1995]. UJNR technical report; no. 22.

Alaska sea grant report; AK-SG-95-03, University of Alaska, Fairbanks, Alaska, pp. 9–16.

Stonich, S., Bort, J.R. and Ovares, L.L. (1997) Globalization of shrimp mariculture: the impact on social justice and environmental quality in Central America. *Society and Natural Resources* 10, 161–179.

Supan, J. (2000) Oyster breeding: the future is now. In: *Book of Abstracts, Aquaculture America 2000*, US Chapter of the World Aquaculture Society, New Orleans, Louisiana. World Aquaculture Society, Baton Rouge, Louisiana, p. 315.

Woods, L.C. III (2000) Morone broodstock domestication II – Present status and future needs. In: *Book of Abstracts, Aquaculture America 2000*, US Chapter World Aquaculture Society, New Orleans, Louisiana. World Aquaculture Society, Baton Rouge, Louisiana, p. 363.

Woods, L.C. III, Hallerman, E.M., Douglass, L. and Harrell, R.M. (1999) Variation in growth rate within and among stocks and families of striped bass. *North American Journal of Aquaculture* 61, 8–12.

Understanding the Interaction of Extractive and Fed Aquaculture Using Ecosystem Modelling

14

Mac V. Rawson, Jr,[1] Changsheng Chen,[2] Rubao Ji,[2] Mingyuan Zhu,[3] Daoru Wang,[4] Lu Wang,[5] Charles Yarish,[6] James B. Sullivan,[2] Thierry Chopin[7] and Raquel Carmona[6]

[1]Director, Georgia Sea Grant College Program, The University of Georgia, 220 Marine Sciences Building, Athens, GA 30602–3636, USA; [2]Department of Marine Sciences, The University of Georgia, 220 Marine Sciences Building, Athens, GA 30602–3636, USA; [3]The First Institute of Oceanography, State Oceanic Administration, PO Box 98, Qingdao 266003, People's Republic of China; [4]Hainan Marine Development, Planning and Design Institute, 10th Yiyuan Building No. 69, Haifu Rd, Haikou, Hainan 570203, People's Republic of China; [5]Marine and Fishery Department of Hainan Province, Haikou, Hainan, People's Republic of China; [6]University of Connecticut, Department of Ecology and Evolutionary Biology, 1 University Place, Stamford, CT 06901–2315, USA; [7]University of New Brunswick, Centre for Coastal Studies and Aquaculture, and Centre for Environmental and Molecular Algal Research, PO Box 5050, Saint John, New Brunswick, E2L 475, Canada

Abstract

One of the most difficult tasks resource managers face is understanding the carrying capacity of coastal waters for aquaculture. Aquaculture, like many other human activities, can threaten coastal waters. Understanding eutrophication and the interaction of two different types of aquaculture is very important to the safe and effective management of coastal aquaculture. The first type of aquaculture, producing shrimp and finfish, depends on supplemental feeding and can contribute to eutrophication. The second type, involving bivalve molluscs and macroalgae, extracts plankton and nutrients from surrounding waters and can have a significant positive impact on moderately eutrophic waters. These species depend on the water's basic productivity and will not grow effectively in water with low nutrient levels. Balancing extractive and fed aquaculture is of obvious importance to maximizing the safety and optimizing the carrying capacity of an embayment.

Ecosystem modelling offers a three-dimensional physical, chemical and biological simulation that can help scientists and managers understand and predict the eutrophic impact of aquaculture for a specific embayment. Such a model is being explored in China in research sponsored by the Sino-US Living Marine Resources Panel. In this study, two projects are using the model to simulate the impact of aquaculture on Jiaozhou Bay, Shangdong Province, and on Xincun Lagoon, Hainan Province. Jiaozhou Bay is in the temperate zone adjacent to the Yellow Sea. There, a major port and industrial city, Qingdao, and scallop and shrimp aquaculture interact with the physical and biological components of the bay. The other modelled environment is very different. Xincun Lagoon is a small embayment (~22 km^2) in southeastern Hainan Island adjacent to the South China Sea. Aquaculture in Xincun Bay includes 6500 fish pens (3 m × 3 m), 100 ha of shrimp ponds, pearl culture rafts and a new macroalgae culture operation that produced 3500 tonnes of *Eucheuma* in 1998–1999. The surrounding area has ~15,000 people and Xincun City is a major offshore fishing port (~500 vessels, > 10 m length) and Monkey Island Wildlife area with > 400,000 visitors annually. Extractive and fed aquaculture, along with the external activities, all have an impact on the carrying capacity of the bay for aquaculture.

These two models show much promise for simulating local eutrophic conditions and for increasing the general understanding of the complex interactions of aquaculture and other human activities. Models that simulate the impact of aquaculture and other human activities and eventually predict carrying capacity should become useful tools for resource managers.

Introduction

Effectively integrating aquaculture into coastal management is an important goal throughout the world. This chapter discusses an ecosystem approach using three-dimensional models. The first step in integrating aquaculture is to determine a body of water's capacity to sustain human activities, with an emphasis on aquaculture. This carrying capacity is the interaction of physical, chemical and biological factors. We have little or no control over the natural processes, but human activities can be altered to optimize our use of water. The key factor over which humans have control is eutrophication. Humans are the source and the solution. The input of organic matter and inorganic nutrients beyond moderate levels reduces a body of water's capacity to support water-related activities. Learning to balance inputs with nutrient extraction is a difficult goal to achieve, but it is critical if aquaculture is to be optimized.

Eutrophication affects aquacultural carrying capacity in three ways. First, eutrophication resulting from pollution and other human activities is added to natural nutrient-transport processes. In excessive amounts, nutrients become a serious threat to coastal waters. The growing number of toxic algal blooms and oxygen hypoxia events are often the result of eutrophication, but these environmental catastrophes are only symptoms of a more pernicious problem. Aquaculture's future success depends on abundant and clean water; thus, the industry is more threatened by extreme eutrophic conditions than by any other water use. In addressing eutrophication, industry needs the kind of broad-based approach that integrated coastal management provides.

Second, extractive aquaculture (bivalve molluscs and seaweeds), which removes plankton and nutrients from surrounding waters, can have a significant positive impact on moderately eutrophic waters. In fact, these species require moderate nutrient levels from natural (Blanton *et al.*, 1987) or human sources to maintain basic planktonic productivity, from which they must filter their food. Without the moderate levels and appropriate types of nutrients and plankton species, the extractive aquaculture species cannot thrive.

Third, fed aquaculture (shrimp and finfish), which depends on supplemental feeding to grow its products, contributes to eutrophication and should be balanced with extractive aquaculture species. The question is how can carrying capacity be managed to create a balance between extractive aquaculture, fed aquaculture systems, and anthropogenic nutrient inputs?

The first step is to determine the quantity of input and extraction of organic matter and nutrients into a body of water. The interactions of the complex physical, chemical and biological processes are just beginning to be understood. The techniques for simulating these interactions in ecosystem models also are improving rapidly (Chen *et al.*, 1999). Three-dimensional ecosystem modelling of coastal estuaries and embayments offers an excellent tool for integrating the natural processes and simulating both negative and positive aspects of aquaculture. The focus of this chapter is to illustrate how three-dimensional models can be used to develop a management strategy and integrate that strategy into coastal management.

Background

The negative aspects of aquaculture, particularly those systems that require supplemental feeding, have received substantial attention. Less emphasis has been placed on the positive aspects of extractive aquaculture. Seaweeds, molluscs and filter-feeding fish extract nutrients, organic particles and plankton. Used in conjunction with fed aquaculture in an integrated system, extractive aquaculture systems will utilize food supplies more efficiently, improve the water quality, reduce costs of production and increase the productivity of the ecosystem (Naylor *et al.*, 2000). How important could integrated aquaculture (polyculture) be to the environment? Naylor *et al.* (2000), in their evaluation of the ecological links between intensive aquaculture and capture fisheries, contend that worldwide, molluscs and seaweeds produced 9 and 8 million tonnes, respectively, in 1997. In the same year, fish cages and fish or shrimp ponds produced 2 and 18 million tonnes, respectively. Thus, world production was within 3 million tonnes, showing a balance between fed and extractive aquaculture; however, the two types of aquaculture are very often geographically disjunctive because of the predominant monoculture approach. Moreover, these numbers do not account for other sources of nutrients or for the difference in species composition. If integrated aquaculture can help balance extractive and fed aquaculture systems, there is hope that regionally integrated polyculture can reduce eutrophication from aquaculture.

The next step is to find a tool that will allow aquaculturists and resource managers to understand the relationships between aquaculture and physical, chemical and biological ecosystem processes. Ecosystem modelling promises to be this tool. Modelling techniques are improving rapidly and three-dimensional models provide excellent visual tools to see the potential impacts of management decisions. In this chapter, two case studies of the use of ecosystem models to evaluate the impacts of aquaculture are discussed. The first example describes conditions in Jiaozhou Bay in northern People's Republic of China, and the second describes Xincun Lagoon on Hainan Island in southern People's Republic of China.

Jiaozhou Bay Coupled Model Experiment

Jiaozhou Bay is a shallow, semi-closed bay adjacent to the Yellow Sea (YS) in the northeastern Province of Shandong, People's Republic of China. The bay's area is ~400 km^2 with an average depth of 7 m and maximum depth over 50 m in the strait to the YS. The bay has an extensive intertidal zone in its northwest quadrant. Tidal amplitude ranges from ~120 cm at the entrance to ~130 cm near the northeast shoreline. Average tidal current is > 15 cm s^{-1} with a maximum of 150 cm s^{-1} at the bay's entrance. The bay is dominated by a southerly or southeasterly wind in the spring and summer and by a northerly or northwesterly wind in the autumn and winter. Average wind velocity is ~5 m s^{-1} in summer and ~7 m s^{-1} in winter (Zhao *et al.*, 1995). Freshwater input is primarily from six major rivers, whose total maximum discharge is 135 m^3 s^{-1}. The largest river, the Dagu River, accounts for > 80% of the freshwater flow and is located on the west side of the bay (Fig. 14.1; Table 14.1; Chen *et al.*, 1999).

Qingdao (120 °22′E, 36°4′N) and surrounding districts have a population of 6.99 million people and the area is a major port and tourist destination. The ecology of Jiaozhou Bay has changed dramatically over the last three decades as a result of increased industries, aquaculture, agriculture and domestic sewage (Liu, 1992). Annual average concentrations of total inorganic nitrogen and phosphate increased from 1.2 and 0.14 μmol l^{-1} in 1962/63 to 10.4 and 0.45 μmol l^{-1}, respectively, in 1992. During the same period, the ratio of total inorganic nitrogen to phosphate shifted from ~10 to 24.2.

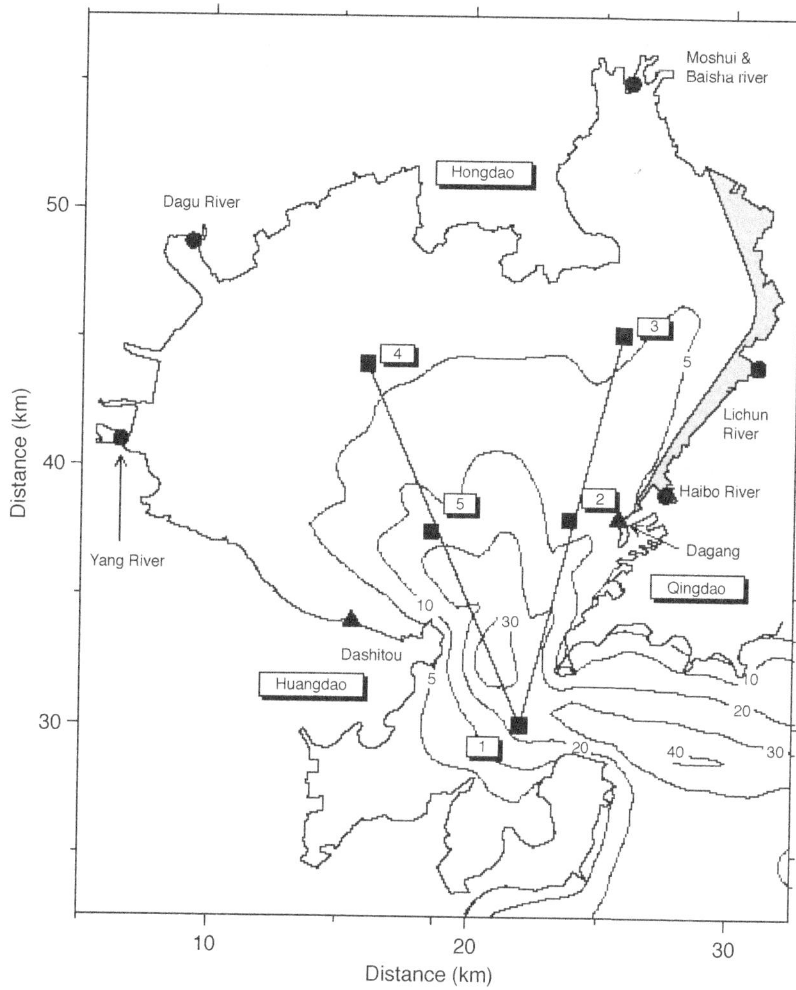

Fig. 14.1. Bathymetry (metres) of Jiazhou Bay, Qingdao, P.R. China. The solid squares are the biological measurement stations. The heavy solid lines are the sections for the model–data comparison. The solid triangles are tidal measurement stations. The light grey area indicates the intertidal zone. Locations of the six major rivers are indicated by solid circles. Dotted area indicates the initial locations of particles released at the 10th model day.

Nitrogen, which was the limiting factor in growth of phytoplankton, has given way to a phosphorus-limited ecosystem (Shen, 1995). Light intensity, water temperature, turbidity, and the continuing input of nutrients from rivers also contribute to phytoplankton productivity, which reaches a peak in summer with chlorophyll *a* readings of 0.37 to 9.5 mg m^{-3}. The highest levels are in the northwestern bay, where the Dagu River is located, and the lowest is at

Table 14.1. River discharge to Jiaozhou Bay during summer 1991 (average based on data from Liu and Wang, 1992).

River	Discharge rate ($m^3\ s^{-1}$)
Dagu	87.42
Yang	6.36
Moshui	3.66
Baisha	2.38
Lichun	0.60
Haibo	0.40

the entrance to the bay. Similarly, total primary productivity ranges from 33.60 to 2145.45 mg C m^{-2} day^{-1} with the highest values near the northern coast during summer.

Aquaculture of two scallop species (*Chlamys farreri* and *Argopecten irradians*) in suspended nets was the dominant production system along with bottom-cultured clams. The three major areas of the bay used for suspended rake culture for scallops are indicated in Fig. 14.2 (Chen *et al.*, 1999). These areas occupied 50 km^3, about one-eighth of the bay (Collaudin, 1996). The annual production of aquaculture scallops during the study period was ~40,000 tonnes (fresh total weight). The major clam species, *Ruditapes philippinarum*, cultured in the intertidal and subtidal northern part of the bay, produces ~70,000 tonnes (fresh total weight) annually (Fig. 14.2; Chen *et al.*, 1999).

The pronounced increase in nutrients in Jiaozhou Bay has caused serious environmental problems. The frequent occurrence of harmful algal blooms and the increased mortality rates and decreased rates of growth of natural and cultured organisms are evidence of the seriousness of the eutrophication problem. A three-dimensional model allows us to simulate the physical and biological processes and predict where eutrophication may become a problem or where aquaculture could have positive or negative impacts on the environment.

The model

The physical model used in this study was a modified version of the coastal ocean circulation model developed by Blumberg and Mellor (1987). The model's forcing functions are: (i) tidal oscillations; (ii) wind-driven features; and (iii) time-variable inputs of rivers. The time-variable river inputs and onshore intake/outflow discharges were used to simulate the buoyancy flow caused by river discharges. By far the largest river discharge (85%) into Jiaozhou Bay during summer was from the Dagu River (87 $m^3\ s^{-1}$; Liu and Wang, 1992). The effects of wind-induced currents and resulting mixing

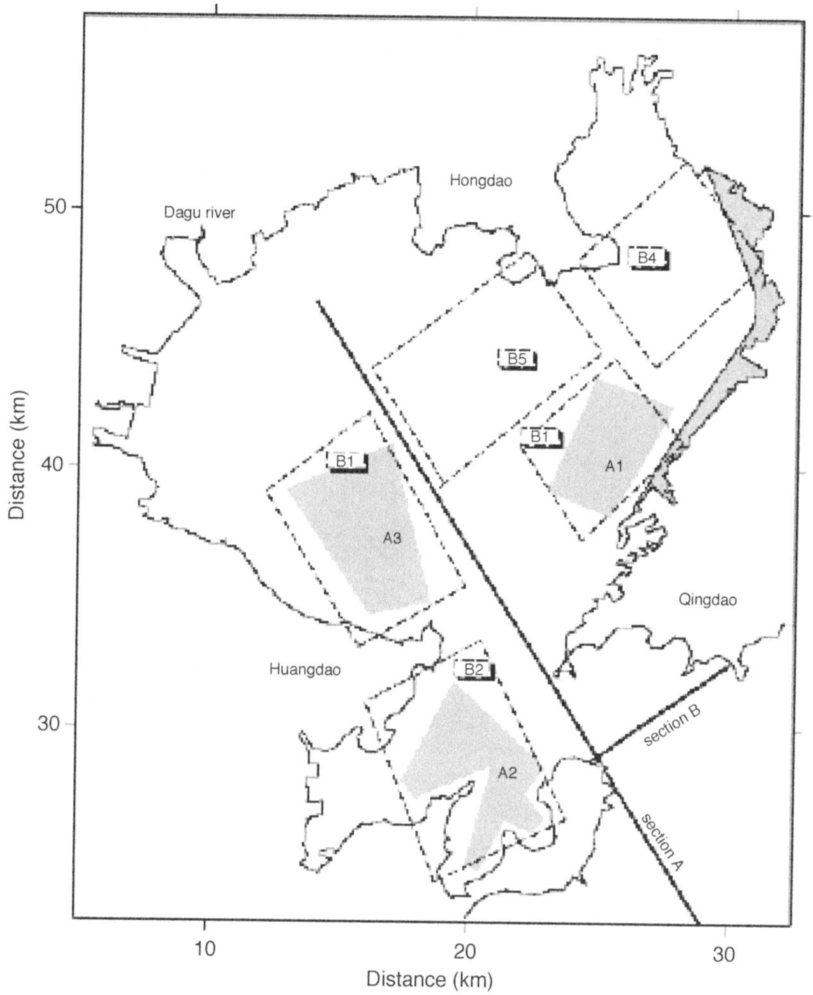

Fig. 14.2. Locations of the shellfish aquaculture sites (shaded areas) and selected regions for the flux estimation of nutrient and phytoplankton (areas enclosed by dashed line). The heavy solid lines indicate two sections used to represent our model results on a cross-bay section (Section A) and flux calculations into or out of the bay (Section B).

influences on the distribution of temperature, salinity, nutrients and phytoplankton were examined in the simulations. The dominant wind force in the summer is from strong southeast winds, but for simplification, a constant southeasterly wind of 5 m s^{-1} was added into the model after 10 model days.

The coupled biological model simulates simple nutrients (N), phytoplankton (P) and zooplankton (Z) using a modified model developed by Franks and Chen (1996). The biological parameters varied widely in time and space. The model was run with an initial set of parameters. Sensitivity analyses were then run over the parameter range. The stock density of shellfish was calculated directly from measurements taken in Jiaozhou Bay in 1996 (Collaudin, 1996). The scallop rafts consist of vertical lines of lantern nets. Based on the average number per lantern net, the standing stock parameter was set at 12 individuals m^{-3}, which is equivalent to 0.012 individuals l^{-1}. The scallop filtration rate varies in a range of 30 to 200 l day^{-1} per individual at around 25°C. This may overestimate the filtration rates since the food availability was not taken into account (Winter, 1978). So, a filtration rate of 100 l day^{-1} per individual was used in the model. The excretion fraction of the filtered food was assumed to be 0.3 or 30%.

Jiaozhou Bay is a phosphorus-limited ecosystem, and in the experiments, phytoplankton and zooplankton were measured in units of carbon (C) and nutrients by units of phosphorus (P). A constant C/P ratio of 100 was used to convert carbon to phosphorus. A 60 C/chlorophyll *a* ratio was used to convert chlorophyll *a* to carbon.

The tidally driven physical model demonstrated that the tidal current turned clockwise as it entered the bay. The tidal amplitude of ~120 cm at the entrance to the bay reached ~130 cm on the northeast coast. It took only 6.2 min for the tidal peak to travel from the YS to the northern coast. The defining physical feature was a residual eddy that developed near the entrance to the main embayment during flood to ebb tides (Fig. 14.3; Chen *et al.*, 1999). The eddies propagated in the vicinity of the scallop rafts and influenced the biological distributions near the outer bay. The velocity was 15–20 cm s^{-1} with a vertically averaged velocity of 10–15 cm s^{-1}. In most regions of the bay where the depth was < 10 m, a weak clockwise current that was relatively stronger on the eastern side (~1.5–2.5 cm s^{-1}) than the western side (< 0.2 cm s^{-1}) was characteristic. River discharges only slightly enhanced the clockwise residual circulation.

The other major physical component was the wind, particularly the southeasterly wind that dominates during the summer months. For example, with a southeasterly wind of 5 m s^{-1}, the near-surface water in the shallow region (< 10 m) moved northwestward and caused surface convergence in the northwest region of the bay – a phenomenon that is important to the biological model.

The results of the coupled model demonstrate that the water temperature and salinity remain vertically well mixed. Summer temperatures remained near 26°C with a tongue of cooler water from the YS intruding into the mouth of the bay. The water injected into the bay has higher salinity than predicted. This was attributed to summer evaporation and the exclusion of the intertidal flats from the model. Nutrients (phosphate) and phytoplankton followed similar trends and were vertically mixed. The maximum values for phosphate and chlorophyll *a* were 1.65 μmol l^{-1} and 1.9 μg l^{-1}, respectively, near the rivers in the inner bay. They dropped to 0.42 μmol P l^{-1} and 0.8 μg chlorophyll *a* l^{-1}

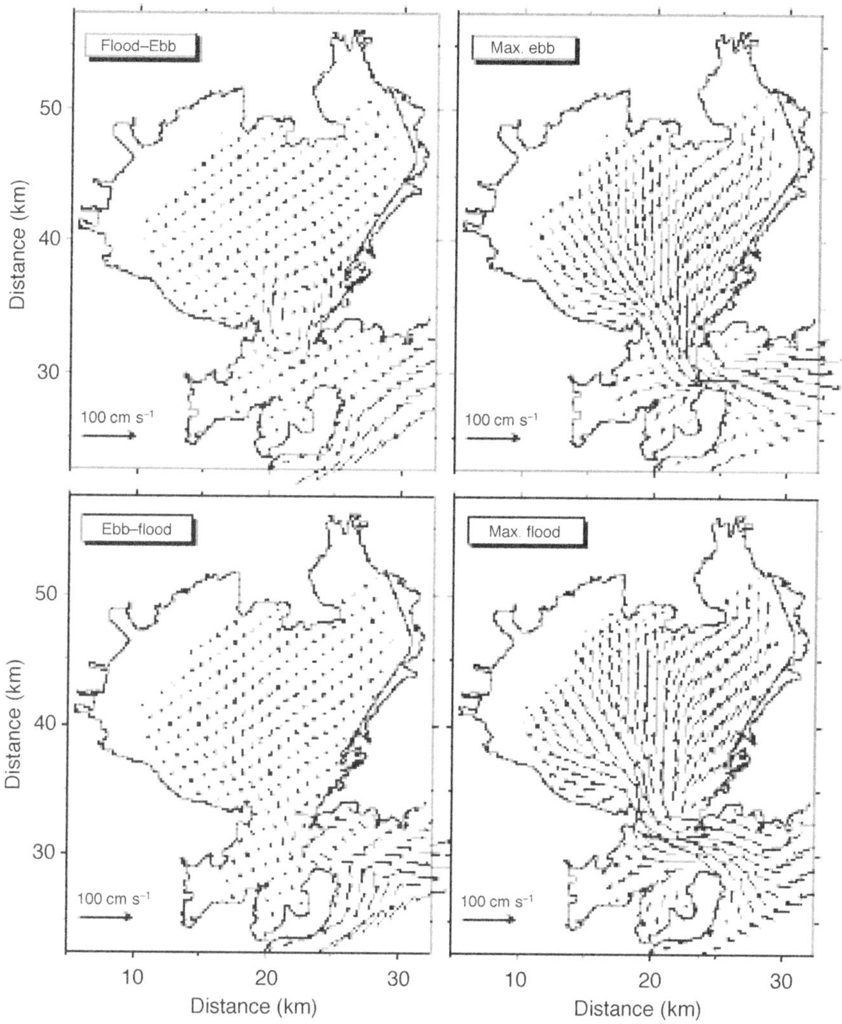

Fig. 14.3. Synoptic distributions of the surface tidal current vectors of the M_2 tide at the times of flood and ebb, with maximum ebb, ebb to flood and maximum flood.

near the southern entrance where the cool-water tongue from the YS was found (Fig. 14.4; Chen *et al.*, 1999).

Effects of aquaculture

Simulation experiments were made with two scallop stocking densities of 12 individuals m^{-3} (0.012 individuals l^{-1}) in the first case and 24 individuals m^{-3}

in the second case. In both cases, scallops dramatically decreased the concentrations of phytoplankton in the culture areas labelled A1–A3 in Fig. 14.2 (Chen *et al.*, 1999). The experiments suggested that the scallops would sharply reduce the concentration of phytoplankton to 0.99 μg chlorophyll *a* l^{-1} in A1, 0.52 μg chlorophyll *a* l^{-1} in A2, and 1.0 μg chlorophyll *a* l^{-1} in A3, about

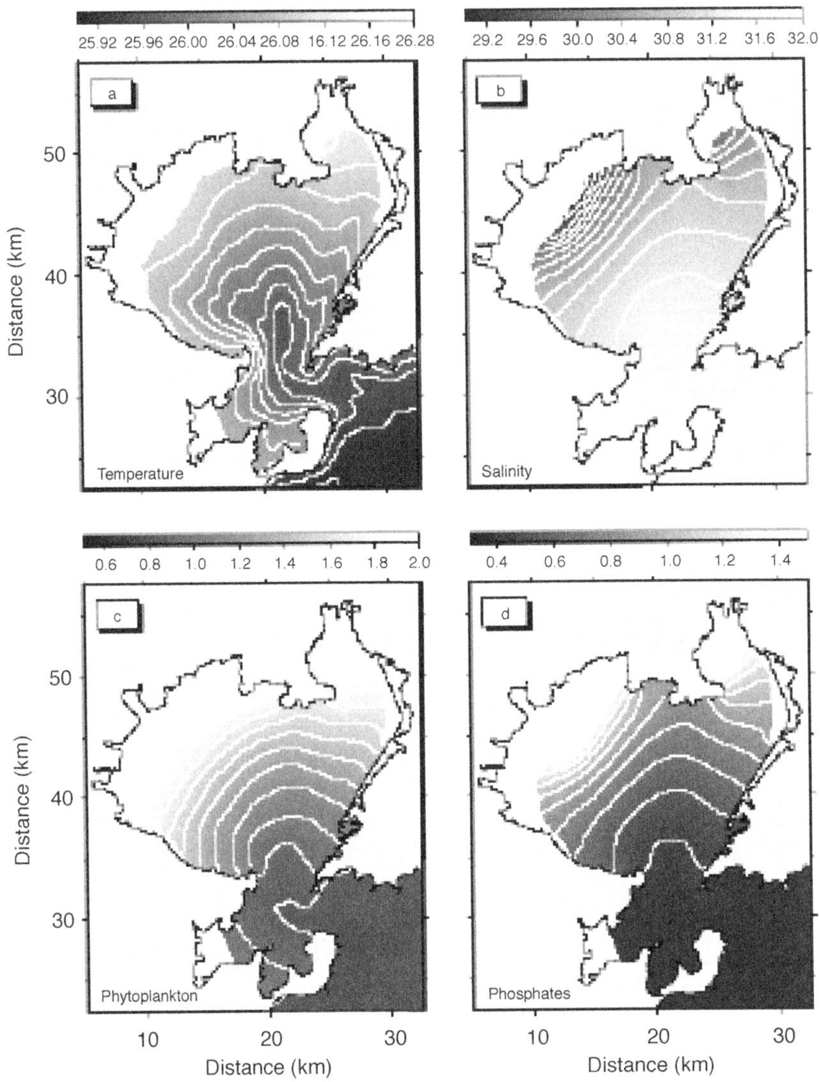

Fig. 14.4. Tidal averaged surface distributions of (a) temperature in °C, (b) salinity in parts per thousand, (c) phytoplankton in μg chlorophyll *a* l^{-1}, and (d) phosphate in μmol l^{-1} at the end of 20 model days for the case with tide and freshwater discharges.

31.8%, 33.3%, and 37.3%, respectively, lower than those in the case without shellfish (Fig. 14.5; Chen *et al.*, 1999). The response of phytoplankton to increased shellfish stocking densities was not linear. When the stocking density was doubled to 24 individuals m^{-3} the concentrations decreased to 0.71 μg chlorophyll *a* l^{-1} in A1, 0.38 μg chlorophyll *a* l^{-1} in A2, and 0.67 μg chlorophyll *a* l^{-1} in A3. These concentrations were ~51.1%, 50.8% and 55.6% lower than the case without shellfish, but only 19%, 18%, and 22% lower than the first case with shellfish (Fig. 14.5b; Chen *et al.*, 1999).

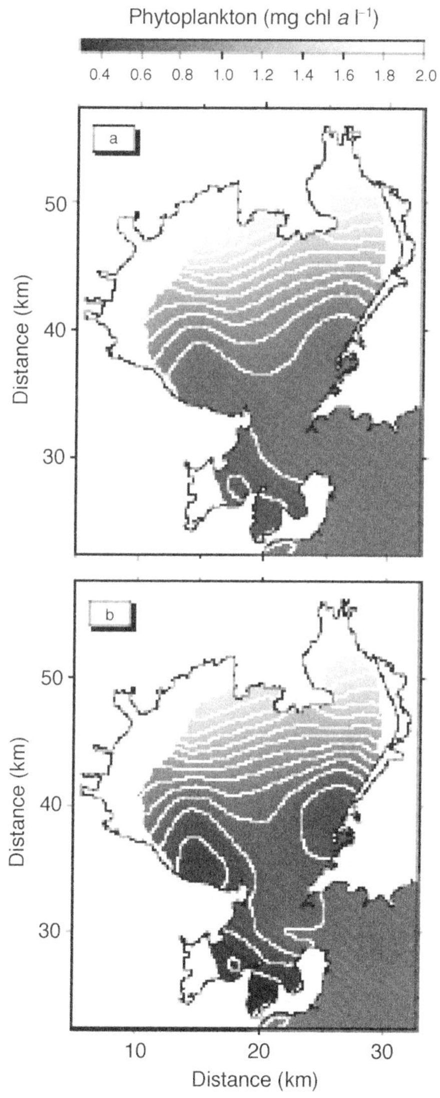

Fig. 14.5. Tidal cycle averaged surface distribution of phytoplankton after 20 model days for the case with shellfish culture densities of (a) 12 and (b) 24 individuals m^{-3}. Physical forcing factors are tide, freshwater discharge, and a southeasterly wind at 5 m s^{-1}.

The impact of scallop culture on the concentration of nutrients was very small, even at the higher stocking density. This is in contrast to previous studies indicating that shellfish have an important role in nutrient cycling and distribution (Dame, 1993). One explanation is that Chen's model did not consider the impact of the biodeposition process of shellfish. The biodeposition process could result in shellfish taking up small particulate organic matter and producing faeces and pseudofaeces that decompose into inorganic nutrients. Our model did include shellfish excretion, which was directly converted to phosphates. If the above explanation is correct, there should be significant modification of nutrient concentrations when the excretion rate or stocking density is increased. However, that was not the case in simulation experiments. Another possible explanation is that most of the phosphates in Jiaozhou Bay were the result of loading from the land and rivers. The recycling of nutrients by shellfish may directly influence the concentration of nitrogen but not of phosphates, or the nutrient regeneration rate may be orders of magnitude smaller than the nutrient loading rate from other sources.

Nutrient uptake and regeneration

To understand the roles of biological processes, the nutrient uptake and regeneration processes in the coupled model must be examined. Nutrient uptake in this modified NPZ model was controlled by the phytoplankton growth rate, incident radiance, the half-saturation constant, phytoplankton biomass and nutrient concentration. Nutrient regeneration was estimated by the sum of zooplankton excretion, death of phytoplankton and zooplankton, and shellfish excretion.

In the case with only tides and freshwater discharges, the model indicated that surface distributions of the uptake and regeneration rates were similar to nutrient distribution, which decreased from the inner to outer bay with the highest values on the northwestern and northern coasts. The maximum regeneration rate in the inner bay was ~2 μmol l^{-1} day^{-1}, which was ~5 times smaller than the maximum uptake rate. This suggests that the physical process associated with river discharge was a major source of nutrients for phytoplankton in the inner bay. Adding a southerly wind did not significantly change the distributions of nutrient uptake or regeneration. The uptake rate was relatively large in the inner bay along the coast as a result of nutrient accumulation by the wind-induced northwestward advection. This phenomenon may explain why a phytoplankton bloom occurs in the innermost bay during a southeasterly wind.

When consumption of phytoplankton by shellfish was included, distribution of nutrient uptake and regeneration were modified, particularly in the aquaculture areas. Nutrient uptake rates in the aquaculture sites dramatically decreased because the phytoplankton decreased. Yet, the nutrient regeneration rate in these sites increased as a result of shellfish metabolism. As a result, the distribution of phosphates was similar in the case with and without shellfish aquaculture, suggesting that the decrease in nutrient uptake rates due to phytoplankton consumption was almost compensated for by the physical processes of advection and diffusion.

One of the main interests in this study was to identify, qualify and quantify the roles of physical and biological processes in maintaining ecosystem health. In the simple NPZ food web, the physical processes included advection and diffusion, M_2 tide, river discharge and wind. Biological processes related to phytoplankton included nutrient uptake and regeneration, phytoplankton grazing and mortality, and shellfish consumption. To examine effects of freshwater discharge and shellfish aquaculture the net flux of nutrients and phytoplankton in five closed regions and across the outer strait were estimated. The flux was calculated based on tidally averaged values over the ten tidal cycles. When we say 'equilibrium state', that means the flux is balanced for a first-order approximation in which the biological field changes slowly but no steady state could be reached over the course of the study period. Sensitivity analysis of the phytoplankton revealed that the spatial distribution of phytoplankton remained unchanged, although the concentration varied remarkably with changes in parameters. This suggests that the model results for phytoplankton were robust.

The model results revealed that physical processes had a direct impact on temporal and spatial distributions of nutrients and phytoplankton as well as on shellfish aquaculture. Tidal mixing caused physical and biological variables to be well mixed vertically. The concentrations of nutrients and phytoplankton were high near the northwestern and northern coast near river sources but decreased from the inner bay to the outer bay. The model results suggested that prevailing river discharges and tidal mixing, and the southeasterly wind in the summer may cause unusual nutrient accumulation and lead to phytoplankton blooms in the innermost bay. The fact that a phytoplankton bloom can occur under a condition of southeasterly wind implies that physical processes may have a direct impact on the occurrence of red tide along the northern coast of Jiaozhou Bay. The overloading of nutrients from inland shrimp aquaculture, industries and other urban human activities caused a high nutrient concentration in the inner bay, which provided favourable conditions for eutrophication. Accumulation of nutrients due to the southeasterly wind speeds up the eutrophication process and causes the red tide.

The estimation of nutrient and phytoplankton fluxes in the five identified sites suggested that the nutrients were maintained by physical processes, while the phytoplankton was controlled predominantly by biological processes. Shellfish aquaculture tended to alter the entire Jiaozhou Bay ecosystem. The loss of phytoplankton in shellfish aquaculture sites was compensated for by nutrients that advected and diffused from surrounding waters. High levels of phytoplankton consumption also caused a net flux of phytoplankton into the bay from the YS, even though nutrients were advected out of the bay. In addition to eutrophication caused by human activities, high densities of suspended, feeding shellfish will alter the lower trophic food web by grazing phytoplankton, excretion and biodeposition. Aquaculture populations of bivalves tended to transfer large quantities of materials from the water column to the sediment, which can dramatically change the content of the organic matter in the benthic layer (Kasper *et al.*, 1985; Kautsky and Evans, 1987). The benthic processes, in turn, may alter nutrient cycling in the bay (Jorgensen, 1990; Barg, 1992; Dame, 1993). It also should be noted that this model did not include the intertidal zone. The direct impact of the intertidal process remains unclear, but a large quantity of nutrients can potentially be advected back to the bay during ebb tide.

Xincun Bay Modelling Experiment

Xincun Lagoon is located on the southeast coast in Lingshui County, Hainan Island (110°E, 18°25′N). The county population is more than 300,000, of which 158,000 are of the Li ethnic minority. The coastline is about 110 km long and about half the people live in the coastal townships (Marine and Fishery Department of Hainan, 1998). Xincun Bay has a gourd-shaped basin 6 km long and 4 km wide, covering an area of 21.97 km^2. Maximum depth is 10.6 m, and the bay is connected to the open sea by a single tidal inlet about 120 m wide (Fig. 14.6). People began to cultivate pearl oysters in the bay in

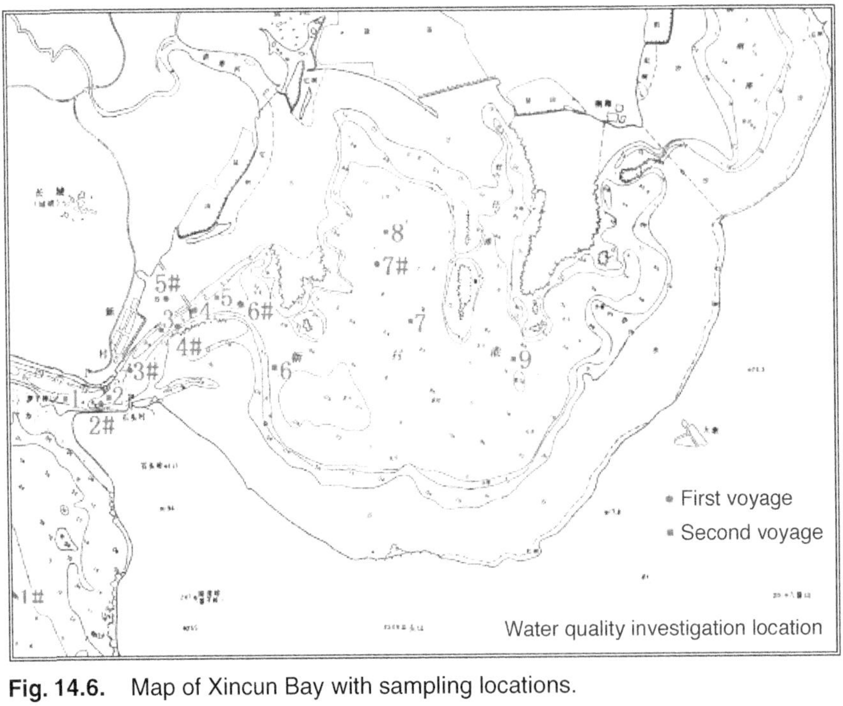

Fig. 14.6. Map of Xincun Bay with sampling locations.

the 1970s, and aquaculture has continued to grow in importance to the local economy. At present fish, shrimp, molluscan shellfish and the macroalga *Kappaphycus* sp. are cultured over an area of 160 ha in the bay. In 1997, the total income of aquaculture was about US$7 million. Lingshui County also has become an important mariculture base.

Aquaculture industry characteristics

The area used for coastal aquaculture in Hainan Province has increased at an annual rate of 9% for the past 10 years. It increased from 3500 ha in 1989 to more than 8200 ha in 1997. Total aquaculture production also increased at a rate of more than 25%. The increase in aquaculture area was at the expense of some coastal forests, wetlands and estuaries, and the intensification of aquaculture resulted in increased coastal pollution (Marine and Fishery Department of Hainan, 1998).

The aquaculture industry in Xincun Bay is dominated by the fish cage culture systems that are located near the bay's entrance and adjacent to the navigation channel. There are approximately 450 floating cage units consisting of 3 m × 3 m net cages, generally configured in a square with three cages per side. Each unit also includes a house with a family (four or five people) and one or two labourers. Several species are grown in the cages from fingerlings. Included are cobia, *Rachycentron canadum* (L.), local grouper and pompano. The fish are fed 2–5% of their body weight daily with ground whole fish from the trawl fishery and nearshore light-net fisheries.

Culture of *Kappaphycus* sp. is relatively new in Xincun Bay. *Kappaphycus* sp. is cultured on suspended lines (rakes) seasonally from October to April. These rakes occupy a significant portion of the middle area of the bay and are often mixed with pearl oyster culture rafts. *Kappaphycus* sp. segments are tied on to long ropes suspended horizontally within a metre of the surface. This culture system is used extensively in the Lian Lagoon, which is adjacent but not connected to Xincun Bay. In Xincun Bay, production of *Kappaphycus* sp. is increasing and in the 1998/99 and 1999/2000 seasons production was 1500 and 2000 tonnes, respectively.

Shrimp are cultured in 85 ha of ponds and production was about 4500 kg in 1999. Some of the aquaculture ponds were constructed at the expense of the coastal forest. To date, the major species produced was the Chinese shrimp, *Fenneropenaeus chinenis*, but initial success with the Pacific white shrimp *Litopenaeus vannamei* has prompted a shift to the latter species. Food conversions of 1.5 kg of feed to 1 kg of shrimp, as compared to a 1.8 : 1 conversion ratio for the former species, and disease outbreaks led to the species change.

Pearl oysters are cultured, but the primary bivalve molluscan production is clams. They are harvested from the intertidal zone out to 3 m depth in the middle portion of the bay. The other major biological components are the beds of seagrass, *Enhalus acoroides*, located in the middle and upper portions of the bay, and mangroves in the uppermost region (Marine and Fishery Department of Hainan, personal communication, 2000).

The history of aquaculture in Xincun Bay is one of boom and bust. Rapid expansion of the fish cage aquaculture industry in the 1990s exceeded the assimilative capacity to maintain the water quality in the bay. In 1997, water quality problems became very serious. As a result the cultured species began to grow slowly and the mortality rate increased dramatically. Fish cage culture declined from 200 and 230 ha in 1995 and 1996, respectively to 20 ha after the 1997 disaster. It increased to 33 ha in 1999. The number of hectares and fish cages was reduced and they are concentrated near the mouth, where tidal exchange is greatest. Shrimp production also collapsed from 100 ha in 1996 to 10 ha in 1997 and remained low (5 ha) through 1999 (Fig. 14.7). The shrimp ponds are in the extreme interior of the bay. Although environmental factors played a role in the mortalities in the shrimp ponds, diseases that ravaged shrimp production throughout China also may have been a major factor in the disastrous decline in production. After the two major aquaculture industries (fish and shrimp) collapsed, production of other aquaculture species increased and emerged as important economic and environmental factors. Pearl oyster culture expanded dramatically in 1998 then declined to 85 ha in 1999. Seaweed culture grew from 52 ha in 1998 to 133 ha in 1999 and increased to 160 ha in 2000.

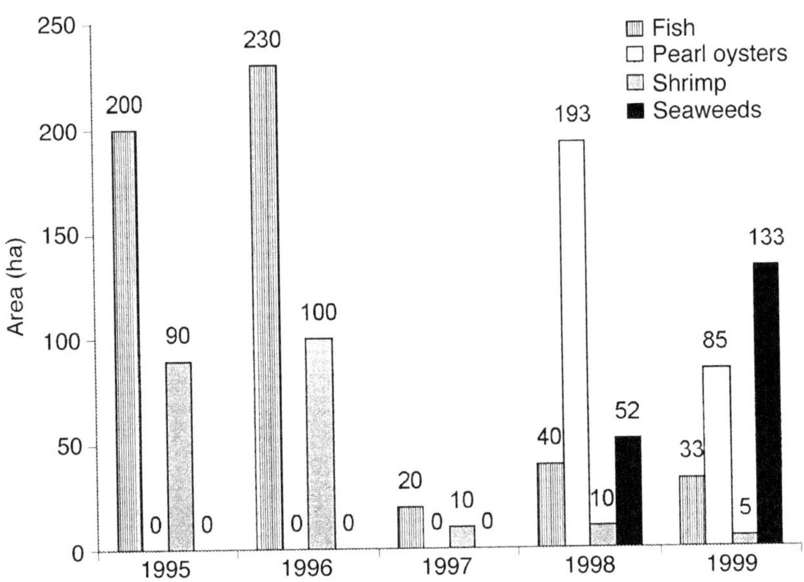

Fig. 14.7. Histogram of area utilized in the 1990s for aquaculture in Xincun Bay.

Ecosystem study and model of Xincun Bay

Observations

Environmental problems severely limit the sustainable development of Xincun Bay mariculture. To determine the severity of the existing environmental problems, two field surveys of Xincun Bay were conducted. From 31 July to 6 August 2000, two continuous stations and nine area stations were sampled (Fig. 14.6). For the continuous stations, the physical parameters monitored included temperature, salinity, DO, DO%, pH, current speed and direction, turbidity, and tidal level. Primary productivity, basic water quality and nutrients, chlorophyll *a*, ATP, chlorophyll *a*/ATP ratio, MPN (most probable number) total coliform organisms, MPN *E. coli* organisms, chemical oxygen demand (COD), suspended solids (SS), pH, turbidity, *ortho*-phosphate, nitrogen, NO_2^-, NO_3^-, NH_4^+, DO, temperature, salinity and bacterial communities were measured at the area stations. During the survey, pollutant sources were evaluated to understand the sewage discharge and its impact on water quality.

Macroalgal and seagrass samples were taken from up to four different stations in the bay. In order to evaluate their role as nutrient sinks in the ecosystem, the tissue carbon, nitrogen and phosphorus contents were analysed with a Perkin-Elmer CHN elemental analyser and by the method of Murphy and Riley (1962). Initial samples were taken in May and August for the macroalgae and seagrasses, respectively, along with the water sampling described above. Another sampling was carried out in November 2000 for both the macroalgae and seagrasses. The locations are shown on Fig. 14.6. Stations I, II and III were relatively close to the area of the bay where the fish cages are located, while station IV was in the upper part of the bay, distant from the cages.

Tidal characteristics and current

Twenty days of water level data analysis showed that the bay's tidal characteristics included an irregular diurnal tide with a small average tide range of 69 cm. The maximum tide range was only 1.55 m, which contributed to the slow tidal current velocity. In addition, the bay connects to the open sea via a 120-m-wide navigation channel, and water exchange takes place within a relatively small area adjacent to the channel.

The measurement results from two continuous current meters showed that tidal current velocity varied within a large range of 2 cm s^{-1} to 155 cm s^{-1} at the surface. At the mouth of the bay (station 1), the maximum surface velocity reached 155 cm s^{-1} during the spring ebb tide period on 31 July. During the neap tide period on 6 August, however, maximum velocity only reached 65 cm s^{-1}. At the upper end of the navigation channel (station 2, 7 m depth), the maximum velocity was relatively stable at 43 cm s^{-1} and 40 cm s^{-1} in the corresponding periods. In the middle of the bay, the velocity dropped off quickly to an average value of 10 cm s^{-1}. Station 2 data also showed that the ebb velocity values were larger than the flood velocities. An estimated water flux of 6×10^7 m^3 flowed in and out of the bay during one spring tidal cycle, which is one-thirteenth of the bay's total volume (8×10^7 m^3). It would seem that the bay's water could exchange completely within half a month, but an earlier study indicated the bay has a 90-day water exchange rate. This slow dynamic exchange leads us to conclude that Xincun Bay has a low capacity for transporting pollutants to the open sea. The present water quality environment is consistent with this fact. The pollutant sources, especially the cage culture, add to the oxygen demand by introducing residual feed and fish waste, but these are not the only sources of organic pollution. Domestic sewage and the fishing fleet also contribute to the eutrophication of the bay.

Dissolved oxygen

The dissolved oxygen concentration (DO) is an important indicator of water quality for marine life. Although it is difficult to determine the DO production and consumption mechanisms precisely, the basic trend can be understood by monitoring the degree to which the water is being polluted. The vertical distribution of DO is particularly important in order to determine the extent to which an area is polluted. The vertical characteristics of DO at different stations located from the outside bay (station 1) to the middle bay (station 7) during the neap tide period are shown in Fig. 14.8. DO was lower than 6 mg l^{-1} at stations 2–6 in all layers. At area stations 2, 4, 5 and 6, the DO was less than 5 mg l^{-1}, and the average value was only 4.8 mg l^{-1}. Vertically, DO declined sharply and reached a minimum at 2–3 m in the fish cage culture region. On the other hand, vertical DO distribution was different at station 1 outside the bay, where it increased with depth. DO at the surface was 5.04 mg l^{-1}, very similar to that in stations 2–6. At station 1 DO was 6 mg l^{-1} at 2 m. We speculate that this phenomenon is the result of the surface water outflow from the bay. At station 7, where there are no fish cages, water quality was better. DO was 6 mg l^{-1} at the surface, but at depths > 3 m, the DO value was below 5 mg l^{-1}. The distribution of DO at all stations was not desirable for fish cage culture. The cages used in Xincun Bay are 3 m × 3 m, and are submerged to a depth of 1–4 m. Where fish are cultured at high density, respiratory difficulties for fish may result when DO is low.

A clear distribution of DO can be seen along the survey profile (Fig. 14.8). DO concentrations were high outside the bay and in the inner bay regions. In the navigational channel region, DO concentrations were low at greater depths. This is the region where fish cage culture is concentrated. From the above analysis, we can draw the basic conclusion that DO concentration was relatively low in fish cage regions.

When the relationship between salinity and DO saturation was analysed, we found that DO saturation has a concave relationship with salinity, reflecting the continuum from the sea to the location of the fish cages (Fig. 14.9). When salinity was less than 31 ppt, DO approached saturation. Planktonic algal populations were larger within a salinity range of 28–31 ppt, but the data

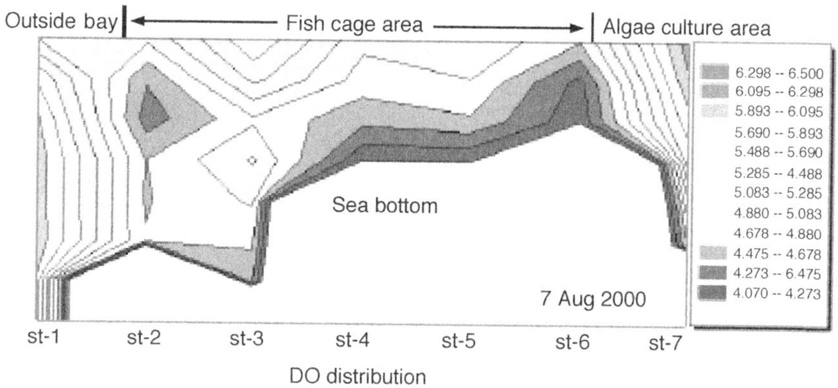

Fig. 14.8. Dissolved oxygen saturation and salinity in Xincun Bay.

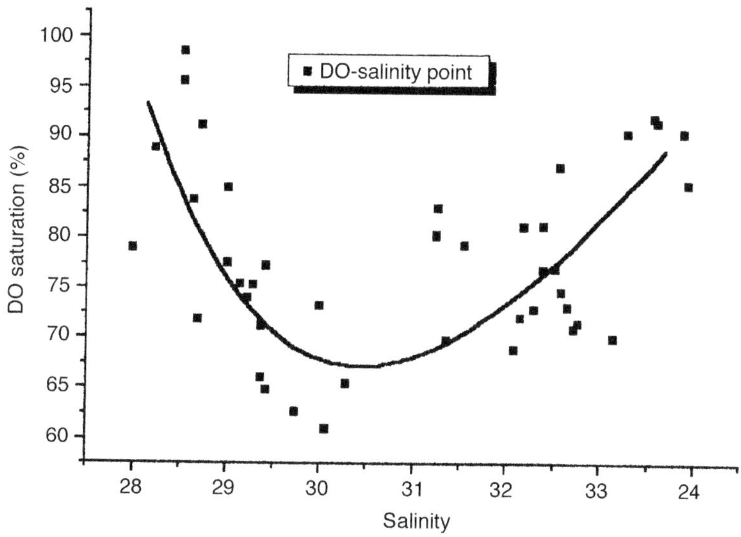

Fig. 14.9. Relationship between oxygen saturation and salinity in Xincun Bay

at station 7 indicate high DO saturation at the surface. It is assumed that larger algal populations in the middle region of the bay produced O_2 in the surface layer and/or the BOD or COD was greater in the cage culture area.

Nutrients

The distribution of nitrates (Fig. 14.10) was characterized by their spatial distribution in the outside bay, tidal channel and inner bay. The nitrate concentrations exhibited a peak of about 5.97 $\mu g\ l^{-1}$ within the channel region near the fish cages, where high levels of organic matter fall to the bottom. In both the outside bay and inner bay regions the nitrate concentration was fairly low with a value of 0.83–0.70 $\mu g\ l^{-1}$.

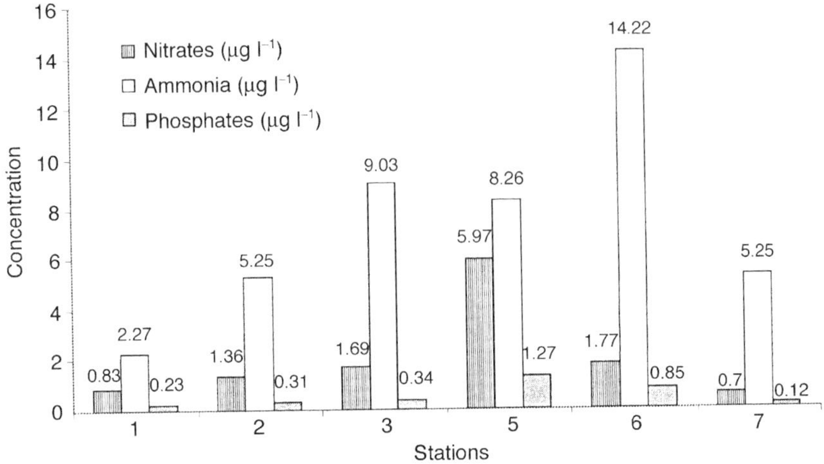

Fig. 14.10. Nutrient distributions in Xincun Bay.

Ammonium concentration had a similar distribution to nitrates in Xincun Bay. It decreased rapidly from the cage fish culture region to the outside bay. Ammonium concentrations exhibited a peak of about 14.22 μg l^{-1} in station 6 in the cage culture region, and declined sharply to 0.09 μg l^{-1} in the inner bay where seagrass beds were dominant.

The phosphorus concentrations were commonly lower than the nitrate and ammonium concentrations in the water column. Phosphorus concentrations were below 0.5 μg l^{-1} at five of the seven stations. The maximum phosphorus concentration reached 1.27 μg l^{-1} at station 5 in the fish cage region.

The spatial and temporal distributions of nutrients are reflected in the DO distribution in the water column. These biochemical characteristics are associated with the aquaculture activities in Xincun Bay. Within the fish cage area, substantial quantities of organic matter accumulate in the sediment owing to uneaten food and fish excrement. Sediment oxygen consumption increases as a result of chemical oxidation, activity of benthic organisms and bacterial decomposition of organic matter. The oxygen in the water column above the sediments can become depleted, leading to anoxic conditions. When oxygen above the sediments is depleted, nitrogen and phosphorus may be released into the water column more readily. Phosphorus is released under anoxic, reducing conditions, whereas it normally complexes with oxidized iron and becomes immobilized. Previous research has examined these relationships between nutrients, dissolved oxygen and sediment oxygen consumption in aquatic systems (Stumm, 1973; Frevert, 1980; Nixon, 1982).

Phytoplankton requirements for nutrients such as inorganic nitrogen (IN) and inorganic phosphate (IP) play important roles in controlling the growth of phytoplankton and primary productivity. The ratio of IN/IP acts as an important indicator of eutrophication. In cultivation areas, the IN/IP changed

Table 14.2. Inorganic nitrogen (IN) in μg l^{-1}, inorganic phosphorus (IP) in μg l^{-1}, IN/IP ratio, and the comprehensive assessment parameter (E) on three dates in 2000.

		Station								
Date	Parameter	1	2	3	4	5	6	7	8	9
June 28	IN	1.68	2.52	2.72	3.78	3.02	3.64	–	–	–
	IP	0.1	0.09	0.14	0.21	0.19	0.19	–	–	–
	IN/IP	16.7	28.1	19.5	18.0	15.9	19.2	–	–	–
	$E = 10^{-3}$	–	–	–	–	–	–	–	–	–
July 31	IN	0.29	1.07	1.8	2.84	2.38	4.53	0.8	0.58	4.27
	IP	0.03	0.13	0.15	0.06	0.04	0.05	0.06	0.06	0.05
	IN/IP	9.7	8.2	12.0	47.3	59.5	90.6	13.3	9.7	85.5
	$E = 10^{-5}$	0.092	1.3	2.5	1.8	1.8	6.2	1.4	0.9	6.3
August 7	IN	3.2	6.81		10.99	15.21	16.3	0.89	–	–
	IP	0.23	0.31		0.34	1.27	0.85	0.12	–	–
	IN/IP	13.9	22.0		32.2	12.0	19.2	7.4	–	–
	$E = 10^{-3}$	0.18	1.1	–	2.2	6.2	4.8	0.06	–	–

due to the effects of cultured species. In scallop culture regions, the ratio of IN/IP was usually high, but the ratio was small in the brown alga, *Laminaria*, culture region (Song Yunli, 1996, personal communication).

Table 14.2 shows that the value of IN/IP always remained high inside the bay (stations 1–6 on 28 June; stations 3–7 and 9 on 31 July; stations 2–6 on 7 August). However, on 31 July 2000, it remained normal in the channel region (stations 1–2), due to the influence of the flood tide. This characteristic distribution of the IN/IP ratio presumably was related to the large area of fish cage culture in the channel region.

The nutrient state of the bay can be assessed by a comprehensive assessment parameter E:

$$\left[E = \frac{\text{COD}(\text{mg}^{-1}) \times \text{IN}(\mu\text{g l}^{-1})\,\text{IP}(\mu\text{g l}^{-1})}{1500} \right]$$

when $E \geq 1$ the water has an over-abundance of nutrients. The results show that the values of E are generally small (Table 14.2), indicating that nutrients do not exceed the critical level of eutrophication in the fish cage region. This is the reason for low chlorophyll *a* (< 0.5 μg l^{-1}) and low primary productivity at all stations sampled in Xincun Bay.

Water quality and sediment assessment

Water quality and sediment survey results are shown in Tables 14.3 and 14.4. Data from 14 stations were collected during four cruises on 10 September, 28 June, 31 July and 7 August 2000. The DO concentration was 21% lower than the Class II water quality standard, while other water quality parameters

Table 14.3. Water quality data.

Parameter	Maximum	Minimum	Average
Temperature (°C)	30.9	23.6	28.9
Salinity (‰)	32.8	28.6	31.8
DO (mg l^{-1})	6.67	3.92	5.12
DO saturation (%)	97	61	87
pH	8.35	8.13	8.26
PO_4^- (µg l^{-1})	3.20	1.60	3.72
NO_2^- (µg l^{-1})	1.12	0.14	0.74
NO_3^- (µg l^{-1})	34.3	4.2	10.4
NH_4^+ (µg l^{-1})	165.2	0.56	28.7
Inorganic N (µg l^{-1})	176.4	4.9	41.0
Silicate (µg l^{-1})	865.2	77.8	278.3
COD (mg l^{-1})	1.02	0.28	0.57
Suspended solids (mg l^{-1})	5.5	0.2	1.84
E. coli MPN (individuals l^{-1})	1700	< 20	156
Bacteria (µg l^{-1})	140	0.10	17.4
Chlorophyll *a* (µg l^{-1})	1.92	0.49	1.24

Table 14.4. Sediment characteristics at four sampling stations.

Station	Sediment type	Sediment colour	Sulphate (mg kg^{-1})	Organic carbon (%)	Organic matter (%)
1	Silt/sand	Black	348.04	4.5	7.8
2	Sand	Grey	330.02	0.3	0.5
3	Silty	Grey black	454.53	1.6	2.8
4	Silty	Grey black	326.99	0.7	1.2

were normal (*GB-99*, National water quality standard, P.R. China 2000). Low DO concentration can be the result of sediment oxygen demand and biological activity in the water column. The sulphate and the organic matter are largely in the sediments. Large organic particles settling in sediment cause a serious water quality problem. The decay of sediment will cause a high demand for oxygen from the overlying water column. This demand may excessively stress the oxygen resources of overlying water and deplete the dissolved oxygen concentration. It is estimated that sediment oxygen demand and the biological decomposition in the water column are the major reasons for low DO concentrations in the Xincun Bay aquatic system.

Macroalgal and seagrass nutrient content

Table 14.5 shows the results for the nutrient content in *Kappaphycus alvarezii* for the May and November sampling periods. The macroalgal samples containing the higher tissue nitrogen and carbon contents were those collected in the

Table 14.5. Tissue nitrogen, carbon and phosphorus content (dry weight percentage) in *Kappaphycus alvarezii* in Xincun Bay during May and November.

Month	Station	Nitrogen	Carbon	Phosphorus
May	III	2.69	28.64	0.11
November	I	1.69	30.64	0.15
	II	1.50	28.61	0.14
	III	1.60	31.69	0.19
	IV	1.21	27.28	0.22

areas close to the fish cages. These areas had an average nitrogen content of 1.6% dry weight (DW) and 30% DW for carbon, compared to the lower values of those two nutrients in specimens from station IV at the head of the bay. On the other hand, internal phosphorus content followed the inverse pattern, since plants at station IV were enriched in that nutrient (0.22% DW). These results reveal that seaweeds are active nutrient scrubbers, which also has an impact on their biomass. Considering the average seaweed production between 1999 and 2000 of 2000 tonnes, the calculated potential nutrient removal by these primary producers would be 28.8 tonnes for nitrogen and 3.66 tonnes for phosphorus for the November sampling period. However, if we based our estimations on the May collections, then the amount of nutrient removal for nitrogen would be 53.8 tonnes and 2.24 tonnes for phosphorus. Obviously, the nutrient removal capacity of *Kappaphycus* is substantial, but does appear to vary seasonally. Other reports have documented how other red seaweed species can act as biofilters; removing nutrients efficiently from fish farm effluents: Kautsky *et al.* (1996) integrated the culture of *Gracilaria* and salmon aquaculture in Chile, reducing the ecological footprint for nitrogen and phosphorus assimilation by 56% and 94%, respectively. Similarly, Chopin *et al.* (1999) reported high values of phosphorus and nitrogen in *Porphyra* grown close to salmon cages in Cobscook Bay, Maine, USA.

Nutrient contents in seagrass tissue (leaves, roots and rhizome) were also determined when material was available, and are reported in Table 14.6. The pattern of tissue nitrogen content indicated different strategies of nitrogen storage and subsequent use, which has been demonstrated in other seagrasses (Kraemer and Mazzella, 1999). Leaf samples collected in August showed higher nitrogen content than those collected in November, but lower carbon and phosphorus levels, coinciding with the peak of fish production in the Bay. It is noteworthy that *Enhalus acoroides* presented high values of nitrogen in leaves and rhizome tissues, even higher than those found in *K. alvarezii* in the same area, as well as higher phosphorus contents in leaves and rhizomes, and higher carbon contents in all three parts. Previous studies showed that the nitrogen uptake and assimilation by leaves can be more important than acquisition by roots (Hemminga *et al.*, 1991; Kraemer and Mazzella, 1996). Therefore, leaves and rhizomes appear to have a nutrient (mainly nitrogen) storage function

Table 14.6. Tissue nutrient content in leaves, roots and rhizomes (dry weight percentage) of *Enhalus acoroides* in Xincun Bay.

		Tissue		
Date	Station	Leaves	Roots	Rhizomes
Nitrogen				
August	III	3.97	0.82	–
November	I	2.13	0.77	2.13
	II	2.42	0.88	1.47
Carbon				
August	II	28.72	39.04	–
November	I	32.43	35.78	32.03
	II	38.22	34.43	34.40
Phosphorus				
August	II	0.17	–	–
November	I	0.48	–	0.91
	II	0.48	–	0.85

and may have a significant role in the ecosystem as a sink/source of nutrients, which must be recognized in any modelling effort.

Dissolved oxygen modelling

A three-dimensional conventional water quality analysis simulation model, which was originally developed by Ambrose *et al.* (1993) and known as WASP5, was modified and used to study the DO in Xincun lagoon. The equations solved by WASP5 are based upon the key principle of conservation of mass. The equations include three major components: the advection and dispersion of transport, the kinetic interaction and transformation, and external loading. The WASP5 eutrophication water quality model considered eight water quality state variables and used the kinetic framework developed by Di Toro *et al.* (1971). In our water quality model, we only consider DO, NH_3, NO_3^- and their major kinetics. This is based on evidence that low DO is the major problem of eutrophication and the nitrate and ammonia concentrations are relatively large within the fish cage region in Xincun lagoon.

The dissolved oxygen cycle

The two major components of the hypolimnetic oxygen depletion are water column oxygen demand (WOD) and sediment oxygen demand (SOD). WOD embraces the biological and chemical oxygen demand primarily due to algal, bacterial and fish respiration. SOD reflects utilization of DO from overlying and interstitial water of the sediments by biological and chemical oxygen demands. As discussed above, DO, which acts as a eutrophication indicator, is predominantly low within the fish culture region, where much of the organic matter deposits to the sediment. With low chlorophyll *a*, low primary productivity, even low nutrients in the water column, but large organic matter in sediments, SOD is the major reason for depletion of DO, rather than WOD. For this reason, we considered SOD to be an important sink of DO in our water quality model. Macroalgal and seagrass photosynthesis and respiration were also considered in the model, owing to the fact that they remove nutrients from the water column. Thus, the DO kinetic interaction and transformation includes processes of reaeration, macroalgal (*Kappaphycus alvarezii*) and seagrass (*Enhalus acoroides*) photosynthesis and respiration, cage fish respiration, nitrification and SOD.

The ammonium and nitrate cycle

The nitrogen cycle (only ammonia and nitrate are considered) is simple in our model. The kinetic processes include the macroalgal and seagrass uptake, nitrification, denitrification and the benthic fluxes.

Determination of the major sinks/sources in the model

Sediment oxygen demand

Factors that affect SOD are rather complex. They include temperature, oxygen concentration, make-up of the biological community, organic and physical characteristics of the sediments, current velocity and chemistry of the sediment–water interface. Velocity effects on SOD were not due to physical resuspension of bottom material to the overlying water. The measurement of SOD showed that it changes in spatial and temporal variation owing to different situations (Hickey, 1984; Whittemore, 1984). In some estuaries, SOD is responsible for 40–50% of the total oxygen uptake (James, 1974).

In water quality models, some studies estimate the SOD as an empirical function of site characteristics such as the sediment depth, sediment chemical or physical properties, biological parameters, or overlying water quality. WASP simulates SOD as a function of the net settling velocity of particulates including algae, first-order reaction rates of particulates and dissolved nutrients and organic material including algal decomposition, and the diffusive exchange rate between dissolved concentration in the sediment interstitial water and the overlying water column (Di Toro *et al.*, 1971). Walker and Snodgrass (1986) defined SOD as the temperature-adjusted rate with DO in linear formula, the temperature-adjusted rate written in Van't Hoff form. In our water quality model, SOD is expressed as:

$$[SOD = K_{Tr}\theta^{(T-Tr)}]$$

where K_T is the rate at a reference temperature, and θ is the temperature-adjusted rate. K_T, determined according to the *in situ* study domain, ranges between 0.05 and 3.4 g O_2 m^{-2} day^{-1} in Xincun lagoon.

Nutrient fluxes

Within aquatic systems sediment nutrient fluxes play an important role in nutrient concentrations in the water column. Studies indicate that nitrogen and phosphorus may be released to the water column more readily when oxygen above the sediments is depleted (Smith and Fisher, 1986). Within the large SOD region, sediments were the dominant sources of ammonium and phosphorus (Fisher *et al.*, 1982; Boynton and Kemp, 1985). The processes of nutrient fluxes in sediments are complex. It is necessary to calibrate by the *in situ* measurement. In our model, we simply consider nutrient fluxes of sediment as the linear relationship with SOD and bottom velocity. The ammonium fluxes and nitrate fluxes ranged between 60 and 600 μmol m^{-2} h^{-1} and 3 and 39 μmol m^{-2} h^{-1}, respectively, in the Xincun water quality model.

The water quality model is coupled with the physical model developed by Blumberg and Mellor (1987). The physical model runs prognostic problems with tidal oscillations, along with wind-driven and rainfall water discharge. When the physical model reaches the equilibrium state, the water quality parameters are added and run with the cases of: (i) 450 fish cages in the channel region (Area 1 in Fig. 14.11) and without the effect of macroalgal culture; (ii) double the fish cages and without the effect of macroalgal culture; and (iii) double the fish cages and consider the macroalgal culture (Area 2 and Area 3 in Fig. 14.11).

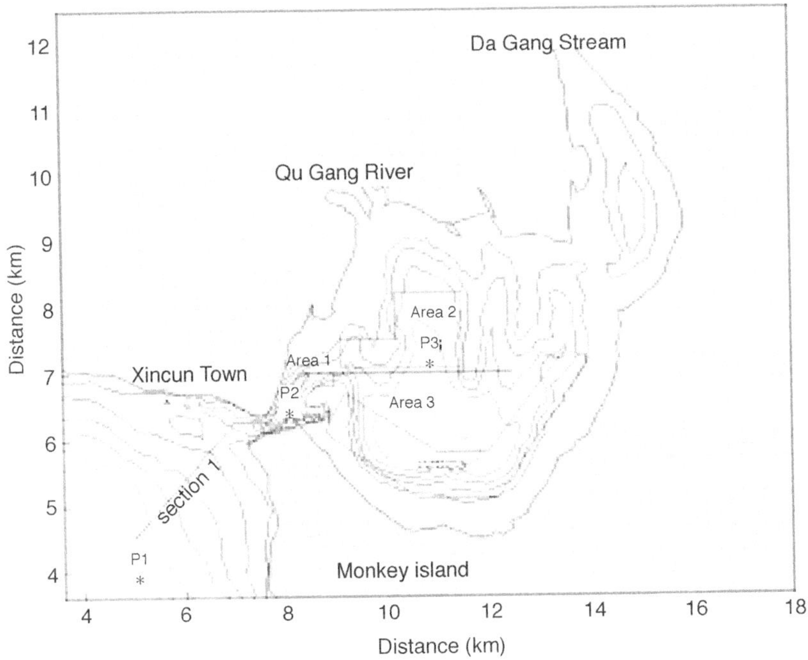

Fig. 14.11. Location of fish cages and macroalgal culture regions at selected points (P1, P2 and P3) in Section 1 in Xincun Bay.

Model results

The model was initialized by the investigation temperature and salinity field on 28 July. The model demonstrated that the temperature and the salinity remain stratified in Xincun lagoon. Tidal cycle average surface temperature and salinity inside the bay is uniform with a magnitude of 30.8°C and 31.8 ppt, but at the bottom, temperature is low and salinity high in the deepest part of the bay. Outside the bay entrance, a strong gradient of temperature and salinity is demonstrated by the effect of flood and ebb tidal current.

A uniform DO, NH_4^+, NO_3^- field was initialized after the temperature and salinity field was adjusted. Figure 14.12 shows the spatial and temporal distribution of DO concentration in cases i and ii (described in the preceding subsection). The distribution of DO is closely related to the tide. During flood tide (upper portion of Fig. 14.12), the minimum DO concentration centre occurs

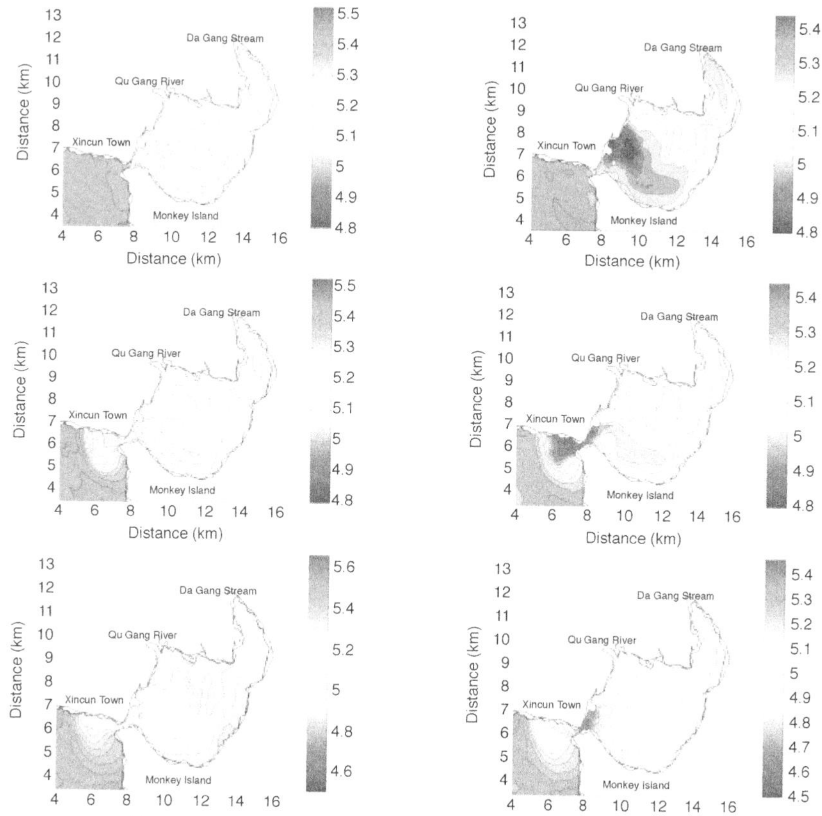

Fig. 14.12. DO concentration distribution during the flood tide (upper pair), ebb tide (middle pair), and flood–ebb tide (lower pair) under existing conditions (left) and with a 50% increase in the number of fish cages (right) in Xincun Bay.

not in the channel fish culture region, but in the inner bay. During ebb tide, the minimum DO occurs outside of the region, near the entrance of the bay only during the middle flood–ebb tide. The lowest DO concentrations occur in the cage culture region. In cases 1 and 2, the lowest DO concentrations are 5.0 mg l^{-1} and 4.5 mg l^{-1}, respectively. The experiments demonstrate that the average DO concentration is reduced by 10% when the fish cages are doubled. It seems that the effect of caged fish is not sensitive to the DO concentration. The response of this process indicates that strong tidal currents in the channel region enhance mixing with outside bay water and increase DO concentration.

Figure 14.13 shows the DO concentration distribution in Area 1 (Fig. 14.11). It clearly demonstrates the effect on DO concentration by SOD, with a stratified structure in the channel region. Large SOD in the fish cage region consumes dissolved oxygen and leads to DO depletion.

Figure 14.14 shows the spatial and temporal distribution of surface DO concentration when macroalgal culture is considered in Areas 2 and 3. Although macroalgae seem not to enhance DO concentration in the fish cage region, within the macroalgal culture regions the DO concentration is improved by macroalgal photosynthesis and nutrient uptake. Figure 14.15 shows the change of DO concentration at selected points outside the bay (P1), in the fish cage region (P2) and in the inner bay (P3) during three tidal

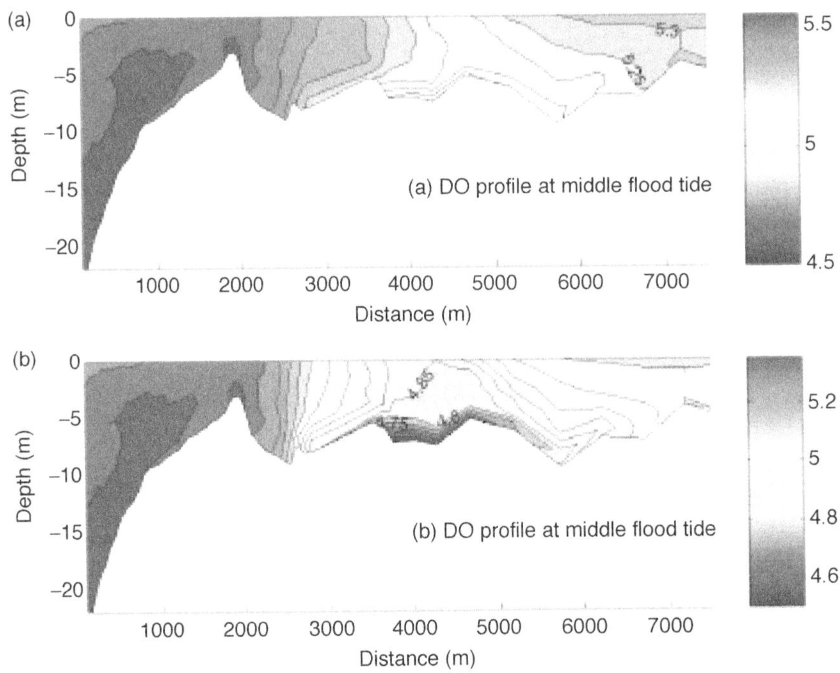

Fig. 14.13. DO profile in a selected section at present (a) and with a doubling of the number of fish cages (b) during mid-ebb tide in Xincun Bay.

cycles. We can see that the change in DO concentration in these three regions coincides with tidal elevation (tidal current). In P2, lowest DO concentration occurs before low tide with a magnitude of 4.2 mg l^{-1} at the bottom; highest DO concentration occurs at the beginning of the flood with a magnitude of 5.5 mg l^{-1}, as at P1. But at P3, the surface DO concentration seems unchanged, while the bottom DO concentration declines to 4.8 mg l^{-1}.

Discussion on the Xincun Bay case

The results showed that the water column in the Xincun Bay aquaculture region was heavily polluted by organic material and, according to sediment sample analyses, was also high in silicate. Except for DO, sampled chemical parameters and nutrients did not exceed the national water quality standards Class I. The data were also evaluated according to Criteria for Surface Water Quality Classifications, Class II (Florida EPA). Although the water generally met the standards, the environmental health of the caged fish area and navigation channel region has declined considerably as evidenced by the mortality of cultured fish and low DO.

Two kinds of pollutants affect water and sediment quality in Xincun Bay. One source is the pollutants produced by four factories, four restaurants, seven gasoline stations and an estimated 481 tonnes of COD from sewage discharge. The other source is a by-product of fish cage operations, which results in

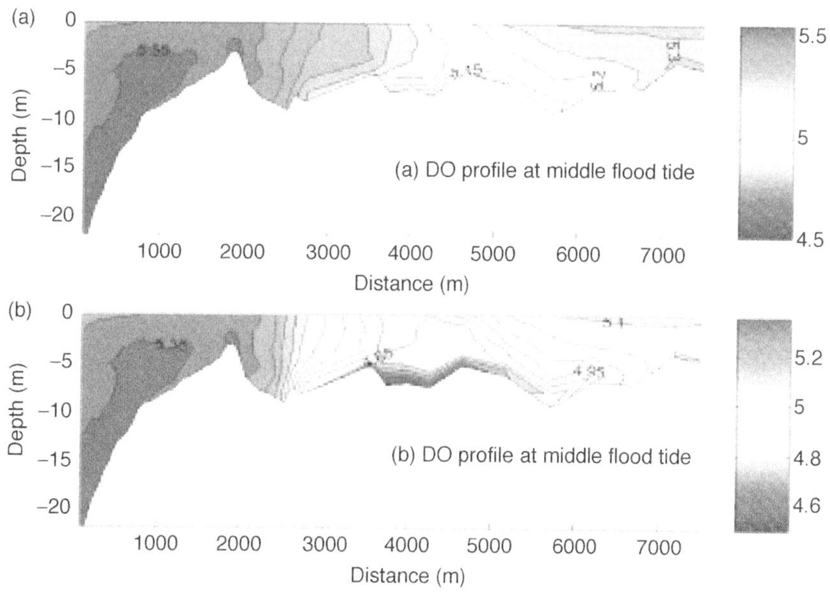

Fig. 14.14. The DO distribution in case iii with the DO at flood tide (left) and at ebb tide (right) in Xincun Bay.

an estimated 5000 tonnes of organic pollutants annually. Water quality sampling indicated that DO is always lower than the value of national water quality standards in fish culture areas. The results show that the main source of pollution is the fish cages. Large amounts of uneaten food and faeces descend to the bottom under the cages. The sediments release NH_3, H_2S and other pollutants by degradation of the organic matter by bacteria. These chemical reactions require large quantities of oxygen and reduce the DO substantially in both the sediments and the water column. The results of DO concentration modelling support that conclusion. The model experiments also show that average DO concentration will decrease by about 10% when fish cages are doubled. This result shows the effect of strong tidal current mixing and transport. Macroalgal culture increases the DO concentration in the culture region, but has less contribution to the DO concentration in the cage fish region.

The assessment of nutrient status showed that inorganic phosphorus is low relative to other nutrients in Xincun Bay, which are also generally present at low levels. This is frequently the case for most aquaculture of *Kappaphycus* sp. cultured in oligotrophic, tropical or subtropical regions. Moreover, nutrient requirements vary greatly between species. The point is that with integrated aquaculture, potential nutrient limitation is alleviated by moving seaweeds closer to the source of nutrients. Also, the internal nitrogen and phosphorus tissue contents in seaweeds and seagrasses in the bay suggest that these producers can act as nutrient scrubbers from water column and sediments,

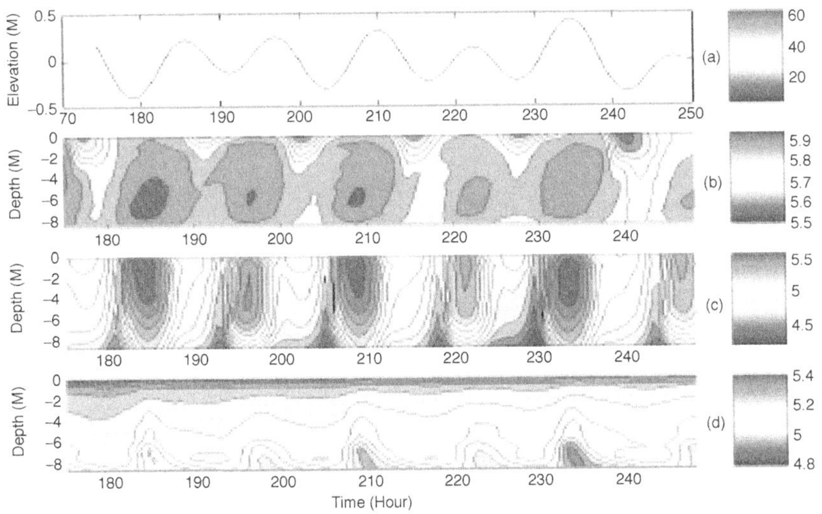

Fig. 14.15. The DO distribution within three tidal cycles at the selected points: (a) is the elevation at the channel; (b) selected point (P1) outside the bay; (c) selected point (P2) in the channel region; (d) selected point (P3) in the inner portion of Xincun Bay.

thus playing an important role in the ecosystem by reducing the risk of other less desirable algal blooms.

Summary

The management of aquaculture in embayments is influenced by numerous physical, chemical and biological factors, many of which cannot be controlled by the aquaculturist alone. It is the responsibility of the resource managers to maintain and improve the environmental situation in the bay for multiple uses, including aquaculture. Balancing the uses of the bay is not a simple task. It calls for advanced technological tools, such as three-dimensional modelling, to inform and integrate the process of coastal management. This chapter presented two examples of how integrated ecosystem models can enhance the understanding of the processes affecting water quality.

Jiaozhou Bay is a relatively large bay (~400 km^2) adjacent to a major city, Qingdao, and has numerous sources of nutrient inputs. In Jiaozhou Bay, tides and southeast winds dominated the summer distribution of nutrients, phytoplankton and zooplankton, but the basic productivity in this phosphorus-limited ecosystem could be altered in the scallop aquaculture areas. Simulation experiments made at scallop stocking densities of 12 and 24 individuals m^{-3} illustrated the potential impact of stocking rate on phytoplankton. In both cases, scallops grazing in suspended rake culture dramatically decreased the concentrations of phytoplankton in the culture areas. However, the potential impact of scallop culture on the concentrations of nutrients was very small, which contrasts with previous studies (Dame, 1993). One explanation is that the model did not consider the impact of biodeposition from the shellfish. Biodeposition could result in shellfish taking up small particulate organic matter and producing faeces and pseudofaeces that decompose into organic nutrients. This model did include shellfish excretion that was converted directly to phosphates. Another explanation is that most of the phosphates in Jiaozhou Bay were the result of loading from the land and rivers. The recycling of nutrients by shellfish may directly influence the concentration of nitrogen but not phosphates, or the nutrient regeneration rate may be orders of magnitude smaller than the nutrient loading rate from other sources. The implication for the coastal manager is that the scallop rakes in Jiaozhou Bay had little impact on the water quality of this large bay, other than to reduce phytoplankton in the aquaculture areas. Importantly, the model did point to the danger of the nutrients accumulating in the shallow northwest quadrant of the bay, where they could contribute to local blooms and red tides.

In Xincun Bay on Hainan Island, the situation is very different. Xincun Bay is a shallow subtropical lagoon of about 22 km^2 with a 120-m-wide inlet. Macroalgae, pearl oysters, shrimp and fish are cultured in the lagoon. The fish pen culture site near the navigation channel is the most economically and environmentally important aquaculture activity in the lagoon. Fish pen culture units that occupied 230 ha of the lagoon in 1996 declined to 33 ha in 1999 after disastrous losses. Results of this study indicated that organic matter heavily polluted the water in the fish pen areas. Still, except for DO, chemical parameters and nutrients did not exceed China's national water quality standard Class I, nor would they have exceeded Florida's (USA) water quality standards. Although the water generally met the standards, the environmental health in the cage culture area and navigation channel is declining as evidenced by the fish deaths and low DO. Two types of pollutants influence water and sediment quality in the lagoon. Organic pollutants from the town of Xincun help reduce the DO, but the main source of organic pollution is from the fish cages. Uneaten food and faeces decay and fall to the bottom. As a result, the sediments have become anaerobic and release NH_3, H_2S and other pollutants. Simulation experiments using the three-dimensional model predicted that the average DO concentration would decrease by 10% if the number of cages was doubled. Macroalgal culture in the middle region of the lagoon can enhance DO concentrations in the macroalgal culture areas but has little impact on DO concentrations in the fish cage areas.

The assessment of nutrient levels shows that inorganic phosphorus is low relative to other nutrients in Xincun Bay. The latter are generally at low levels, except in the fish culture region of the bay: this is beneficial for most aquaculture. As *Kappaphycus* sp. is usually cultured in oligotrophic, tropical or subtropical regions, its productivity is limited by nutrient-depleted waters. Moreover, nutrient requirements vary greatly between species. The point is that with integrated aquaculture, moving commercially important seaweeds closer to the source of nutrients alleviates potential nutrient limitations for seaweed production. Also, from looking at the internal nitrogen and phosphorus tissue contents in seaweeds and seagrasses in the bay, our data suggest that these primary producers can act as key nutrient scrubbers from the water column and sediments. These organisms may play a key competitive role in the ecosystem by reducing the risk of other less desirable, potentially harmful micoalgal blooms.

The case studies presented here illustrate three-dimensional integrated hydrodynamic models that can be used to assess water quality issues. In these cases, we focused on aquaculture. The impacts of extractive aquaculture may be benign or positive. Fed aquaculture, as is the case in Xincun Bay, can have serious impacts on water quality if it is concentrated in small or poorly flushed areas. A solution may be to integrate extractive and fed aquaculture in each embayment. This is not a simple task, but ecosystem modelling offers promise of a management tool to evaluate the consequences of management alternatives. The same ecosystem modelling approach is a valuable tool to identify potential problem areas. For example, in Jiazhou Bay, the model reveals the tendency for nutrient accumulation in the northwest quadrant, thus identifying likely areas of blooms.

References

Ambrose, R.B. Jr, Wool, T.A. and Martin, J.L. (1993) *The Water Quality Analysis Simulation Program, WASP5, Part A: Model Documentation.* US Environmental Protection Agency, Athens, Georgia, 202pp.

Barg, U.C. (1992) *Guidelines for the Promotion of Environmental Management of Coastal Aquaculture Development.* FAO Fisheries Technical Report. No. 328. Food and Agriculture Organization of the United Nations, Rome, 328pp.

Blanton, J.O., Kinner, K.R., Castillejo, F., Atkinson, L.T., Schwing, F.B. and Lavin, A. (1987) The relationship of upwelling to mussel production in the Rias on the western coast of Spain. *Journal of Marine Research* 45, 497–511.

Blumberg, A.F. and Mellor, G.L. (1987) A description of a three-dimensional coastal ocean circulation model. In: Heaps, N.S. (ed.) *Three-Dimensional Coastal Ocean, Coastal Estuarine Science*, Vol. 4. American Geophysical Union, Washington, DC, pp. 1–16.

Boynton, W.R. and Kemp, W.M. (1985) Nutrient regeneration and oxygen consumption by sediments along an estuarine salinity gradient. *Marine Ecology Progress Series* 23, 45–55.

Chen, C., Ji, R., Zheng, L., Zhu, M. and Rawson, M. (1999) Influences of physical processes on the ecosystem in Jiaozhou Bay: a coupled physical and biological model experiment. *Journal of Geophysical Research* 104(C12), 29,925–29,949.

Chopin, T., Yarish, C., Wilkes, R., Belyea, E., Lu, S. and Mathieson, A. (1999) Developing *Porphyra*/salmon integrated aquaculture for bioremediation and diversification of the aquaculture industry. *Journal of Applied Phycology* 11, 463–472.

Collaudin, B.A. (1996) *A First Approach of Aquaculture Development in Jiaozhou Bay (PR China).* Investigation Report, IFREMER, Brest, France.

Dame, R.F. (1993) The role of bivalve filter-feeder material fluxes in estuarine ecosystems. In: Dame, R.F. (ed.) *Bivalve Filter-Feeders in Estuarine and Coastal Ecosystem Processes. NATO Advanced Study Institute Series* 33, 235–270.

Di Toro, D.M., O'Connor, D.J. and Thomann, R.V. (1971) A dynamic model of phytoplankton population in the Sacramento–San Joaquin Delta. In: Hem, J.D. (ed.) *Nonequilibrium Systems in Natural Water Chemistry.* American Chemistry Association, Washington, DC, pp. 131–180.

Fisher, T.R., Carlson, P.R. and Barber, R.T. (1982) Sediment nutrient regeneration in three North Carolina estuaries. *Estuarine, Coast and Shelf Science* 14, 101–106.

Franks, P.J.S. and Chen, C. (1996) Plankton production in tidal fronts: a model of Georges Bank in summer. *Journal of Marine Research* 54, 631–651.

Frevert, T. (1980) Oxygen uptake rates in the sediment of the profundal zone of lake of Constance (Obersee) under oxic and anoxic conditions. *Journal of Hydrology* 42, 56–64.

Hemminga, M.A., Harrison, P.G. and Van Lent, F. (1991) The balance of nutrient losses and gains in seagrass meadows. *Marine Ecology Progress Series* 71, 85–96.

Hickey, C.W. (1984) Chamber studies of benthic oxygen uptake kinetics in the Waioyapu River, New Zealand. In: Hatcher, K. (ed.) *Sediment Oxygen Demand.* Institute of Natural Resources, University of Georgia, Athens, Georgia, pp. 37–62.

James, A. (1974) The measurement of benthal respiration. *Water Research* 8, 955–959.

Jorgensen, C.B. (1990) *Bivalve Filter Feeding: Hydrodynamics, Bioenergetics, Physiology, and Ecology.* Olsen and Olsen, Fredensborg, Denmark, 140pp.

Kasper, H.F., Gillespie, P.A., Boyer, I.C. and Mckenzie, L. (1985) Effects of mussel aquaculture on the nitrogen cycle and benthic communities in Kenepru Sounds. *New Zealand Marine Biology* 85, 127–136.

Kautsky, N. and Evans, S. (1987) Role of biodeposition by *Mytilus edulis* in circulation of matter and nutrients in Baltic coastal ecosystem. *Marine Ecology Progress Series* 38, 201–212.

Kautsky N., Troell, M. and Folke, C. (1996) Ecological engineering for increased production and environmental improvement in open sea aquaculture. In: Etnier, C. and Guterstam, B. (eds) *Ecological Engineering for Waste Water Treatment.* Lewis Publishers, Boca Raton, Florida.

Kraemer, G. and Mazzella, L. (1996) Nitrogen assimilation and growth dynamics of the Mediterranean seagrasses *Posidonia oceanica, Cymodocea nodosa,* and *Zostera noltii.* In: *Proceedings of the Rottnest Island International Seagrass Symposium,* Perth, Australia, pp 181–190.

Kraemer, G. and Mazzella, L. (1999) Nitrogen acquisition, storage, and use by the co-occurring Mediterranean seagrasses *Cymodocea nodosa* and *Zostera noltii. Marine Ecology Progress Series* 183, 95–103.

Liu, F. and Wang, K. (1992) The rivers along Jiaozhou Bay coastline and their geological effects. *Marine Science* [in Chinese] 1, 25–28.

Liu, R. (1992) The natural environmental characteristics of Jiaozhou Bay. In: Liu, R. (ed.) *Ecology and Living Resources of Jiaozhou Bay.* Science Press, Beijing, pp. 2–3.

Marine and Fishery Department of Hainan (1998) Coast and Marine Environment Plan. Asia Development Bank Report ADB-5712. Haikou, Hainan, PR China.

Murphy, J. and Riley, J.P. (1962) A modified single solution approach for the determination of phosphate in natural waters. *Analytica Chimica Acta* 27, 31–36.

Naylor, R.L., Goldberg, R.J., Primavera, J.H., Kautsky, N., Beveridge, M.C.M., Clay, J., Folke, C., Lubchenco, J., Mooney, H. and Troell, M. (2000) Effect of aquaculture on world fish supplies. *Nature* 405, 1017–1024.

Nixon, S.W. (1982) Nutrient dynamics, primary production and fishery yields in lagoons. *Oceanology Acta* 5, 357–371.

Shen, Z. (1995) The variations of nutrients in Jiaozhou Bay. In: Dong, J. and Jiao, N. (eds) *Ecological Study of Jiaozhou Bay.* Science Press, Beijing, pp. 47–52.

Smith, L.K. and Fisher, T.R. (1986) Nutrient fluxes and sediment oxygen demand associated with the sediment–water interface of two aquatic systems. In: Hatcher, K. (ed.) *Sediment Oxygen Demand.* Institute of Natural Resources, University of Georgia, Athens, Georgia, pp. 343–366.

Stumm, W. (1973) The acceleration of biogeochemical cycles in phosphorus. *Progress in Water Technology* 2, 133–144.

Walker, R.R. and Snodgrass, W.J. (1986) Modeling for sediment oxygen demand in lakes. *Journal of Environmental Engineering* 112, 25–43.

Whittemore, R.C. (1984) Implementation of *in-situ* and laboratory SOD measurements in water quality modeling. Water Pollution Control Federation 56th Annual Conference, Atlanta, Georgia, 2–6 October 1983.

Winter, J.E. (1978) A review on the knowledge of suspension-feeding in lamellibrachiate bivalves, with special reference to artificial aquaculture systems. *Aquaculture* 13, 1–33.

Zhao, Y., Chen, Y. and Lin, Z. (1995) The climate around Jiaozhou Bay. In: Dong, J. and Jiao, N. (eds) *Ecological Study of Jiaozhou Bay.* Science Press, Beijing, pp. 8–24.

Shrimp Farm Effluents

15

Granvil D. Treece

Texas Sea Grant College Program, 2700 Earl Rudder Highway South, Suite 1800, College Station, TX 77845, USA

Abstract

In the past, intensive shrimp farm effluents have been characterized as having an environmental impact, but with improvements in technology to control the factors that contribute to the impact, this situation has improved rapidly. In 1994, Texas shrimp farmers began to retrofit their farms and to reuse culture water. Farmers have found that they could reuse far more water than ever thought possible, and the process cleans the renewable natural resource. The management teams have led the way to sustainable, environmentally friendly shrimp farming in the United States by recirculating water and have cut their water use over time to filling ponds and evaporation replacement. In the process, farms have cut their stocking densities, use lower protein feeds, feed more frequently, have widened and deepened their discharge canals, have placed weirs or baffles in discharge canals to help settle the solids, have increased mechanical aeration in those canals and the ponds, and have increased production as a result. Recirculated water has been monitored on the farms and comparisons of intake versus average recirculation-system water have been made. Results show marked improvement in water quality when recirculation is employed. Because of these successes, farms are stocking more shrimp ponds in anticipation of equally successful future crops using the modified procedures.

The Move to Water Reuse

Effluents from intensive shrimp farms have generally been characterized as having an environmental impact (Fig. 15.1). However, this situation is changing with improvements in technology to control factors that contribute

Fig. 15.1. Shrimp farm effluent.

to the reported impact. In the early 1990s, necessity compelled Thailand to lead the way in practising the reuse or recirculation of water on intensive culture shrimp farms to control diseases and improve water quality (Fig. 15.2).

In the early 1990s, Texas (USA) shrimp farming expanded (Fig. 15.3, Table 15.1), and because of more stringent regulations on water quality and discharges, the farms began to reuse water (Figs 15.4–15.8). Farms found that water could be reused after implementing water recirculation practices far longer than ever thought possible. The process also improved the quality of the renewable natural resource. In 2000, Texas shrimp farms produced over 2.54 million kg (5,638,498 lb) of shrimp (Table 15.2) on 879 ha (2171 acres) or approximately 2912 kg ha^{-1} (2600 lb acre^{-1}), using semi-closed systems. In 2000, The Arroyo Aquaculture Association (AAA) stocked 162 ha (400 acres) and produced 638,000 kg (1,406,450 lb), or approximately 4000 kg ha^{-1}. As presented in Table 15.2, there were 11 operating farms within the AAA and shrimp survivals ranged from 22% to 83% during the growout period in 2000. Survival throughout the state averaged 46.99% and ranged

Fig. 15.2. Water recirculation on shrimp farm in Thailand.

Table 15.1. Texas annual farm-raised shrimp production from 1987 to 2000.

Year	Production (kg)
1987	582,000
1988	540,000
1989	499,000
1990	662,000
1991	753,000
1992	1,720,000
1993	1,910,000
1994	1,670,000
1995	635,000
1996	907,000
1997	1,160,000
1998	1,440,000
1999	2,400,000
2000	2,540,000

from a low of 6.49% to a high of 83.33% from a total of 25 facilities producing shrimp (including commercial and research facilities). The loss of 712,546 kg (1,567,602 lb) of shrimp on Texas shrimp farms due to cold weather in 2000 can also be seen in Table 15.2.

Other farms have followed AAA's lead in Texas and now all the farms practise water reuse and recirculation techniques. The methodology may vary from farm to farm, but the principal techniques used are very similar. The management team at AAA* has led the way toward sustainable, environmentally friendly shrimp farming in Texas by recirculating water and by cleaning up that water before discharging. Over time, water use has been reduced from

Fig. 15.3. Southern Star shrimp farm and Arroyo Aquaculture Association (AAA) farm in Texas.

Fig. 15.4. Water intake system at AAA.

* Data were provided by the Harlingen Shrimp Farms, Inc. management group, which manages 45 ponds of 2–3 ha at the Arroyo Aquaculture Association Farm.

37,620 l water per kilogram of shrimp produced in 1994 to 2938 l of water per kilogram of shrimp produced in 1999 and 2000 (Table 15.3). Most of the water is used to fill the ponds and offset evaporation. The stocking density of shrimp was reduced from 50 m^{-2} to 36 m^{-2}. At the same time aeration in each pond was increased from 20 to 25 hp ha^{-1}. The discharge canal at AAA was widened and deepened and weirs or baffles were positioned within the canal

Fig. 15.5. Water reuse system at AAA.

Fig. 15.6. Recirculation pump at AAA.

to aid in the settlement of solids. This practice was also implemented at the Harlingen Shrimp Farm, Inc. farm. Mechanical aeration was also added to the AAA farm canal and the Harlingen discharge canal as shown in Fig. 15.9. Screens were placed on all intakes to eliminate predators and virus pathways (Fig. 15.10) and on the shrimp farm effluents to eliminate shrimp escape, and to control bubbles and foam as the water was discharged. Using precise feeding

Fig. 15.7. Recirculation pumps at Harlingen Shrimp Farms, Inc.

Fig. 15.8. Reusing water at Harlingen Shrimp Farms, Inc.

techniques and lower protein feeds is helping Texas farms maintain good water quality (Fig. 15.11).

Table 15.2. Texas farm-raised shrimp production detail in 2000 from 25 individual facilities (including commercial and research).

Farm	production (kg)	Area (ha)	Lost to cold (kg)	Loss* (%)	Number stocked	Number harvested	Survival (%)
AAA-1	7,484	4.0	6,804	90.9	1,600,000	350,000	21.9
AAA-2	14,700	6.1	3,629	24.6	1,800,000	720,000	40.0
AAA-3	54,400	14.2	1,588	2.8	5,600,000	2,870,789	51.3
AAA-4	322,000	74.9	12,800	3.8	24,000,000	15,000,000	62.5
AAA-5	14,000	6.1	13,600	49.3	2,025,000	810,000	40.0
AAA-6	44,000	10.1	0	0.0	3,350,000	1,990,000	59.4
AAA-7	32,400	6.1	0	0.0	2,100,000	1,749,830	83.3
AAA-8	8,392	2.0	0	0.0	788,000	365,170	46.3
AAA-9	82,600	22.3	19,800	19.3	7,800,000	4,002,760	51.3
AAA-10	27,900	8.1	4,082	12.8	2,700,000	1,280,000	47.7
AAA-11	29,900	8.1	6,804	18.5	2,700,000	1,450,000	53.7
AAA-12	66,300	32.4	72,600	52.3	13,800,000	3,688,000	26.7
13	726,000	159.4	31,800	4.2	73,000,000	46,000,000	63.0
14	680	0.2	227	25.0	60,000	33,000	55.0
15	79,400	167.9	261,000	76.7	30,500,000	4,700,000	15.4
16	5,760	8.1	19,500	77.5	2,000,000	246,700	12.3
17	0	0.1	0	100.0	440,000	0	0.0
18	132,000	36.8	48,800	27.0	25,000,000	9,280,000	35.7
19	462,000	180.1	156,000	25.2	45,000,000	25,125,000	55.8
20	236,000	80.9	34,000	12.6	40,000,000	13,024,000	32.6
21	2,595	1.1	0	0.0	170,000	140,000	82.4
22	1,361	3.2	9,526	87.5	1,050,000	68,100	6.5
23	82,600	22.7	4,536	5.2	8,000,000	4,000,000	50.0
24	115,000	20.2	5,443	4.5	10,190,000	5,838,340	57.3
25	10,400	3.2	9	0.1	1,200,000	800,000	66.7
Total	2,557,781	878.4	712,546	27.8	305,473,000	143,540,689	47.0

*Percentage of total potential harvest lost to cold weather.

Table 15.3. Litres of water used to produce 1 kg of shrimp.

Year	Water usage (l kg^{-1})
1994	37,620
1995	33,220
1996	8,637
1997	2,341
1998	2,701
1999	2,938
2000	2,938

AAA's recirculated water is monitored from July to November. All water samples are analysed in the laboratory at the farm using standard methods approved by the United States Environmental Protection Agency. Each water

Fig. 15.9. Mechanical aeration in discharge canal at Harlingen Shrimp Farms, Inc.

Fig. 15.10. Screens are used on both intake and discharge canals.

sample is collected and analysed in duplicate. All results are compared against a standard variance, and if the variance between duplicates is too large, the sample is discarded or the analysis is repeated. Results of several water analyses over a growing season are presented in Table 15.4. Salinity and total suspended solids readings decreased slightly during the year of sampling shown and all the other measured parameters (pH, carbonaceous biochemical oxygen demand (CBOD), nitrate, nitrite, ammonia, total phosphorus, reactive phosphorus and volatile suspended solids) increased slightly during the 1 July to 5 November sampling period. Water was allowed to settle and was aerated before discharge.

Results of the water analyses and comparisons between intake and average recirculation system water can be seen in Table 15.5. Suspended solids levels were higher in recirculated water than in the intake water, but all the other parameters were equal or lower in the recirculated water than in the intake water.

Other Changes in Management

Additional management practices that have been implemented on Texas shrimp farms include multiple feedings (up to four times daily), use of feeds with reduced protein content (25% protein during the final growout stages), an increase in the frequency of water quality monitoring, and maintenance of

Fig. 15.11. Computer-operated shrimp feed blowers help Texas shrimp farmers provide feed more accurately.

the pond bottoms. In 1997 and continuing to 2000 (Table 15.6), the AAA farm has equalled or surpassed its 1994 production of 4000 kg ha^{-1} (approx. 4000 lb acre^{-1}).

These additional management practices have significantly reduced all the discharge parameters when compared over time. For example, in 1994, the AAA generated 3.5 kg of TSS kg^{-1} of shrimp produced. By 1998 (continuing to 2000), only 0.06 kg of TSS was generated for every kilogram of shrimp

Table 15.4. Mean values for several parameters from duplicate water samples taken from a recirculated shrimp pond water system from July to early November.

Date	Ammonia (mg l^{-1})	Nitrite nitrogen (mg l^{-1})	Nitrate nitrogen (mg l^{-1})	Total nitrogen (mg l^{-1})	Total suspended phosphorus (mg l^{-1})	Salinity solids (ppt)
1 Jul	0.10	< 0.01	0.8	0.27	113	21
2 Jul	0.03	0.01	0.9	0.17	58	23.5
8 Jul	0.03	0.05	1.2	0.18	28	23.5
9 Jul	0.03	< 0.01	1.0	0.19	31	23.5
15 Jul	0.05	0.01	0.9	0.19	277	23.5
16 Jul	0.12	0.20	1.5	0.58	44	22.5
22 Jul	0.03	< 0.01	1.6	0.21	56	25.5
23 Jul	0.07	< 0.01	1.1	0.29	42	23.5
29 Jul	0.08	0.02	1.3	0.49	73	22.5
30 Jul	0.10	0.06	0.9	0.49	93	21
5 Aug	0.28	0.11	1.2	0.44	74	19.5
6 Aug	0.09	0.07	1.7	0.40	61	23
12 Aug	0.78	0.06	1.0	0.41	78	24
13 Aug	0.66	0.21	0.8	0.51	87	24
19 Aug	0.07	0.25	1.1	0.42	35	25
20 Aug	0.46	0.06	1.4	0.52	17	25
26 Aug	0.41	0.10	1.4	0.52	67	27
27 Aug	0.87	0.09	1.3	0.45	56	27
3 Sep	0.41	0.07	1.5	0.58	72	27
4 Sep	0.18	0.07	1.7	0.65	154	27
10 Sep	2.04	0.20	1.2	0.63	40	20
11 Sep	2.61	0.23	1.1	0.43	50	22.5
18 Sep	1.40	0.23	1.2	0.45	45	20
24 Sep	0.45	0.08	1.1	0.58	40	20
2 Oct	4.39	0.15	1.8	0.79	137	20.5
9 Oct	0.14	0.11	1.4	0.50	54	19
15 Oct	2.75	0.05	1.0	0.90	99	22
16 Oct	3.98	0.06	1.1	1.01	99	22
22 Oct	2.18	0.12	1.6	0.78	134	21
23 Oct	2.02	0.02	1.4	1.17	61	21
28 Oct	2.62	0.12	0.8	0.83	39	20
5 Nov	2.26	0.28	1.2	0.52	42	20

produced (Table 15.7). In 1994, AAA discharged almost 50 g of ammonia (NH_3) for every kilogram of shrimp, but by 1998 (continuing to 2000) only 0.4 g of NH_3 was discharged for every kilogram of shrimp (Table 15.7). During 1994 and 1995, the AAA farm discharged between 119 g and 179 g of CBOD for every kilogram of shrimp produced. By 1998 (continuing to 2000) discharge of CBOD had been reduced to 11 g kg^{-1} of shrimp produced (Table 15.7).

Table 15.5. Comparison of selected mean water quality parameters between intake water and recirculated water at the AAA shrimp farms.

Parameter	Intake	Recirculating
Ammonia nitrogen	0.19 mg l^{-1}	0.06 mg l^{-1}
Nitrite	0.02 mg l^{-1}	0.02 mg l^{-1}
Nitrate	1.16 mg l^{-1}	1.12 mg l^{-1}
Total phosphorus	0.44 mg l^{-1}	0.30 mg l^{-1}
Total suspended solids	42 mg l^{-1}	118 mg l^{-1}
CBOD	3.5 mg l^{-1}	2.9 mg l^{-1}
pH	8.4	7.8

Table 15.6. Average production (kg) per 2 ha pond at AAA from 1994 to 2000.

Year	Average production (kg)
1994	9185
1995	4309
1996	7258
1997	9662
1998	9526
1999	9526
2000	9526

Table 15.7. Total ammonia, suspended solids, and CBOD per kilogram of shrimp produced at AAA from 1994 to 2000.

Year	Ammonia (mg)	Suspended solids (mg)	CBOD (mg)
1994	0.0490	3.49	0.119
1995	0.0410	1.31	0.179
1996	0.0017	0.34	0.060
1997	0.0010	0.07	0.009
1998	0.0004	0.06	0.011
1999	0.0004	0.06	0.011
2000	0.0004	0.06	0.011

Despite an estimated 0.71 million kg (1.57 million lb) of shrimp lost to a 100-year weather event in October 2000, during which the water temperature in ponds fell rapidly to 15°C (59°F), Texas still produced a state record. This goes a long way in offsetting one of our country's largest trade deficits (seafood). Next to oil, automobiles and electronics, seafood represents the nation's largest trade deficit. Disease-resistant Pacific white shrimp (*Litopenaeus vannamei*), developed for domestication by the US Marine Shrimp farming programme, largely contributed to the industry's success in recent years.

With improved, automated harvest techniques (Fig. 15.12) used on Texas farms, the farms harvest a clean seafood product, often in the 18–20 g size range (45–55 shrimps per kilogram) and some are even larger.

The recirculating practices at Texas shrimp farms are highly successful approaches to a wise use of renewable natural resources to produce food for the growing population of the world. In the future, Texas shrimp farmers would like to further clean the water by having their discharge permit period extended into the winter months when water temperatures drop below 20°C (68°F). After the first cold front, the algae (volatile suspended solids or VSS) will fall out and die and the farms can discharge water with lower levels of VSS and maintain their discharge limits.

Fig. 15.12. Automated harvest device for shrimp.

Additional considerations by farms are feed management, possibly using more feeding trays. This would be a means of helping water quality and might be used in conjunction with additional aeration and passage of water through constructed wetlands. The wetlands used thus far on Texas shrimp farms have shown that increased aeration is necessary to offset the decaying plant material and organic material, which act as an oxygen sump. Wetlands may lead to the removal of all oxygen from the water as it passes through. However, benefical effects are seen in sediment removal and in reduced ammonia levels. Waterfowl add fertilizer and ammonia to the wetlands during winter months. At one farm a created wetland is used for duck hunting.

The AAA farm in Texas has experienced the following discharge averages:

pH = 7.97 (permit limit 6–9 units),
NH_3-N = 0.14 mg l^{-1} (permit limit 1.0 mg l^{-1}),
TSS = 21.9 mg l^{-1} (permit limit 30 mg l^{-1}),
CBOD 3.95 mg l^{-1} (permit limit 4 mg l^{-1}).

Conclusions are that sustainable, economically viable production levels can be maintained in recirculation systems, with the result being environmentally friendly effluents. The quality of reuse water showed only slight increases in constituents of concern, and discharge water quality averages for November and December on Texas farms can be significantly lower than permitted values if regulatory agencies will allow farms to discharge during those months. Texas shrimp farms have vastly improved their effluents with the addition of recirculation techniques.

Fish Meal: Historical Uses, Production Trends and Future Outlook for Sustainable Supplies

16

Ronald W. Hardy[1] and Albert G.J. Tacon[2]

[1]University of Idaho, Hagerman Fish Culture Experiment Station, 3059 National Fish Hatchery Road, Hagerman, ID 83332, USA; [2]Aquatic Farms Ltd, 49–139 Kamehameha Highway, Kaneohe, HI 96744, USA

Abstract

Fish meal in some form has been used as a component of animal feeds for centuries, but it is only in the past 50 years that fish meal production has become a global enterprise. Fish meal is typically produced from species of fish not used for direct human consumption, or from the by-products of seafood processing. Fish meal is by far the most valuable non-edible commodity produced from fishing, and, over the past decade, annual global production has ranged between 5.5 and 7.5 million tonnes (Mt). Approximately 30% of annual global fisheries harvest is used to produce fish meal; yields from landed fish (wet) to fish meal (dry) and fish oil average 26%. The wet reduction method of processing is the most widely used production method, and improvements in production technology have led to a higher proportion of fish meal production being classified as premium grade.

Although annual global production has been relatively constant over decades, during El Niño years, production in Peru and Chile is substantially reduced. Those countries account for about one-third of global production, but up to 65% of the fish meal traded internationally; thus changes in their production of fish meal greatly affect global supplies and prices. The largest single use of fish meal is as a constituent of poultry feeds. Aquaculture feeds utilized less than 10% of annual fish meal production until 1990, but the proportion of annual production used in fish feeds has tripled over the past decade. Increasing use of fish meal in fish feeds has come primarily at the expense of its use in poultry feeds.

Fish meal is the protein source of choice in feeds for fry of many species, and in feeds for carnivorous fish species. The amino acid profile of fish meal combines favourably with plant protein concentrates to produce blended products that support rapid and economical fish growth. Increasing concerns over the presence of organic contaminants in fish meal from certain areas may result in restrictions in its use in some

 Responsible Marine Aquaculture (eds R.R. Stickney and J.P. McVey)

aquaculture applications. Nevertheless, for the foreseeable future, fish meal will be used as a constituent of feeds for many farmed fish species.

Fish meal use is concentrated in a small proportion of global aquaculture production; nearly 70% of global use is in salmon, trout and shrimp feeds. Predictions of future use of fish meal in these sectors are for the amount to remain more or less constant, and for the proportion of fish meal used in feed formulations to decrease. Increasing efforts to reclaim protein from seafood processing by-product will increase the supply of fish meal by as much as 10%, enough to offset decreases in production associated with natural variation in landings and with cessation of fishing for stocks that have been depleted by overharvesting.

Introduction

Fish meal is a dry powdered material produced from species of pelagic fish that are captured primarily for the purpose of producing fish meal and fish oil. The whole fish is subjected to the wet rendering process, and factories producing fish meal run continuously when fish are being landed. Fish meal is also produced from seafood processing by-product, which is either filleting waste, waste trimmed during canning, or parts of the fish left over from the production of surumi. Most of the world's fish meal is produced from whole fish.

Fish meal is a finite global resource, meaning that the amount produced every year is more or less constant, except in El Niño years, and that the stocks of fish used in fish meal production are fully exploited. As the human population grows, so does the demand for seafood. Given that seafood production from capture fisheries is also a finite global resource, and that stocks of fish used for seafood production are also fully, or, in some cases, overexploited, it is the consensus of all concerned that seafood production from capture fisheries will be unable to meet the expected demand for seafood in the future.

At present, aquaculture supplies 33% of total landed foodfish supplies, and just to keep up with with current world per capita seafood (fisheries products) consumption of 16 kg year^{-1} will require that aquaculture production expand by 50 Mt by the year 2050 (Tacon and Forster, 2001). This will require greater input of pelleted feeds, which will in turn require higher quantities of various feed ingredients, including protein sources. Since fish meal is a finite global resource, and if current fish meal use patterns in feeds within various segments of aquaculture are extrapolated into the future, it is clear that sooner or later the demand for fish meal for use in fish feeds will exceed the annual global production of the commodity. The exact point at which demand for fish meal will exceed supply is debatable, and may occur in the distant future, but the realization that this situation awaits us has been at the core of efforts at research centres throughout the world for the past 15 years or more to develop and test alternative protein sources to replace fish meal in the diets of farmed fish.

More recently, those not involved in research to find alternatives to fish meal have embraced the issue of fish meal use by the aquaculture industry for a variety of reasons, and expanded awareness beyond the segment of the scientific community that deals with these issues every day (Naylor *et al.*, 2000). While all agree that fish meal use patterns by certain segments of the aquaculture feed industry need to be changed, not all agree on how critical the issue is, nor do all agree on how the issue is portrayed in the scientific literature or popular press. Like all issues, the fish meal/aquaculture issue is complex, and it is made more so by growing public concerns about food safety, over-exploitation of resources, and a general sense that the scientific community ignores the larger environmental, social and ethical issues that need to be addressed in a rational evaluation of global resource utilization. In the current environment of sound-bite journalism, it is relatively easy to construct a compelling story suggesting that the expected growth of aquaculture has the potential to ruin the environment or cause wild fish stocks to collapse. A more responsible (and difficult) approach is to examine the facts in detail and develop constructive solutions based upon historical facts and realistic projections for the future.

History of Fish Meal Use

Dried fish have been used as an animal feed and as fertilizer for thousands of years. Historical records show that in Norway fish oil was being produced as early as AD 800. Fish oil was used as a dietary constituent and for producing oil for lamps and candles. The material remaining after oil was pressed from the fish was used as animal feed and fertilizer. Marco Polo reported seeing sun-dried fish used to supplement the feed of animals during his exploration of the Far East in the 14th century. Colonial writings in North America mentioned the use of menhaden as animal feed and fertilizer, both as wet and dried products. Menhaden were caught from shore in haul seines. In the early 1800s in the United States, menhaden were fished primarily for oil production. Oil was pressed from cooked fish using a rock-weighted process, and the residue (called presscake) was then used mainly for fertilizer. By the 1850s, mechanical screw presses were developed to press oil from the cooked fish; this was followed by hydraulic presses several decades later. Fish oil was a major industrial commodity, and was used in paints, lubricants, soaps, printing ink, and in the tanning of hides.

During the first half of the 20th century, animal and poultry production expanded greatly, driven by research that established the essentiality of specific nutrients in the diets of animals and humans. Discovery of the vitamins and demonstration of the requirement for certain amino acids and minerals in the diet led to the development of formulated feeds for raising livestock. United States President Woodrow Wilson directed the National Academy of Sciences (NAS) to ensure adequate food supplies for the nation's population during World War I, leading to the creation of the National Research Council (NRC), a part of the NAS. The NRC created the Committee on Animal Nutrition (CAN) in 1928, and since then, the CAN has been the national authority on animal nutrition, evaluating published research findings from academia and other sources and developing recommendations for the nutritional requirements of domestic animals, companion animals and zoo animals (NRC, 1999). Those requirements are published as NRC bulletins in the series, *Nutritional Requirements of Domestic Animals* (available from National Academy Press, Washington, DC, USA). The bulletins formed the basis for expansion of animal and poultry production by providing the best available knowledge to the animal feed industry. It is difficult to imagine, but in the early years of the 20th century, poultry was a luxury food; annual per capita consumption of poultry in the United States was 0.45 kg. Today, per capita consumption is 37.3 kg, and poultry is considered a dietary staple in many parts of the world.

This expansion created a demand for high-quality protein sources, and fish meal became the premier protein source in animal feeds, especially for poultry and swine. Fish meal was not only a rich source of essential amino acids, but it also contained relatively high levels of trace minerals, the essentiality of which was not known at first. Fish meal was said to contain 'unidentified growth factors', without which poultry and swine would not flourish. Identification of essential dietary trace minerals in fish meal and their supplementation to diets led to more sophisticated feed formulations and eventually to lower and lower levels of fish meal in the diets of poultry and swine. Today, in the United States those feeds are primarily based upon soybean meal and maize. Nevertheless, in many parts of the world where soybeans and maize are not grown, fish meal remains a major protein source for poultry and swine feeds.

Fish Meal Production Trends

Prior to World War II, global production of fish meal was about 1 Mt per year, but following the war years production began to increase, reaching 2 Mt by 1960. The next 20 years saw rapid expansion that resulted in production by the mid-1980s of 6–7 Mt year^{-1}. Construction of fish meal production plants in Peru, Chile, Norway, Iceland, Denmark and several other countries increased global production capacity during the period. Production since the mid-1980s has been relatively constant, except in years when the El Niño phenomenon occurred in the eastern Pacific ocean and lowered landings of anchoveta in Peru and northern Chile. By comparison, annual production of soybean meal since 1960 has increased from 18 Mt to 100 Mt, over a fivefold increase.

World landings of fish have varied between 84 Mt and 94 Mt over the past decade, and of this, between 19 Mt and 28 Mt were used to produce fish meal and fish oil. Fish captured for fish meal and oil production are sometimes termed industrial fish to distinguish them from food fish. While the species of fish harvested to produce fish meal are edible, they are normally not used for direct human consumption due to their small size, undesirable taste, or some other reason. Yields of fish meal and fish oil from landings of industrial fish are typically 6–7 Mt of fish meal, and 1.2 Mt of fish oil, except, as noted above, in El Niño years, when landings are reduced in some areas.

Fish meal is traded around the world between fish meal producing countries and fish meal consuming countries, but the amount that is traded is lower than total annual global production. Exports average about 3.6 Mt per year, which is about 60% of total annual production. Four countries dominate the global trade in fish meal: Peru, Chile, Denmark and Iceland. Norway, Japan, Thailand, the United States and the nations making up the former USSR are major producers of fish meal, but those countries either consume most of what they produce, or export slightly more than they import, as is the case for the United States. Of the fish meal exporting countries, Peru and Chile account for about 80% of the international trade in fish meal, a fact that has important implications for the market price of fish meal. During El Niño years such as 1972, 1977, 1987, 1992 and, most recently, 1998, fish landings and consequently production of fish meal from Peru and Chile has been substantially reduced, causing global production to fall by as much as 1.75 Mt. This cuts the amount of fish meal traded throughout the world by as much as 50%, resulting in a near doubling of fish meal prices during those periods. The price can reach over US$600 per tonne, FOB Peru, compared to as low as US$330 per tonne during periods of abundance. Many users of fish meal are very price sensitive, providing a buffer in the supply and demand relationship, and making it impossible to make a linear correlation between annual production of fish meal and average price per ton.

Trends in Aquaculture Production

Total production of all aquaculture products increased from 10 Mt to over 30 Mt between 1964 and 1997 (FAO, 1999). Removing aquatic plants and molluscs (which are not fed prepared diets) from production, yields values of about 24 Mt of fish and crustaceans (mainly shrimp), up from 7 Mt in 1994. Fish production in 1998 was nearly 19 Mt, of which more than 70% was attributed to the production of cyprinids, i.e. various species of carp (FAO, 2000). Most of the production of carp was in Asia, and until recently, most Asian carp production depended on natural food in ponds stimulated by fertilization and other low technology forms of nutrient input, rather than on pelleted feeds added to ponds to feed the fish directly. However, yields of carp can be increased by three- to fivefold by using pelleted feeds compared with extensive culture techniques based upon pond fertilization and enrichment, and it is likely that the proportion of carp reared using pelleted feeds will increase in Asia (Cremer *et al.*, 1999).

The remaining 30% of aquaculture production in 1998 was comprised of species groups of fish that are fed pelleted diets, e.g. eels, catfish, marine fish, tilapia, salmon, trout and various other freshwater fish species. Total fish feed production in 2000 was estimated to be slighly over 13 Mt, of which 53% was used to rear carp, 12.5% for salmon and trout, 12% for shrimp, and the remainder for eels, catfish, marine species and tilapia (Table 16.1). Of these species groups, the largest increase between 1995 and 2000 was in the carp group, which grew from about 4 Mt to nearly 7 Mt (Fig. 16.1). Growth in this sector over the 5-year period was nearly equal to the total feed produced in 2000 for salmon, trout and shrimp. Despite the rapid increase in the production of pelleted feeds for farmed aquatic species, fish and shrimp feed production is still a small proportion of the total amount of animal feed produced worldwide, accounting for less than 5% of global feed production.

Table 16.1. Estimated world fish feed production in 2000 by species group.

Species group	Feed produced (tonnes)	Per cent of total
Salmon/trout	1,636,000	13.1
Catfish	505,000	4.0
Tilapia	776,000	6.2
Cyprinids (carp)	6,991,000	55.8
Marine fish	1,049,000	8.4
Shrimp	1,570,000	12.5
Total[a]	12,527,000	100.0

[a]Includes minor species.

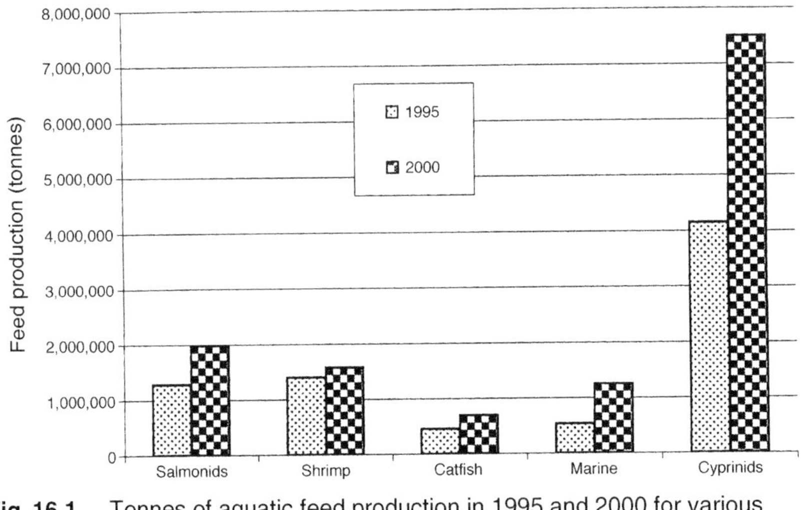

Fig. 16.1. Tonnes of aquatic feed production in 1995 and 2000 for various species groups.

Fish Meal Use in Feeds

Global fish meal production has averaged about 6.2 Mt per year over the past 15 years, except in El Niño years when it decreases by 1–2 Mt. If one uses the 15-year average of fish meal production, the largest agricultural user of fish meal in the world is the poultry industry, which uses about 46% of annual production to produce poultry feeds. Aquaculture is the next largest industry segment user of fish meal, accounting for about 34%, followed by swine at 9–10%, and others, such as pet foods and minor farmed species, total 10–11%. The aquaculture feed industry's use of fish meal has increased greatly over the past decade, and predictions are for further increases in the present decade (Fig. 16.2). The increased share of global fish meal production being consumed by the aquaculture feed industry has come at the expense of use in poultry and swine feeds. In other words, there has not been an increase in fish meal production (and the quantity of fish harvested from the sea to produce fish meal) to supply the growing needs of the aquaculture feed industry. Rather, fish meal use patterns have shifted among industry segments, in part because the aquaculture feed industry can pay higher prices for fish meal than can the poultry or swine industries, and in part because there are well-established alternative feed ingredients, such as soybean meal and maize, that can be substituted for fish meal in poultry and swine feeds when the price differential between fish meal and soybean meal so dictates. Thus, economics controls fish meal use patterns in animal and fish feeds.

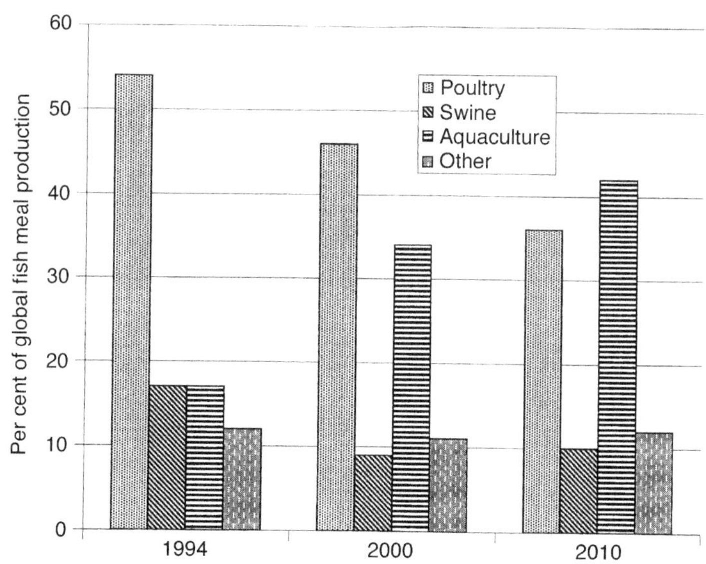

Fig. 16.2. Percentage of fish meal used by various feed sectors in 1995, 2000 and estimated for 2010.

Fish meal use by various sectors of the aquaculture industry is related to the nutritional requirements of the species groups, and the fact that among protein sources available for use in producing fish feeds, fish meal has a number of properties that make it the protein source of choice. Many of the species of fish for which strong consumer markets exist are piscivorous, meaning that in nature they are high in the food chain and consume other fish. For example, salmon consume herring, capelin, squid and other such prey; trout consume insects, fish and invertebrates; and marine fish being raised in aquaculture, such as sea bream, sea bass, yellowtail, grouper and halibut, also consume other fish in nature. In contrast, tilapia and catfish are omnivores, and while they consume other fish if the opportunity arises, they also consume vegetation, molluscs, insects, zooplankton and a variety of other foods. Fish meal has a high protein content compared to many proteins of plant origin, such as soybean meal. In addition, the balance of essential amino acids in fish meal is well matched to the dietary requirements of fish. Protein and amino acid digestibility is relatively high in fish meal, and fish meal-based feeds are highly palatable to most species of fish, in contrast to feeds in which oilseed meals are the main protein source. Finally, although fish meal varies in price depending upon supply and demand, usually fish meal is the most economical protein source that can be used in feeds for carnivorous fish species, especially when it is compared on a protein-unit basis.

Estimates for use by various fish species groups show that 2.115 Mt of fish meal were used to produce feeds for farmed fish in 2000. Salmon feed producers were the largest species group user of fish meal, followed by marine fish (sea bream, sea bass, yellowtail and other species), shrimp and carp (Table 16.2). The highest use among the farmed fish groups, based on average percentage of fish meal used in feed formulations, was 55% for marine flatfish (turbot, halibut, sole), followed by marine fish (45%), salmon (40%), trout (30%) and shrimp (25%) (Table 16.3). Species groups using the lowest percentage of fish meal in feed formulations were catfish (3%) and carp (5%).

Table 16.2. Estimated fish meal use in aquatic feeds in 2000.

Species group	Fish meal used (tonnes)	Per cent of total
Salmon	454,000	21.3
Trout	176,000	8.3
Catfish	15,000	0.7
Carp	350,000	16.4
Eels	173,000	8.1
Flatfish	69,000	3.2
Other marine fish	415,000	19.5
Other fish	106,000	5.0
Shrimp	372,000	17.5
Total	2,130,000	100.0

Table 16.3. Estimated average percentage and tonnage of fish meal in feeds for various fish species groups for 2000 and 2010.

	Fish meal used (tonnes)		Per cent of total	
Species group	2000	2010	2000	2010
Salmon	454,000	377,000	40	30
Trout	176,000	147,000	30	25
Catfish	15,000	0	3	0
Carp	350,000	675,000	5	2.5
Flatfish	69,000	263,000	55	45
Other marine fish	415,000	688,000	45	40
Shrimp	372,000	485,000	25	20
Total[a]	2,115,000	2,831,000		

[a]Includes minor species.

Carp, by virtue of being widely farmed, was the third largest industry segment in terms of tonnes of fish meal used, despite the low percentage of fish meal used in the diets formulated for these fish. The actual percentage of fish meal used in feeds for some species groups varies somewhat depending on the price of fish meal. For example, when fish meal is expensive, the percentage of fish meal in trout feeds drops to 25% or less, but when the price of fish meal is inexpensive, the percentage of fish meal in feeds approaches or even exceeds 40%. Other species are less flexible in terms of the percentage of fish meal used in their feeds; examples being eels, marine flatfish and some other species of marine fish. Predictions for fish meal use patterns in 2010 call for lower percentages in feeds for all species groups and complete elimination of fish meal from the diets of catfish (Barlow, 2000). These predicted reductions will not, however, lower the absolute amount of fish meal used by the aquaculture feed industry because aquaculture production is expected to increase between now and 2010. However, the predicted reductions will lower (by approximately 1.25 Mt) the amount of fish meal used in aquaculture feeds in 2010 compared to what would be used if the percentages of fish meal in feeds of various species groups remained the same as in 2000 (Table 16.4). The dietary protein needs of the fish will not change, however, and alternative proteins will have to be used to replace the dietary protein that would have been supplied by the 1.25 Mt of fish meal that will probably not be used.

Alternative Protein Sources to Fish Meal

Alternative protein sources to fish meal for use in feeds for fish and shrimp must have certain properties to be serious candidates. These properties include: price, which must be competitive with fish meal, both on a protein-unit basis and in terms of fish performance; effect on product quality (there cannot be a negative effect); pollution potential (cannot increase pollution by adding fibre or indigestible nitrogen or phosphorus); they must be legitimate articles of commerce that are available in large quantities; familiar; and easy to ship, handle and use at the feed production facility. The list of suitable alternatives that meet the criteria listed above is quite short, consisting of protein concentrates made from grains, oilseeds and pulses (peas and beans); rendered products; and meals produced from fisheries bycatch and seafood processing waste. Alternatives that are not articles of commerce, but may become so in the future, include single-cell proteins (e.g. bacteria and yeast grown on industrial by-products, such as products of fermentation) krill products and other, plant-derived materials.

When alternative protein sources are examined, it soon becomes apparent why those feed ingredients are alternative rather than principal protein sources for use in fish feeds. All have some deficiency that limits their use. For example, all plant-derived protein sources, whether from grains, oilseeds or pulses, are deficient in one or more essential amino acids compared with fish meal (Table 16.5). Further, the storage form of phosphorus in seeds is phytate, which is indigestible by monogastric animals, including fish and humans. Only ruminant animals can utilize the phosphorus in phytate. Thus, plant-derived protein sources contain indigestible phosphorus, which passes through the gut into the rearing water and contributes to nutrient enrichment. Plant protein sources also may contain indigestible fibre or other carbohydrates that affect faecal solids production (volume) and the moisture content of faeces,

Table 16.4. Predicted use of fish meal in aquatic feeds in 2010 compared with the amount needed if the percentage of fish meal in those feeds remained the same as in 2000 (Barlow, 2000).

Year	Aquatic feed (tonnes)	Fish meal (tonnes)
2000 (estimated)	1,309,800	2,115,000
2010 (at today's use levels)	3,722,600	4,081,000
2010 (estimated)		2,831,000
Difference		1,250,000

Table 16.5. Nutritional problems in rainbow trout associated with various plant protein sources (Sugiura *et al.*, 2000).

Feed ingredient	Limiting amino acid	Phosphorus digestibility (%)	Fibre (%)
Corn gluten meal	Lysine	8.5	4.4
Wheat gluten meal	Arginine	74.7	6.7
Soybean meal	Methionine	22.0	3.4

which in turn affects the efficiency of faecal solids collection systems designed to remove solids from farm discharge water or recirculation systems.

Nevertheless, advances are being made in two areas pertaining to the use of plant-derived protein sources in fish feeds. First, efforts to determine maximum and optimum dietary use levels of plant protein sources in fish feeds for various species are widespread and encouraging. Second, research on ways to modify specific negative characteristics of some plant protein sources through selection of desirable cultivars, ingredient processing, and dietary supplementation is also yielding results. Thus, in the near future, it is likely that higher use levels of plant protein sources in fish feeds will be common, assuming that economic factors are favourable for their use.

Rendered products are derived from the by-products of animal and poultry processing, and they have been valuable ingredients in feed formulations for fish for decades. Their value stems from their protein content and balance of essential amino acids, which complements the amino acid deficiencies of plant proteins. The major rendered products are meat and bone meal, poultry by-product meal, blood meal and hydrolysed feather meal, and just in the United States and the European Union, the total amount of these products is in excess of 8 Mt.

However, rendered products have several drawbacks that limit their use in feeds for fish. First, the quality of rendered products varies among suppliers, especially in ash content and in the digestibility of protein (and therefore amino acids). Second, owing to the high level of ash in meat and bone meal and poultry by-product meal, there are limitations on their use due mainly to the levels of calcium and phosphorus they contain. Third, concerns about contamination of rendered products in the European Union have led to a ban on their use in animal (and fish) feeds as well as a temporary ban on exports. Consumer concerns about bovine spongiform encephalitis (BSE) have further restricted the use of these products in non-European Union countries, especially those that export poultry, fish and other animal products to the European Union. Taken in total, it is clear that rendered products of bovine origin are not promising candidates to replace a portion of fish meal in the diets of farmed fish and shrimp; however, poultry by-product meals remain promising products.

Outlook for Sustainable Supplies of Fish Meal

For the most part, the stocks of fish that are harvested to produce fish meal and oil are fully exploited, but not overexploited. Stocks are monitored closely by fishery managers before, during and after fishing seasons. Fishing closures are imposed to protect stocks if deemed necessary by agencies responsible for stock assessment and management. However, stocks are known to decline precipitously from time to time. Such changes in stock strength are thought to be the result of undesirable ocean conditions during spawning and/or the early life history of the fish. This results in one or more year-classes of fish from a specific stock being several orders of magnitude less abundant than normal, leading to very low recruitment to the fisheries. Depending upon the species of fish, such weak year-classes can affect the strength of the stock, and consequently the size of the harvest, for years. Naturally, overfishing during periods of low recruitment exacerbates the problem of stock strength, and can extend the period of stock recovery. An example of this phenomenon was the Japanese sardine stock collapse of 1938, which resulted in a very weak year-class, and affected sardine abundance and fisheries catches for many years until later year-classes became old enough to enter the fishery. The opposite sometimes occurs as well; a very strong year-class can dominate a fishery for years, at least in herring, which live for about 10 years. Other species of fish with shorter life cycles are affected for shorter periods of time when a year-class fails to reach normal abundance.

The major species of fish comprising the collective term industrial fish include anchovies, anchoveta, capelin, eel pouts, herring, horse mackerel, menhaden, pilchard and sardines. Anchoveta are the dominant species captured in Peru, although horse mackerel are also caught there and in Chile. Capelin and herring dominate catches in Norway, Iceland and Canada. Eel pouts are found in the North Sea and used in Danish production, pilchard in South Africa, and sardines in Japan and Central America. When one stock decreases in abundance, as has been the case with capelin in the 1980s and sardines in Japan in the 1990s, total world landings of industrial fish decrease somewhat. However, during El Niño years, when warm currents make Peruvian anchoveta move to areas not accessible to fishermen, total world landings decrease significantly. Despite the ups and downs of various stocks of fish landed to produce fish meal, world landings have averaged slightly over 6 Mt year^{-1} over the past 15 years, except, as noted, in El Niño years, and the outlook for the future is for landings to remain at that level.

The prospects for increasing fish meal production from traditional sources of fish are slight, but the prospects are more encouraging for increasing fish meal production from non-traditional sources, namely fisheries bycatch and seafood processing waste. The total quantity of raw material from these two sources is difficult to measure, but has been estimated to be at least 90% of the average world harvest of industrial fish (New, 1996). Two complicating issues prevent recovery and use of this material: (i) logistic issues associated with the unpredictable supply of material with respect to amount, location, and seasonal abundance; and (ii) technology to deal with small or fluctuating amounts of material which, in the case of processing waste, is high in bone content. Both of these issues limit the extent to which conventional fish meal processing equipment can be used to convert the raw material into fish meal. Thus, alternative technologies are required. Progress has being made in both areas, especially in approaches to dealing with fish bone in seafood processing waste (Babbitt *et al.*, 1994; Rathbone *et al.*, 2001).

Unexpected Factors That Could Affect Fish Meal Abundance

The most obvious threat to predictions of future fish meal abundance concerns contamination of fish meals arising from the use of contaminated raw material. The most visible concern with respect to contamination is polychlorinated biphenyls (PCBs), followed by dioxin. Recently, press reports have suggested that farmed salmon contain higher levels of PCBs than do wild salmon, but those reports have been based on two studies, one involving relatively few samples and the other conducted by an advocacy group against salmon farming. Neither study has been subject to peer-review, and a complete description of sampling protocols, sample size, analytical methods used and so on has not been made available. In fact, according to the European Union, there are no differences in PCB content between farmed and wild salmon. Given the uncertainties about sampling and analytical methods, the variety of locations where fish are farmed, and the variability in feed (and fish meal) inputs to farmed fish, it is not rational to make sweeping generalizations about the contaminant load in farmed or wild fish without widespread, scientific testing conducted by independent, certified analytical laboratories in a transparent manner.

The implication of press reports is that contamination of farmed fish is via the feed, and, further, that among the constituents of fish feed, fish meal is the likely source of contamination. Fish meals are produced in many countries from many raw materials, making it likely that there are differences in contaminant load in fish meals produced from different species and different regions. Given that the key countries supplying the bulk of fish meal that is traded throughout the world are Peru and Chile, followed by Iceland and Denmark, it is logical that fish meal from those countries be the focus of study.

Apart from the issue of contaminants in fish meal, another factor that could affect the abundance of fish used to produce fish meal is global warming. The effects of global warming on deep ocean water currents, and on coastal shelf upwelling are unknown, but we must be aware of the potential for such changes to affect the stock strength of forage fishes.

Challenges for the Future

Given that global fish meal supplies are finite and that stocks of fish used to produce fish meal are fully exploited, emphasis must be placed on wise use of global fish meal supplies and on efforts to recover and utilize fish protein from fisheries bycatch and seafood processing waste, which is currently discarded. To ensure that at least 1.25 Mt of protein is available from various sources to replace a portion of fish meal in feeds of farmed fish species, expanded research and development are needed on protein sources derived from grains and oilseeds, specifically to produce products designed for use in fish feeds. It is also critical to continue research on feed ingredients and fish feed formulations designed to lower the environmental impacts of fish farming on the aquatic environment. These challenges seem overwhelming, but such challenges faced the poultry and swine industries decades ago, and as a result of intensive research, poultry and swine products today are much less expensive than they were several generations ago. Similar opportunities await the aquaculture industry.

Growing global concerns for food safety have created a greater awareness of the need to prevent human exposure to contaminants through food. In the case of food from animal and fish production, one clear source of potential contamination is via the feeds fed to livestock, and this has resulted in increasing scrutiny of the feed ingredients used to manufacture animal and fish feeds. These examinations are focused on all ingredients, not just fish meal or fish oil. They cover chlorinated hydrocarbons, pesticides, herbicides, heavy metals and microbial toxins. In these examinations, we must remember that the oceans are the recipient of many industrial pollutants, and if we look, we will find them, not only in sediments but also in fisheries resources, especially in light of the incredibly sensitive analytical tests available to detect trace amounts of some contaminants. Whether specific contaminants are present in seafood products at levels that threaten human health is yet to be determined, but we must remember that we have few opportunities to prevent contaminants from being present in wild-caught seafood, whereas we can prevent them from being present in the feeds of farmed fish. Thus, through vigilant testing of feed ingredients, including fish meal and oil, and the modification, if necessary, of their production methods, we at least have the potential to ensure the safe production of food fish for a growing human population through aquaculture production.

References

Babbitt, J.K., Hardy, R.W., Reppond, K.D. and Scott, T.M. (1994) Processes for improving the quality of whitefish meal. *Journal of Aquatic Food Product Technology* 3(3), 59–68.

Barlow, S. (2000) Fish meal and oil. *Global Aquaculture Advocate* 3(2), 85–88.

Cremer, M.C., Baoxin, Z., Schmittou, R. and Zhang, J. (1999) In: *Status of Forecast for the Freshwater Aquaculture Production and Feed Industries of China. International Aquafeed Directory and Buyer's Guide, 1999.* Turret RAI PLC, Uxbridge, UK.

FAO (1999) *Aquaculture Production Statistics 1988–1997.* FAO Fisheries Circular 815, Rev.11. Food and Agriculture Organization of the United Nations, Rome, 197pp.

FAO (2000) *FAO Yearbook. Fisheries Statistics: Aquaculture Production 1998*, Vol. 86/2. FAO Fisheries Series No. 56, FAO Statistics Series 154. Food and Agriculture Organization of the United Nations, Rome, 169pp.

Naylor, R.L., Goldberg, R.J., Primavera, J.H., Kautsky, N., Beveridge, M.C.M., Clay, J., Folke, C., Lubchenco, J., Mooney, H. and Troell, M. (2000) Effect of aquaculture on world fish supplies. *Nature* 405, 1017–1024.

New, M.B. (1996) Responsible uses of aquaculture feeds. *Aquaculture Asia* 1(1), 3–15.

NRC (1999) *The First Seventy Years: Committee on Animal Nutrition.* National Academy Press, Washington, DC.

Rathbone, C.K., Babbitt, J.K., Dong, F.M. and Hardy, R.W. (2001) Performance of juvenile coho salmon *Oncorhynchus kisutch* fed diets containing meals from fish wastes, deboned fish wastes, or skin-and-bone by-product as the protein ingredient. *Journal of the World Aquaculture Society* 32, 21–29.

Sugiura, S.H., Babbitt, J.K., Dong, F.M. and Hardy, R.W. (2000) Utilization of fish and animal by-product meals in low-pollution feeds for rainbow trout, *Oncorhynchus mykiss* (Walbaum). *Aquaculture Research* 31, 585–593.

Tacon, A.G.J. and Forster, I.P. (2001) In: *Global trends and challenges to aquaculture and aquafeed development in the new millennium. International Aqua Feed Directory and Buyer's Guide.* Turret RAI PLC, Uxbridge, UK, pp. 4–25.

17

The Use of Wild-caught Juveniles in Coastal Aquaculture and its Application to Coral Reef Fishes

Cathy Hair,[1] Johann Bell[1] and Peter Doherty[2]

[1]ICLARM – The World Fish Center, PO Box 500, GPO 10670, Penang, Malaysia; [2]Australian Institute of Marine Science, PMB No. 3, Townsville, Queensland, Australia

Abstract

Worldwide, there are many substantial coastal aquaculture and stock enhancement operations based on collection of wild juveniles. These include: growout of shrimp (Penaeidae), milkfish (*Chanos chanos*), eels (*Anguilla* spp.), yellowtail (*Seriola quinqueradiata*), southern bluefin tuna (*Thunnus maccoyii*), edible oysters (Ostreidae) and mussels (Mytilidae); stock enhancement of scallops (Pectinidae); and the culture of pearls in farmed blacklip pearl oysters (*Pinctada margaritifera*). The growout of wild puerulus larvae of spiny lobsters (Palinuridae) is also developing rapidly.

The advantages of using wild-caught juveniles for aquaculture are: (i) low costs of obtaining animals for stocking as compared with hatchery production; (ii) availability of individuals fit for growout in the sea; (iii) no risks of 'genetic pollution' from deliberate or accidental releases; (iv) reduced likelihood of transferring diseases; and (v) a broader range of economic benefits, including opportunities for coastal dwellers in developing countries to sell stock to larger enterprises. In addition, responsible capture and culture of wild juveniles can improve overall fisheries productivity for target species by circumventing the high rates of natural mortality associated with settlement of postlarvae from the plankton. Careful management of this process is needed, however, to ensure that replenishment of the stock, and fisheries targeting adults, are not affected. Where large numbers of postlarvae are taken, or where aquaculture is based on larger juveniles, these goals can be met by returning a proportion of the cultured juveniles to the wild, or through the transfer of fishing effort from adults to juveniles. The disadvantages of using wild juveniles for aquaculture are: (i) the number of animals available for growout can be limited and variable; (ii) there is no scope for increased productivity through selective breeding; and (iii) potential effects on the ecosystem stemming from mortality of bycatch and removal of prey from the food chain.

On balance, the advantages outweigh the disadvantages and new applications for the use of wild-caught juveniles are under investigation. In particular, there has been interest in using aquaculture to supply the trade in ornamental and live food fish from coral reefs to overcome problems resulting from overfishing of adults and the use of destructive fishing techniques. However, it is technically difficult and expensive to propagate postlarvae of many coral reef fishes so cost-effective hatchery production of juveniles for aquaculture is likely to develop only for a minority of target species. As an alternative, the feasibility of harvesting pre-settlement coral reef fishes from the plankton in numbers that do not affect the replenishment of natural populations, and rearing them for a short period before sale to the ornamental trade or as juveniles for growout for the live fish market, is being assessed. Two sampling techniques, light traps and crest nets, have proved suitable for the capture of live pre-settlement fishes and substantial progress has now been made in applying these methods to the development of artisanal fisheries for ornamental species. Although the capture and culture of postlarvae is unlikely to meet the demand for all tropical marine fish required by the ornamental trade, it has the potential to create important niche markets, e.g. for ecolabelled specimens, and provide sustainable economic benefits from coral reef resources for coastal villagers.

Introduction

The responsible development of aquaculture, including its use in rearing or propagating juveniles for release in restocking and stock enhancement programmes, holds the promise of supplying increased yields of fish and shellfish without reducing production from wild stocks. The responsible practices we refer to are described in other chapters of this book (see Chapters 6 and 7, in particular) and address many of the potential concerns about coastal aquaculture raised by Naylor *et al.* (2000).

In this chapter, we focus on the simplest form of coastal aquaculture – the growout of juveniles caught from the wild. We believe that there is a need to highlight the importance and potential of this type of aquaculture because, with the recent advances in knowledge of larval biology and aquacultural engineering, there is a tendency to assume that further development of aquaculture will be predicated on the mass production of juveniles in hatcheries. While the use of hatchery technology may be the only way to produce sufficient numbers of juveniles for stocking or increase the supply of them beyond current levels for many species, much of the world's coastal aquaculture production can still be expected to come from wild-caught juveniles.

A particularly attractive feature of aquaculture based on wild-caught juveniles is that many of the environmental concerns associated with the growout of juveniles produced in hatcheries, e.g. transfer of diseases and 'genetic pollution' of wild stocks (Munro and Bell, 1997, and references therein), are not inherent in the process. Collection of juveniles from the wild does, however, come with its own set of responsibilities. Chief among these is the need to ensure that the increased production from the culture of juveniles more than offsets any loss in yield from the wild stock. In particular, collections should not adversely affect replenishment of the wild population or disadvantage other users of the resource.

Here, we briefly describe the range of coastal aquaculture activities based on the culture of wild juveniles and appraise the advantages and disadvantages of the general process for increasing productivity in a sustainable way. We then outline the recent emergence of a fishery based on the collection and growout of postlarval coral reef fishes as a responsible way of supplying the growing trade in live reef fish for the aquarium and food trades.

The current practice of exploiting mostly adult coral reef fishes, often using destructive methods, is widely considered to be biologically and ecologically unsustainable (Sadovy and Vincent, 2002). The alternative we describe reduces fishing effort on the spawning biomass, protects the habitat and creates new opportunities for coastal communities to earn income.

Types of Coastal Aquaculture Based on Wild-caught Juveniles

The variety of aquaculture enterprises based on wild-caught animals (Table 17.1) shows that harvest occurs at life history stages ranging from planktonic (pre-settlement) postlarvae to large juveniles. To provide an overview of the types of coastal aquaculture based on wild-caught juveniles, we have classified the collection process into three broad groups: (i) juveniles caught at the end of the pelagic phase, or immediately after settlement; (ii) juveniles collected several weeks after settlement; and (iii) large juveniles captured before they have entered the adult fishery. We do not consider the collection of gravid females, e.g. shrimp, as a source of wild stock because they are used for spawning in hatcheries in lieu of captive broodstock.

Collection of pre-settlement and settlement stages

Bivalve spat

The collection of the final larval stage of bivalve mollusc, by providing settlement substrates in the water column, is the most widespread use of wild-caught juveniles in coastal aquaculture. Although many species of bivalves are routinely produced in hatcheries, the scale of wild spat collection often dwarfs hatchery production and, in some cases, is so efficient that it eliminates the need for hatchery production altogether. The main groups of bivalves involved are the mussels (Mytilidae), edible oysters (Ostreidae), scallops (Pectenidae) and pearl oysters (Pteridae). The farming of mussels and edible oysters is the most widespread and has the longest history, dating back 700 years in the case of mussels in Europe (Jeffs *et al.*, 1999) and 400 years for edible oysters in Japan. The blue mussel (*Mytilus edulis*) is collected using

Table 17.1. Examples of species groups harvested as wild juveniles, the countries where the aquaculture is based and how they are harvested.

Group	Target species	Regions involved	Collection and culture methods
Shrimp	Family Penaeidae	S America, Indochina, Asia, India	Postlarvae move naturally into coastal ponds via tidal pulses or are collected by hand and transferred to ponds
Spiny lobsters	*Panulirus* and *Jasus* spp. (Family Palinuridae)	Australia, NZ, Asia, Indochina	Pueruli harvested from collectors made of artificial seaweed, mesh, etc., set at various depths usually inshore. In Vietnam 'lampara' nets used with lights. Post-settlers also collected, from submerged objects or using trunks drilled with holes as artificial habitat. Tank and cage growout
Scallops	Family Pectinidae	Australia, NZ, Europe, N and S America, UK, Japan, China, Korea	Collection of spat from the water column using monofilament net in onion bags, juveniles grown-out in lantern cages, reseeding of natural habitat to enhance wild fishery
Oysters	Family Ostreidae	Australia, NZ, N and S America, Japan, China, Europe, UK	Spat collection on sticks in estuaries. Subsequent growout on trays or sticks
Mussels	*Mytilus* and *Perna* spp. (Family Mytilidae)	NZ, Europe, N and S America, Asia, Australia, Africa, Pacific Islands, India	Spat collection from plankton on to fibrous rope. In NZ settled spat harvested mainly from stranded macroalgae

Blacklip pearl oysters	*Pinctada margaritifera*	Pacific Islands	Spat collection on bundles of local hardwood, nylon rope, shademesh or plastic sheets hung on subsurface lines. Young oysters protected from predation for some time, then hung on longlines
Milkfish	*Chanos chanos*	Philippines, Sri Lanka, Pacific Islands, Indonesia	*Bangus* fry collected with push nets or various kinds of set or lift nets. Grown-out in ponds
Eels	*Anguilla* spp.	Japan, Asia, Europe, Australia, UK, China, N Africa, N America	Collection of glass eels and elvers in estuaries during upstream migration, growout in ponds or released into natural waterbodies to enhance wild stocks
Coral reef fish I	*Epinephelus* and *Plectropomus* spp. (Family Serranidae)	Asia (in particular Indonesia, Philippines, Vietnam)	Postlarvae, fry and fingerlings collected by various methods including: gangos (fish nest), miracle holes, fish shelters (with and without lights), nets, hook and line. Growout in fish cages or ponds
Coral reef fish II	Many families and > 1000 species	Pacific Islands	Postlarval coral reef fish and invertebrates are harvested with light traps and crest nets, then reared in tanks, raceways or sea cages using low or high technology fish husbandry methods
Yellowtail	*Seriola* spp. (Family Carangidae)	Japan, Indochina	Pelagic juveniles, associated with floating seaweed, caught with encircling nets and on-grown in cages for 2–3 years
Southern bluefin tuna	*Thunnus maccoyii*	Australia	Immature fish of 2–4 years old are caught by line (part of the adult wild fishery quota) and then fattened for up to 6 months in high-tech sea cages

fibrous rope and other natural and artificial substrates, whereas oysters are usually collected on harder surfaces, including wooden and plastic sticks. A unique source of wild-caught juveniles is found in New Zealand, where there are still no suitable hatchery techniques for the green mussel, *Perna canaliculus*. Instead, the industry obtains 80% of the required wild spat from detached macroalgae stranded on Ninety Mile Beach in the Marlborough Sounds (Jeffs *et al.*, 1999). In this case, the spat would perish if they were not harnessed for aquaculture.

Scallops are in a different category to edible oysters and mussels in that spat collected from the wild are used widely to enhance fisheries, e.g. in Japan, New Zealand and Europe (Ventilla, 1982; Dao *et al.*, 1999; Arbuckle and Metzger, 2000). The spat are collected in onion bags containing nylon netting suspended from longlines, reared to a size where they can escape most predation and then released in truly outstanding numbers (Ventilla, 1982). Releases in Hokkaido, Japan, alone exceed 3 billion per annum and have improved yields in the capture fishery from 5000 t annually in the 1960s to 200,000 t in the 1990s (Masuda and Tsukamoto, 1998; Imamura, 1999). Another 200,000 t per annum are produced in hanging culture (Kitada, 1999).

The development of the black pearl industry in French Polynesia is also an excellent example of the importance of aquaculture based on wild spat. Initially, the industry relied on amassing and implanting sufficient adults of the blacklip pearl oyster, *Pinctada margaritifera*, to establish viable pearl farms. However, large-scale collection of spat commenced in the late 1960s to prevent overexploitation of adult stocks (Tisdell and Poirine, 2000). The industry still relies on spat collection and grew to an annual value of US$200 million in 2000, second only to tourism as a source of foreign earnings. Cook Islands have also applied the same culture system with great success, although the scale of the industry there is much smaller than in French Polynesia and is currently worth US$6 million per annum. Several other countries in the Pacific (e.g. Federated States of Micronesia, Fiji and the Solomon Islands) have now assessed the potential to collect spat of *P. margaritifera*, identified productive areas and established farms. Interestingly, the larger species used for farming the white 'South Seas' pearls in Australia, Indonesia and the Philippines, *P. maxima*, cannot be collected reliably as spat and the pearl industries there depend on wild adult shell and hatchery-reared juveniles.

Vagile postlarvae

This life history stage is used for at least three forms of high-value coastal aquaculture in the tropics: shrimp, spiny lobsters and grouper. The farming of shrimp based on vagile pre-settlers has the longest history. Traditionally, postlarval shrimp were introduced into ponds in India and Indonesia by tidal action or hand collection from backwaters at low tide (Phillips *et al.*, 1993). These early extensive farming systems progressed to more active collection of postlarvae and growout in semi-intensive and intensive operations, and finally to the use of wild broodstock to produce hatchery-bred juveniles to enhance the supply. Although hatchery production of postlarval shrimp is now routine worldwide (Rosenberry, 2000), many small-scale farmers continue to practise extensive culture using wild larvae and some larger-scale farmers prefer wild-caught shrimp because they regard hatchery-produced seed as inferior in quality (Phillips *et al.*, 1993). Shrimp farming in Ecuador illustrates the significance of wild-caught juveniles to the industry. The abundance of wild postlarvae and the ease of growing them to market size resulted in a boom in the sector which comprised the collectors (*larveros*), middlemen, shrimp farms, processing plants and exporters (Ortiz, 1991). The industry produced 70,000 tonnes in 1988 (Aiken, 1990), predominantly from growout of wild postlarvae. However, there has since been a steady decline in supply of wild-caught juveniles, leading to greater reliance on collection and spawning of gravid females and postlarvae produced in hatcheries (Ortiz, 1991; Rosenberry, 2000).

The relatively new aquaculture industry for spiny lobsters (mainly *Panulirus ornatus*) in many Asian countries, including Vietnam, the Philippines, India, Thailand, Burma, China, Taiwan, Malaysia and Singapore, is based entirely on the capture of wild juveniles. Much of the stock is collected as postlarvae (pueruli), although settled juveniles are also collected from floating structures and from artificial and natural benthic habitats. The puerulus is the final stage larvae. It is 6–12 mm in carapace length and looks essentially like a small transparent lobster. The culture of pueruli collected using fibre-based materials from the wild for farming and stock enhancement is also under active consideration in Australia and New Zealand (Booth and Kittaka, 2000; Walker, 2001).

The industry in Vietnam provides an example of the benefits of this form of aquaculture to a developing country. There, specialized fishermen can collect in excess of 200 puerulus of *Panulirus ornatus* per boat per night during the settlement season (L. Tuan, personal communication). The pueruli are sold at US$4–5 each to farmers, who construct pens in shallow bays. The spiny lobsters attain market size of ~1 kg within 2 years on a diet of crustaceans, molluscs and fish. Survival during growout is reputed to approximate 90% and the price obtained by farmers is about five times greater for live animals, at \$25–30 kg^{-1} (Tuan *et al.*, 2000). In 2000, approximately 1000 t of live lobster were exported from Vietnam (L. Tuan, personal communication). The spiny lobsters usually spawn at least once during the growout period but as yet there is insufficient information to assess whether this industry is sustainable (see section below headed 'Advantages of using wild juveniles for aquaculture'). The capture and culture of pueruli, however, is destined to be the mainstay of spiny lobster farming worldwide for quite some time owing to the problems facing economically viable production of postlarvae in hatcheries. In particular, the numerous oceanic pelagic larval stages in the development of spiny lobsters, which take 4 months for tropical species and at least a year for those in temperate areas, are difficult to rear in captivity.

The collection of postlarval grouper (known as 'tinies' owing to their small size), to meet the demands of the live reef fish industry, is now widespread in many Asian countries, particularly Indonesia, the Philippines and Vietnam (Sadovy, 2000). Although some hatchery-reared groupers are available, wild-caught juveniles remain the primary source of fish for aquaculture of these species (Sadovy and Vincent, 2002). Due to their high value, potential impact on fish stocks and failure to meet growing demand, fisheries for postlarval grouper in Asia have been studied in detail (e.g. see Ogburn and Johannes, 1999; Sadovy, 2000; Tuan and Hambrey, 2000). The tinies are harvested mainly using fish shelters (deployed or suspended from boats sometimes in conjunction with lights), lift nets, and various kinds of set nets. Some of these methods can yield thousands of postlarval grouper per operator per night during the high season. The postlarvae are transferred to ponds or sea cages for rearing. Mortality is often high for these fish, due either to damage during capture or to poor fish husbandry (e.g. overcrowding, overfeeding and poor water quality) during culture (Ogburn and Johannes, 1999; Sadovy, 2000).

Collection of juveniles well after settlement

The large-scale use of juveniles in this category is restricted primarily to finfish. Milkfish (*Chanos chanos*), freshwater eels (*Anguilla* spp.) and yellowtail (*Seriola quinqueradiata*) are prime examples of species harvested several weeks after settlement.

Milkfish are one of the most important species cultured in Southeast Asia (Lee *et al.*, 1997). In the Philippines, for example, 'bangus' fry are collected using various netting techniques from many coastal areas by thousands of villagers, then sold to larger farms for growout (Ahmed *et al.*, 2001). The importance of the industry in the Philippines is reflected in the prohibition on the capture of adults. Approximately 165 million wild fry were harvested in 1995, but demand is greater than supply, and it is likely that some of the 150,000 t of production in 1996 may have come from imported fry (Ahmed *et al.*, 2001).

Taiwan leads Asia in hatchery production, which is becoming more common in the Philippines and Indonesia in order to augment supply from the wild (Lee *et al.*, 1997). Milkfish are farmed on a much smaller scale in the Pacific for food and as bait for tuna fishing (Fitzgerald, 1996).

Eels inhabit freshwater as adults but breed in the ocean and undergo an extended pelagic, oceanic larval phase. After metamorphosis from glass eel to elver in estuaries, juveniles make their way upstream to freshwater. During these migrations, the bulk of the cohort can be intercepted using fixed nets across river mouths, then stocked in ponds. Eel culture using wild-caught juveniles has been underway for several decades in Japan, Korea, China and Taiwan using locally caught juveniles, and in Malaysia using imported elvers (Tzeng, 1997). To gain an idea of the scale of the industry, there are > 5000 elvers to a kilogram and Japanese production peaked in 1979 with a harvest of 130 tonnes. It has since declined to around 50 t year^{-1} (Tzeng, 1997). Europe also has an eel aquaculture industry, which includes exports to Asia to make up for shortfalls in local supply (Wood, 1999), and there is a small but growing industry on Australia's eastern seaboard, with growout of elvers in ponds prior to release into lakes and impoundments for stock enhancement (L. McKinnon, personal communication).

There has been cage culture of yellowtail (predominantly *Seriola quinqueradiata* but also *S. dumerili*) in Japan since the middle of the 20th century with rapid expansion occurring in the 1980s (Anonymous, 1997). Current annual production is about 150,000 t (FAO, 2001). Yellowtail are caught around the country but culture is concentrated off southern Japan (Anonymous, 1997). The pre-settlement fish (1.5–10 cm in length) are associated with drift algae after hatching and then adopt a truly pelagic lifestyle. The fish are harvested with encircling nets and grown in cages until they are 2–3 years old and 70–80 cm in length (T. Hara, personal communication). Vietnam's rapidly expanding mariculture operations also include the related *S. dumerili* and *S. nigrofasciata* (Tuan *et al.*, 2000), principally for export to Japan, where an estimated 450,000 fry were sent in 1995 (Dao, 1999).

Collection of large juveniles

An extreme example of the use of wild-caught juveniles for aquaculture is the southern bluefin tuna (*Thunnus maccoyii*). This species commands premium prices as sashimi and the sea cage farming of large juveniles has been underway in South Australia since the early 1990s. Two- to four-year-old fish are caught by net or line, and transferred to large sea cages for fattening over periods of 3–10 months. The fish are sold when Japanese market prices are optimal. Over 95% of Australia's catch quota for this species is farmed in this manner.

Advantages of Using Wild Juveniles for Aquaculture

The use of wild juveniles has distinct advantages for aquaculture and potential advantages for the overall productivity of the species involved.

Bypassing the need to propagate juveniles

A major advantage of using wild-caught juveniles for coastal aquaculture is that it enables farmers, and managers of restocking and stock enhancement programmes, to bypass the often expensive, technically complex, process of mass-producing that life stage in hatcheries. This process is acknowledged as one of the major bottlenecks in aquaculture (Rimmer, 1999). The advantages of wild-caught juveniles in this regard are not limited to cheaper production costs; they also extend to the availability of juveniles that are already fit for rearing in coastal waters. This is not always true for hatchery-reared juveniles, particularly those released in stock enhancement programmes, where morphological and behavioural deficits can increase vulnerability to predation and other sources of mortality (Munro and Bell, 1997).

A related advantage is that wild juveniles can be used to assess the potential for aquaculture of coastal species, even when they are relatively difficult to collect. Provided sufficient numbers can be obtained for meaningful experiments, problems associated with keeping the species in captivity, and factors that affect rates of growth and survival, can be determined without the need to invest in hatchery technology. Bell *et al.* (1991) and Ramofafia *et al.* (1997) provide some relevant examples from the Pacific.

Reduced impacts on wild stocks

Wild juveniles have special advantages over hatchery-reared animals for stock enhancement programmes. For this process, it is important that the released juveniles have comparable gene frequencies to the wild stock (Munro and Bell, 1997, and references therein). To achieve this, hatcheries need to collect large numbers of broodstock from the area where the juveniles are to be released, and replace broodstock regularly to ensure that selective breeding and inbreeding do not occur. The costs involved in managing broodstock this way are substantial and the protocols are not always successful, with the result that multiple cohorts need to be released to maintain gene frequencies (Bartley and Kent, 1990). All these problems are avoided by using juveniles collected from the wild.

Juveniles caught from the wild also generally have a much lower incidence of disease than those propagated in hatcheries. Once again, this is an important consideration when releasing juveniles in stock enhancement programmes as diseases are not only capable of affecting conspecifics, they are often more virulent in atypical hosts (Langdon, 1989). Low risk of disease is also an obvious advantage to mariculture ventures. Although the use of wild-caught juveniles should reduce the chances of spreading diseases compared to the use of hatchery-reared juveniles, particularly when wild-caught juveniles are not held at high density prior to growout or release, there are still risks involved. These risks are greatest when animals are moved from one place to another (Munro and Bell, 1997, and references therein).

Socioeconomic benefits

In many situations, the collection and growout of juveniles confers a greater range of socioeconomic benefits than aquaculture based on animals supplied from hatcheries, e.g. collection and sale of juveniles to growout enterprises can provide income for sectors of the population otherwise excluded from aquaculture operations. Such considerations are especially important in developing countries, where advanced technology and capital intensive operations are restricted and where traditional marine tenure operates. Ogburn and Johannes (1999) report the positive effects of collection of grouper juveniles in the Philippines, where fewer people now practise destructive fishing with dynamite and poison, and where there has been a reduction in fishing pressure on wild-caught adults and less targeting of spawning aggregations, which otherwise leads to overexploitation (Johannes and Riepen, 1995; Birkeland, 1997).

In one group of islands in French Polynesia, Tisdell and Poirine (2000) estimated that a quarter of all families earn a living from the pearl industry by selling spat to larger farms. This reliance on wild spat has led to conservation of adults to ensure continued supply of oysters, and provided a model for other nations in the Pacific, which have begun to conserve wild stocks of blacklip pearl oysters to pave the way for development of their own pearl industries (Friedman, 1999). In response to McAllister's (1999) concerns about aquaculture depriving local fishers of their livelihood, the capture and culture of wild juveniles should actually increase the opportunities to earn income, provided the growout of the animals occurs in the country of harvest.

Increasing overall productivity

Perhaps the most exciting advantage of using wild-caught juveniles for aquaculture is the scope for increasing the overall productivity of the species involved. This possibility arises because the natural mortality of juveniles at or shortly after settlement is often extreme, and because 'nursing' during aquaculture can quarantine juveniles against such mortality so that they survive in consistently higher proportions. As described above, these juveniles can then either be grown-out for sale (see also Carleton and Doherty, 1999; Dufour, 1999) or used in restocking or stock enhancement programmes (see also Doherty, 1994; Maroz and Fishelson, 1997; Munro and Bell, 1997; Watson *et al.*, 2001).

The literature on early mortality of coral reef fish and invertebrates supports this contention. Although coral reef fish are highly fecund (Sale, 1980), they experience great mortality during their pelagic development phase, which lasts from weeks to months (Leis, 1991), and during settlement to shallow-water habitats (Carr and Hixon, 1995; P. Doherty, unpublished data, but see Koenig and Coleman, 1998). Recent studies on settlement mortality of an acanthurid (surgeonfish) in Moorea Lagoon found that more than 50% did not survive the first night (P. Doherty, personal communication). Survivorship generally improves with age (Sale and Ferrell, 1988; Hixon, 1991) but fewer than 1% of larvae ever reach adulthood (Doherty, 1991). Coral reef invertebrates also experience massive mortality at settlement. For example, the numbers of pueruli of the spiny lobster, *Panulirus argus*, entering nursery habitats are much larger than the available shelter can sustain (Smith and Herrnkind, 1992) and less than 5% of settlers are estimated to survive the first benthic year (Herrnkind *et al.*, 1997). The estimate that 90% of pueruli of *P. ornatus* collected in Vietnam survive to market size during growout shows the enormous gains that can be made through capturing and culturing juvenile spiny lobsters. Similarly, Friedman and Bell (2000) demonstrated that the very high mortality rates for wild spat of the blacklip pearl oyster, *Pinctada margaritifera*, described by Sims (1993), can be transformed into survival rates of around 80% by using spat collectors and then rearing the juveniles in intermediate culture until they reach a size where they escape predation.

Despite the fact that there is scope for increasing productivity by salvaging a high proportion of settling larvae from natural mortality, there are concerns that the use of wild-caught juveniles simply constitutes another capture fishery which may lead to overexploitation if fishing pressure is excessive (Sadovy and Pet, 1998). Commercial and recreational fishers who target adult stocks wish to be assured that early removal of individuals will not have an impact on the number of adults that they are permitted to catch. Although there is a lack of dedicated research into the effects of removing juveniles from wild populations for aquaculture, there is already evidence that some of the fisheries for wild-caught juveniles are in decline (e.g. Phillips *et al.*, 1993; Sadovy, 2000; Ahmed *et al.*, 2001). The main questions revolve around the relative contribution of overfishing and degradation of essential habitat to the decline. For example, the decline in abundance of postlarval shrimp in Ecuador was preceded by extensive mangrove deforestation during farm construction (Ortiz, 1991; Phillips *et al.*, 1993). Similarly, reduced catches of juvenile eels in Taiwan and Japan and declining numbers of adults have been associated with habitat destruction, river impoundment and changes in currents, in addition to many years of heavy exploitation of elvers (Tzeng, 1997). Significantly, reduced juvenile grouper catches in the Philippines have only been reported from areas where habitat has been damaged (Ogburn and Johannes, 1999).

At the high rates of mortality associated with settlement, the harvest of juveniles for aquaculture will have minor effects on the abundance of adults compared to the increased production from farming. Nevertheless, managers will need to achieve a responsible balance between maintaining spawning biomass and increasing productivity. From the data available for spiny lobsters presented earlier, a hypothetical example, based on 100 million pueruli at settlement, 5% survival after the first benthic year and 90% survival during culture, indicates that removal of 2 million pueruli for growout would reduce recruitment to the fishery after 1 year from 5 million to 4.9 million individuals. However, this modest reduction would be more than offset by the additional 1.8 million market-size animals produced from the 2 million pueruli reared in aquaculture, giving a net gain of 1.7 million individuals. Even if survival in aquaculture decreased to 50% the net gain would be 0.9 million individuals, and survival of 20% would still yield an additional 0.3 million lobsters.

We hasten to point out that these calculations are hypothetical and that the potential for aquaculture to add to production will depend on the numbers of animals settling, and their rates of survival in the wild and in culture. To apply this approach to the management of *P. ornatus* in Vietnam, for example, the following accurate and precise data are needed: (i) population size and catch rates of wild adults; (ii) spatial and temporal variation in abundance of larvae; (iii) survival of pueruli after settlement; (iv) proportion of pueruli collected for growout; (v) survival of the spiny lobsters during growout; and (vi) the fate of larvae emanating from stock spawning in aquaculture cages – there is the possibility that these larvae may be entrained too far inshore to complete their complex oceanic pelagic phase.

The potential increases in productivity described above need not disadvantage fishers who target the adults. In Australia and New Zealand, where there is much interest in collection and growout of pueruli, proposals under consideration include releasing part of the 'bonus' catch nursed through the period of high mortality for stock enhancement so that there is a net increase in numbers of animals in the wild, or at least biological neutrality. In Tasmania, Australia, there are plans to release 5% of the animals taken out of the sea as postlarvae plus 20% of the survivors (Walker, 2001). In New Zealand, quotas to fish for pueruli are granted on expiration of adult licences (Booth, 2000).

We can also foresee situations where managers of depleted fisheries will be able to restore stocks, and maintain a level of productivity, by transferring fishing effort from adults to juveniles. The prerequisites for such management would be a low spawning biomass and the existence of relatively easy systems for collecting and growing the juveniles. The levels of fishing permitted for adults, and the proportions of juveniles held for growout and release to the wild in restocking programmes, would depend on the status of the population. For example, in severely depressed fisheries there would need to be a moratorium on the capture of adults, and all juveniles should be released after a period of growout to rebuild the population. On the other hand, where stocks are not so depressed, harvesting of reduced numbers of adults, combined with farming of juveniles and releases of a portion of the cultured young in the wild, should increase spawning biomass and pave the way for increased productivity. Elements of this process can be found in the Tasmanian proposal for lobster pueruli aquaculture, where a proportion of reared juveniles are to be introduced into prime habitats with low larval supplies, thus enhancing natural productivity (Gardner *et al.*, 2000).

A key assumption in such developments is that the juveniles are taken prior to the mass mortality associated with settlement. Strict controls on the size of individuals, and perhaps timing of operations, is needed to fulfil this important assumption. The consequences of mismanagement of this part of the process are evident from the work of Dufour (1999), who hypothesized that if juveniles of coral reef fish were collected 50 days rather than 5 days after settlement, there would be a subsequent reduction of ~40% in the future adult population. Thus, the benefits of increased production from aquaculture of wild-caught juveniles will diminish as the age of the animals increases and there will need to be persuasive reasons why individuals should be diverted from wild fisheries into aquaculture for large juveniles that have already run the gauntlet of high natural mortality. The farming of southern bluefin tuna depends on 2- to 4-year-old fish; however, in this case, the quota previously taken by the tuna fishery has been transferred to the aquaculture operations so there is no additive negative effect on replenishment.

Disadvantages of Using Wild Juveniles in Aquaculture

The use of wild-caught juveniles for aquaculture is not without its limitations and pitfalls. The various constraints inherent in the process, and some ways in which this form of aquaculture could affect the ecosystem, are summarized below.

Limitation to the supply and quality of captured juveniles

The availability of animals for aquaculture based on wild-caught juveniles is constrained by the natural supply of larvae, which can be expected to change significantly from year to year (Doherty, 1999). This is proving to be a problem for the milkfish farming industry in the Philippines, where the supply of wild-caught juveniles cannot meet the current annual demand of 1.65 billion fry for aquaculture (Ahmed *et al.*, 2001). In this case, simple hatchery technology for milkfish is available (Tamaru *et al.*, 1995) and elsewhere, e.g. Taiwan, the industry relies heavily on hatchery-reared juveniles so there is no impediment to meeting increasing demand. This is not true for eels, however, where demand exceeds the supply of glass eels or elvers in some countries and there is little prospect of closing the life cycle of these animals in captivity, let alone producing juveniles cost effectively in hatcheries. Spatial patchiness in the availability of wild juveniles and uncertainty about the timing of arrival, e.g. green mussels in New Zealand, can make such industries vulnerable if there are few alternative sites for cost-effective spat collection. To illustrate this point, the discovery of toxic dinoflagellate cysts on the macroalgae that support most of New Zealand's mussel spat led to restrictions on the movement of spat to prevent the spread of the algae (Smith and Rhodes, 1994), affecting farms reliant on that source of spat. Variable quality is also an issue for aquaculture based on wild-caught juveniles. Poor quality animals, often a result of the mode of harvest, have been reported in the lobster pueruli industry in Vietnam (Tuan *et al.*, 2000), mussel aquaculture in New Zealand (Jeffs *et al.*, 1999) and grouper collections in Asia (Sadovy, 2000).

Lack of selective breeding

Reliance on wild juveniles prevents any form of genetic improvement (e.g. hybridization, selective breeding) to the seed for aquaculture. While this is desirable for stock enhancement programmes (see the preceding section), it is a distinct disadvantage for farming operations because > 80% gains in growth rate have been made through selective breeding within five to six generations (Gjerde and Korsvoll, 1999; Dey and Gupta, 2000). Without the ability to maintain captive broodstock, there is also no scope for development of a genetic tag to assess the contribution of cultured wild juveniles to subsequent generations in restocking programmes. An appropriate tag for wild juveniles in put-and-take stock enhancement programmes has been available in some situations, however. For example, the stress band laid down in scallop shells during transfer from intermediate culture to the sea bed clearly identifies the released animals (Dao *et al.*, 1999).

Effects on the ecosystem

The collection of wild juveniles for coastal aquaculture has at least two potentially undesirable effects on the ecosystem: the capture of non-target species as bycatch, and removal of animals from the food chain. The amount of bycatch associated with the collection of wild juveniles for aquaculture is not well documented, but Phillips *et al.* (1993) reported that the bycatch of finfish and shrimp of lesser value can be an order of magnitude greater than the target species during harvest of juvenile *Penaeus monodon* in India. The severity of the bycatch problem depends on the survival of the animals, which can be expected to vary with the species involved, fishing methods, and the time needed to sort the catch. For example, large numbers of *Siganus* spp. (rabbitfish) removed from fish attractors in the Philippines, and deemed to be unsuitable for culture, are released unharmed (Sadovy, 2000). On the other hand, the use of fyke nets for harvesting grouper tinies often results in high mortality of bycatch (Ogburn and Johannes, 1999). Similarly, low-value shrimp sorted from the catch of juvenile *Penaeus vannamei* in Ecuador were usually killed during the process (Ortiz, 1991). Clearly, collectors need to be made aware of the effects of their activities on other species and should be trained in the use of methods that minimize bycatch.

In coral reef ecosystems, the fate of new settlers has been colourfully referred to as making a 'suicide drop onto the reef' (Kaufman *et al.*, 1992) or having to contend with 'a wall of mouths' (Hamner *et al.*, 1988). The large-scale mortality of juveniles at or shortly after settlement, and evidence from dietary studies, indicate that postlarvae are an important source of food for piscivorous species higher up the food chain (Roberts, 1996). The impacts of removing a proportion of this food source are unknown. However, they should not be any greater than for other accepted forms of fishing. The only difference is that removal of large fish can spare prey species, whereas the removal of prey may have an effect on the condition, and therefore egg production, of adult stocks.

Capture and culture of postlarval coral reef fishes for the live reef fish trade

An emerging fishery based on capture and culture of postlarval coral reef fishes aims to supply the live food and ornamental markets, which have burgeoned since the 1970s and late 1980s, respectively (Wood, 2001). The former market is currently estimated to be worth US$830 million (Sadovy and Vincent, 2002). The industry supplies, live to the table, a few dozen species of larger fish, predominantly from the families Serranidae (cod, groupers) and Lutjanidae (snappers) as well as the Napoleon wrasse (*Cheilinus undulatus*). The major markets for the live food trade are Hong Kong, China and Taiwan, where diners are prepared to pay inordinate amounts for fresh gourmet fish meals. A prodigious effort is being made to breed species in the groups mentioned and rear the juveniles in hatcheries, but collection of adult fish and growout of wild-caught juveniles continues to sustain the industry (Rimmer, 1999; Sadovy, 2000).

In contrast, hundreds of small, brightly coloured or otherwise interesting species are targeted for the rapidly growing marine ornamental trade (Table 17.2; Fig. 17.1). The value of this market is difficult to assess – current estimates of retail and FOB values range from US$90 to 300 million and US$28 to 44 million, respectively (Wood, 2001). The vast majority of the fish are collected from the wild and there is little commercial hatchery production, despite promising results in closing the breeding cycle of a small number of species (Wood, 2001; Sadovy and Vincent, 2002).

The widespread use of destructive fishing techniques and overfishing seriously threaten the sustainability of these two trades in live coral reef fishes. In particular, sodium cyanide solution is often used in Southeast Asia to stun fish during collection, often resulting in death or such severe internal damage that they die shortly after purchase by hobbyists. This method of fishing not only taints the reputation of the industry, it also causes collateral damage to the habitat, i.e. death of corals and other non-target species of fish and invertebrates (Jones, 1997; McManus *et al.*, 1997; Mous *et al.*, 2000). Further destruction occurs when stunned fish retreat into holes in the reef and coral is broken to retrieve them, although collectors may also do this even when the poison is not used. Several organizations (e.g. The International Marinelife Alliance and The Marine Aquarium Council) have alerted coastal communities to the threat posed by destructive fishing (Barber and Pratt, 1997), and established a system of certification that promotes sustainable trade in aquarium fishes by rewarding good practices at all points along the supply chain (Holthus, 1999; Wood, 2001).

Table 17.2. Comparative numbers of species in the main ornamental and live food fish families cultured from postlarvae in Solomon Islands (Hair, unpublished data) and French Polynesia (source: AquaFish Technology product list).

Family	Solomon Islands	French Polynesia
Fish		
Muraenidae (moray eels)	na	> 3
Holocentridae (squirrelfish, soldierfish)	1	> 11
Serranidae (groupers, coral trouts, soapfish)	3	8
Cirrhitidae (hawkfish)	na	4
Apogonidae (cardinalfish)	> 5	> 8
Lutjanidae (snapper)	4	4
Chaetodontidae (butterflyfish)	3	18
Pomacanthidae (angelfish)	1	5
Pomacentridae (damselfish)	> 18	> 15
Labridae (wrasses)	na	16
Blenniidae (blennies)	na	7
Gobiidae (gobies)	na	> 7
Antennaridae (frogfish)	na	3
Ephippidae (batfish)	2	1
Siganidae (rabbitfish)	2	2
Zanclidae (Moorish idol)	na	1
Acanthuridae (surgeonfish, tangs)	> 10	19
Balistidae (triggerfish)	6	5
Ostraciidae (boxfish)	1	3
Tetraodontidae (pufferfish, tobies)	5	6
Invertebrates		
Assorted shrimps	> 2	7
Palinuridae (spiny lobster)	2	

na, no attempt to culture this group yet although they may occur in catches.

Overfishing is a major concern for species in the live fish trade and is exacerbated by two features of the market and fishery. First, rarity confers greater value, with the result that poor fishers continue to target high-value species far beyond the stage of economic viability in other fisheries (Sadovy and Vincent, 2002). Second, the groupers form spawning aggregations, which makes them highly vulnerable to capture, even at low population densities (Johannes and Riepen, 1995). Consequently, groupers are in decline throughout Asia (Sadovy and Vincent, 2002) and buyers are shifting their operations further from the markets as stocks are depleted. Overexploitation was previously considered to be a lower risk for ornamental fishes, with the exception of rare or highly endemic species. More recently, however, there are signs that ornamental fishes are not as resilient to heavy fishing pressure as was once believed (Wood, 2001; Sadovy and Vincent, 2002).

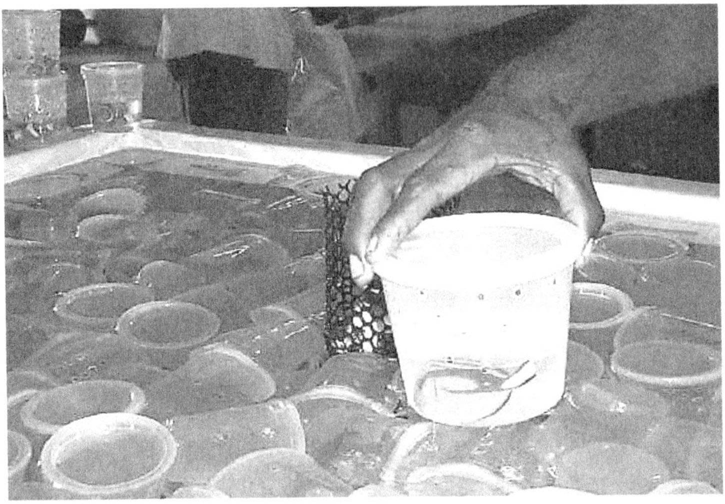

Fig. 17.1. A tropical marine aquarium products export facility in the Solomon Islands, showing how fish are prepared for international air freight to the United States. The species displayed is *Paracanthurus hepatus*.

An environmentally friendly and biologically sustainable way to address the twin problems of destructive fishing methods and overexploitation facing the live coral reef fish industries is the responsible capture of juveniles during their transition from the planktonic to the demersal stage (see preceding sections; Doherty, 1994; Carleton and Doherty, 1999). Various methods are available to harvest wild juveniles at settlement (Choat *et al.*, 1993); however, light traps and crest nets appear to be particularly suitable as they target postlarval fishes and then retain them in good condition. Light traps are passive devices that attract and concentrate photopositive zooplankton including pre-settlement fishes (Doherty, 1987). The crest net, which rarely exceeds 3 m in mouth gape, was developed by Dufour and Galzin (1993) in French Polynesia and has also been used in Australia (Doherty and McIlwain, 1996) and the Solomon Islands (Hair *et al.*, 2002). Crest nets are deployed on shallow reef tops with consistent unidirectional water flow and capture the small fish as they move from deeper waters to settle on reefs at night. The deployment of relatively small nets that sample only a very minor proportion of the total reef crest has a built-in check against overharvesting because the removal of large numbers of potential settlers is not possible.

In French Polynesia and the Solomon Islands, two groups are now using these sampling techniques to explore the feasibility of a new fishery targeting pre-settlement reef fish (Dufour *et al.*, 1999; Whitty, 2001; Hair *et al.*, 2002). The practice of harvesting the fish and keeping them from several weeks up to 6 months reduces the stress normally associated with the rapid succession of capture, enclosure, transport and introduction to new surroundings that is typical of current industry practice (Baquero, 1995; Dufour *et al.*, 1999; Wood, 2001).

The project in the Solomon Islands is concentrating on development of low technology capture and growout methods for local communities. Fish captured using crest nets and light traps are reared in land-based tanks supplied with unfiltered seawater, where the fish are fed fish roe and finely minced fish or shellfish. Rocks encrusted with algae and invertebrates (live rock) are used to supplement the diet of the grazers and provide shelter for most species. More than 60 species from 15 families have been reared in this experimental project (Table 17.2) for trial shipments by a local exporter of marine aquarium products. The next stage of the project involves growout experiments in simple, small (1–2 m^3) sea cages in lagoonal areas adjacent to participating villages. On-shore facilities are likely to be beyond the means of most local communities, particularly in the initial stages of establishing this artisanal fishery.

In contrast, the operation in French Polynesia is more intensive and uses current aquarium technology. A specially formulated pelleted feed promotes rapid growth rates (V. Dufour, personal communication). More than 150 species from 21 families have been reared in French Polynesia (Table 17.2); however, relatively few of those species make up the bulk of individuals marketed. Initial sales from the French Polynesian company are promising due to the 'captive bred' status of the fish, i.e. they are accustomed to captivity, are well fed, and comparatively stress-free (V. Dufour, personal communiction).

The competitiveness of these enterprises will be enhanced greatly by the planned introduction of eco-labelling for aquarium specimens, which will raise the value of fish caught and reared in an environmentally sustainable manner (Holthus, 1999; Wood, 2001). Another advantage of this emerging form of aquaculture is the prospect of supplying new, that is rare, species for the market. This is particularly true for crest nets, which capture the juveniles of a number of cryptic species, many of which are not usually collected for the aquarium trade. For example, pueruli of the spiny lobster, *Panulirus versicolor*, are captured frequently in crest nets in the Solomon Islands. After the first moult, the juveniles develop attractive coloration (Fig. 17.2), are easy to maintain and have been well accepted by the trade. Although it is sometimes more difficult to keep younger specimens (Wood, 2001), it is also true that some species that do not readily make the transition from the wild to an aquarium as adults may adapt to captivity if caught early (Dufour *et al.*, 1999).

The main constraints to this fishery are that light traps and crest nets are selective, and that not all species caught will be amenable to culture, e.g. in the Solomon Islands, wrasses (Labridae) are generally captured at a size too small to be reared successfully. Further, in French Polynesia the supply of Picasso triggerfish (*Rhinecanthus assasi*) exceeds demand while the opposite is true for other species (V. Dufour, personal communication). Thus, the small-scale fisheries will not be able to supply the full range of species sought by the market and will need to identify places where pre-settlement fish of sufficient species in high demand can be collected reliably in economically viable numbers. Such

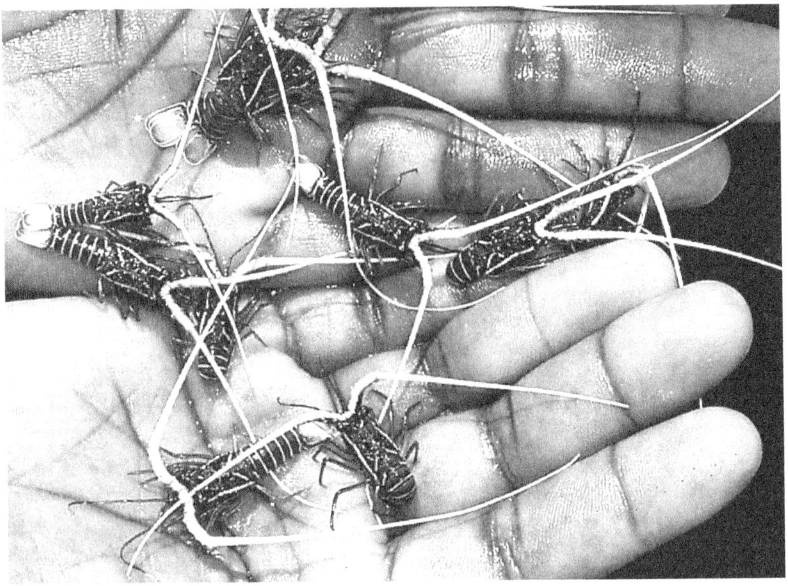

Fig. 17.2. Juvenile spiny lobsters (*Panulirus versicolor*) reared for 3 weeks after capture by crest nets in the Solomon Islands and ready for sale to an exporter of marine aquarium products.

settlement hotspots are known for coral reef invertebrates, such as pearl oysters (Friedman *et al.*, 1998). Catch rates of postlarval fishes in French Polynesia have certainly been in this category (Dufour *et al.*, 1996) but the identification of premium fishing sites involves time and expense (V. Dufour, personal communication). Numbers of ornamental fishes taken from the few locations where crest nets and light traps have been used in the Solomon Islands have so far been relatively low (Hair *et al.*, 2002). Extensive collections using light traps on the Great Barrier Reef have also generally yielded low catches of valuable species (Carleton and Doherty, 1999). This indicates that considerable research may be needed to find locations for commercial operations.

Another potential impediment to the development of this activity is that very few species suitable for the live food market have been caught in large numbers in the South Pacific, although large numbers of an *Epinephelus* species have been caught sporadically in French Polynesia (Dufour *et al.*, 1996). The results from the South Pacific are in contrast to Asia, where very large numbers of groupers are harvested (Sadovy, 2000, and references therein).

A quick appraisal of input costs suggests that adaptation of the crest net technique will be more suited than light traps to development of low technology enterprises for the capture and culture of postlarval coral reef fish at the village level. However, sites with shallow reef tops and unidirectional water flow are not as common as sites where light traps can be used. Thus, low cost, low maintenance night lights, combined with collection by simple nets, as practised for the collection of spiny lobster pueruli in Vietnam, may yet prove to be suitable for villagers interested in catching and growing juvenile coral reef fishes for the aquarium trade.

Conclusions

The fact that only a small percentage of marine species that settle into nursery habitats survive to become breeding adults is a persuasive argument for using some of the settling cohort to increase productivity through growout in aquaculture or in stock enhancement programmes. The wide range of aquaculture based on collection of small animals demonstrates that this process has potential for any species that can be collected *en masse* at an early stage. However, responsible application of aquaculture based on animals captured from the wild will depend on capturing juveniles before they experience the severe mortality associated with settlement, limiting the catch to ensure replenishment of spawning biomass, returning sufficient juveniles to the wild to compensate fisheries targeting adults, and use of capture methods that minimize bycatch of non-target species and do not damage supporting habitats. Not all forms of aquaculture that use wild juveniles are being implemented in this way and particular attention is needed in the areas of overharvesting, the life history stage exploited, and the effects on other users of the resource. On balance, the potential advantages of using wild juveniles for aquaculture, which also include lower operating costs, the maintenance of genetic diversity, and the socioeconomic benefits that accrue from allowing more people to participate in the production of the species, outweigh the inherent disadvantages.

The potential for new aquaculture enterprises based on wild juveniles is embodied in the recent developments in the South Pacific designed to supply live coral reef fishes and invertebrates mainly for the aquarium trade. Although the light traps and crest nets used to harvest the settling juveniles are unable to catch all species demanded by the market, the early results indicate that these methods are: (i) an environmentally friendly source of supply for many species; (ii) capable of catching economically viable numbers of animals once areas with consistently good settlement of larvae are identified; and (iii) able to create niche markets for species that are normally difficult to collect, or to transfer from the wild to captivity as adults. In addition, postlarval coral reef species nursed through the survival bottleneck at settlement can be released to enhance stocks in overfished areas.

Management of this emerging form of aquaculture will have the added task of ensuring that practices comply with requirements for eco-labelling. These controls may take the form of limits on fishing gear, areas closed to fishing, closed seasons, and prohibited species if it is found that post-capture mortality rates are unacceptably high (Sadovy, 2000). As a developing fishery, these controls will need to evolve and appropriate research should be undertaken to adapt management measures to meet the needs of the producers, marketplace and stocks of fish.

If artisanal fisheries for the capture and culture of pre-settlement fish can be established in a responsible manner, they should enhance the employment and economic opportunities for coastal communities in developing countries through novel and sustainable use of inshore marine resources. They may also be able to compensate for some of the damage incurred to coral reef habitats through current and past practices.

Acknowledgements

We are grateful to a number of people for supplying information, including Vincent Dufour, Le Anh Tuan, Takashi Hara and Kevin Williams. The project in the Solomon Islands on the capture and culture of postlarval fish is supported by funding from the Australian Centre for International Agricultural Research. This manuscript is ICLARM contribution no. 1633.

References

Ahmed, M., Magnayon-Umali, G.A., Valmonte-Santos, R.A., Toledo, J., Lopez, N. and Torres, F. Jr (2001) *Bangus Fry Resource Assessment in the Philippines*. ICLARM Technical Report 58. International Center for Living Aquatic Marine Resources Management, Manila, Philippines, 38pp.

Aiken, D. (1990) Shrimp farming in Ecuador – whither the future? *World Aquaculture* 21, 26–30.

Anonymous (1997) Marine product industry of Japan, 'yellowtail'. Japan Fisheries Resource Conservation Association, Tokyo, Japan, 10pp. [in Japanese].

Arbuckle, M. and Metzger, M. (2000) *Food for Thought – A Brief History of the Future of Fisheries' Management*. Challenger Scallop Enhancement Company Limited, New Zealand, 25pp.

Baquero, J. (1995) The stressful journey of ornamental marine fish. *Sea Wind* 9(1), 19–20.

Barber, C.V. and Pratt, V.R. (1997) Policy reform and community-based programmes to combat cyanide fishing in Philippines. *South Pacific Commission Live Reef Fish Information Bulletin* 3, 26–35.

Bartley, D.M. and Kent, D.B. (1990) Genetic structure of white seabass populations from the southern California Bight region: applications to hatchery enhancement. *Californian Co-operative Fisheries Investigation Report* 31, 97–105.

Bell, J.D., Quartararo, N. and Henry, G.W. (1991) Growth of snapper, *Pagrus auratus*, from south-eastern Australia in captivity. *New Zealand Journal of Marine and Freshwater Research* 25, 117–121.

Birkeland, C. (1997) Disposable income in Asia – a new and powerful external pressure against sustainability of coral reef resources in Pacific Islands. *Reef Encounter* 22, 9–13.

Booth, J.D. (2000) New Zealand's rock lobster fisheries. In: Phillips, B.F. and Kittaka, J. (eds) *Spiny Lobsters: Fisheries and Culture*. Fishing News Books, Oxford, UK, pp. 78–89.

Booth, J.D. and Kittaka, J. (2000) Spiny lobster growout. In: Phillips, B.F. and Kittaka, J. (eds) *Spiny Lobsters: Fisheries and Culture*. Fishing News Books, Oxford, UK, pp. 556–585.

Carleton, J.H. and Doherty, P.J. (1999) The potential for collecting tropical marine fish fry by light attraction. In: Cabanban, A.S. and Phillips, M. (eds) *The Aquaculture of Coral Reef Fishes*. Proceedings of the workshop on aquaculture of coral reef fishes and sustainable reef fisheries held in Kota Kinabalu, Sabah, Malaysia, December 1996, pp. 184–197

Carr, M.H. and Hixon, M.A. (1995) Predation effects on early post-settlement survivorship of coral reef fishes. *Marine Ecology Progress Series* 124, 31–42.

Choat, J.H., Doherty, P.J., Kerrigan, B.A. and Leis, J.M. (1993) Sampling of larvae and pelagic stages of coral reef fishes: a comparison of towed nets, purse seine and light-aggregation devices. *Fisheries Bulletin* 91, 195–201.

Dao, J-C., Fleury, P-G. and Barret, J. (1999) Scallop sea bed culture in Europe. In: Howell, B.R., Moksness, E. and Svasand, T. (eds) *Stock Enhancement and Sea Ranching*. Fishing News Books, Oxford, UK, pp. 423–436.

Dao, M.S. (1999) Marine fish culture in Vietnam. In: Cabanban, A.S. and Phillips, M. (eds) *The Aquaculture of Coral Reef Fishes*. Proceedings of the workshop on aquaculture of coral reef fishes and sustainable reef fisheries held in Kota Kinabalu, Sabah, Malaysia, December 1996, pp. 144–152.

Dey, M.M. and Gupta, M.V. (2000) Socioeconomic aspects of disseminating genetically improved Nile tilapia (*Oreochromis niloticus*) in Asia: introduction. *Aquaculture Economics and Management* 4, 5–12.

Doherty, P.J. (1987) Light-traps: selective but useful devices for quantifying the distributions and abundances of larval fishes. *Bulletin of Marine Science* 41, 423–431.

Doherty, P.J. (1991) Spatial and temporal patterns in recruitment. In: Sale, P.F. (ed.) *The Ecology of Fishes on Coral Reefs*. Academic Press, London, pp. 261–293

Doherty, P.J. (1994) A potential role for light-traps in enhancement of coral reef fisheries. In: Munro, J.L. and Munro, P.E. (eds) *The Management of Coral Reef Resource Systems*. ICLARM Conference Proceedings No. 44. International Center for Living Aquatic Resources Management, Manila, Philippines, pp. 92–93.

Doherty, P.J. (1999) Recruitment limitation is the theoretical basis for stock enhancement in marine populations. In: Howell, B.R., Moksness, E. and Svasand, T. (eds) *Stock Enhancement and Sea Ranching*. Fishing News Books, Oxford, UK, pp. 9–21.

Doherty, P.J. and McIlwain, J. (1996) Monitoring larval fluxes through the surf zones of Australian coral reefs. *Marine and Freshwater Research* 47, 383–390.

Dufour, V. (1999) Population dynamics of coral reef fishes and the relative abundance of their early life history stage – an example from French Polynesia. In: Cabanban, A.S. and Phillips, M. (eds) *The Aquaculture of Coral Reef Fishes*. Proceedings of the workshop on aquaculture of coral reef fishes and sustainable reef fisheries held in Kota Kinabalu, Sabah, Malaysia, December 1996, pp. 198–204.

Dufour, V. and Galzin, R. (1993) Colonization patterns of reef fish larvae to the lagoon at Moorea Island, French Polynesia. *Marine Ecology Progress Series* 102, 143–152.

Dufour, V., Riclet, E. and Lo-Yat, A. (1996) Colonization of reef fishes at Moorea Island, French Polynesia: temporal and spatial variation of the larval flux. *Marine and Freshwater Research* 47, 413–422.

Dufour, V., Jouvenel, J.-Y. and Lo, C. (1999) Collecting marine fish and invertebrate larvae: a sustainable technology. 1st International Conference of Marine Ornamentals. Hawaii. [Abstract only.]

FAO (Food and Agriculture Organization of the United Nations) (2001) *Aquaculture Production Statistics 1970–1999*. Fishstat Plus Universal Software for fishery statistical time series, Version 2.30. CD-Rom.

Fitzgerald, W.J. (1996) *Preliminary Report on the Use of Milkfish (*Chanos chanos*) for Tuna Longline Bait*. South Pacific Commission 26th Regional Technical Meeting on Fisheries Information Paper, New Caledonia, 23pp.

Friedman, K.J. (1999) Pearl culture using wild-caught spat of blacklip oysters (*Pinctada margaritifera*) in Solomon Islands. PhD thesis, James Cook University, Queensland, Australia, 238pp.

Friedman, K.J. and Bell, J.D. (2000) Shorter immersion times increase yields of the blacklip pearl oyster, *Pinctada margaritifera* (Linne.), from spat collectors in Solomon Islands. *Aquaculture* 187, 299–313.

Friedman, K.J., Bell, J.D. and Tiroba, G. (1998) Availability of wild spat of the blacklip pearl oyster, *Pinctada margaritifera*, from open reef systems in Solomon Islands. *Aquaculture* 167, 283–299.

Gardner, C., Mills, D., Ibbott, S., Wilcox, S. and Crear, B. (2000) *Preliminary Investigation Towards Ongrowing Puerulus to Enhance Rock Lobster Stocks While Providing Animals for Commercial Culture*. The Marine Research Laboratories, Tasmanian Aquaculture and Fisheries Institute, University of Tasmania. Technical Report Series No. 13, 67pp.

Gjerde, B. and Korsvoll, S.A. (1999) Realized selection differentials for growth rate and early sexual maturity in Atlantic salmon. Abstracts, *Aquaculture Europe 99*, Trondheim, Norway. European Aquaculture Society Special Publication No. 27, 73–74.

Hair, C.A., Doherty, P.J., Bell, J.D. and Lam, M. (2002) *Capture and Culture of Presettlement Coral Reef Fishes in the Solomon Islands. Proceedings of the 9th International Coral Reef Symposium*, Bali, Indonesia, October 2000.

Hamner, W.M., Jones, M.S., Carleton, J.H., Hauri, I.R. and Williams, D.McB. (1988) Zooplankton, planktivorous fish, and water currents on a windward reef face: Great Barrier Reef, Australia. *Bulletin of Marine Science* 42, 459–479.

Herrnkind, W.F., Butler, M.J. and Hunt, J.H. (1997) Can artificial habitats that mimic natural structures enhance recruitment of Caribbean spiny lobsters? *Fisheries* 22(4), 24–27.

Hixon, M.A. (1991) Predation as a process structuring coral-reef fish communities. In: Sale, P.F. (ed.) *The Ecology of Fishes on Coral Reefs*. Academic Press, San Diego, California, pp. 475–508.

Holthus, P. (1999) The Marine Aquarium Council, certifying quality and sustainability in the marine aquarium industry. *South Pacific Commission Live Reef Fish Bulletin* 5, 34–35.

Imamura, K. (1999) The organization and development of sea farming in Japan. In: Howell, B.R., Moksness, E. and Svasand, T. (eds) *Stock Enhancement and Sea Ranching*. Fishing News Books, Oxford, UK, pp. 91–102.

Jeffs, A.G., Holland, R.C., Hooker, S.H. and Hayden, B.J. (1999) Overview and bibliography of research on the greenshell mussel, *Perna canaliculus*, from New Zealand waters. *Journal of Shellfish Research* 18, 347–360.

Johannes R.E. and Riepen, M. (1995) *Environmental, Economic and Social Implications of the Live Reef Fish Trade in Asia and the Western Pacific*. The Nature Conservancy and the South Pacific Forum Fisheries Agency, Honiara, Solomon Islands, 81pp.

Jones, R.J. (1997) Effects of cyanide on coral. *South Pacific Commission Live Reef Fish Information Bulletin* 3, 3–8

Kaufman, L., Ebersole, J., Beets, J. and McIvor, C.C. (1992) A key phase in the recruitment dynamics of coral reef fishes: post-settlement transition. *Environmental Biology of Fish* 34, 109–118.

Kitada, S. (1999) Effectiveness of Japan's stock enhancement programmes: current perspectives. In: Howell, B.R., Moksness, E. and Svasand, T. (eds) *Stock Enhancement and Sea Ranching*. Fishing News Books, Oxford, UK, pp. 103–131.

Koenig, C.C. and Coleman, F.C. (1998) Absolute abundance and survival of juvenile gags in sea grass beds of the northeastern Gulf of Mexico. *Transactions of the American Fisheries Society* 127, 44–55.

Langdon, J.S. (1989) Disease risks in fish introduction and translocations. In: Pollard, D.A. (ed.) *Introduced and Translocated Fishes and their Ecological Effects*. Australian Society for Fish Biology Workshop, 24–25 August 1989. Proceedings No. 8. Bureau of Rural Resources, Canberra, Australia, pp. 98–107.

Lee, C.S., Leung, P.S. and Su, M.S. (1997) Bioeconomic evaluation of different fry production systems for milkfish (*Chanos chanos*). *Aquaculture* 155, 371–380.

Leis, J.M. (1991) The pelagic stage of reef fishes: the larval biology of coral reef fishes. In: Sale, P.F. (ed.) *The Ecology of Fishes on Coral Reefs*. Academic Press, San Diego, California, pp. 183–230.

Maroz, A. and Fishelson, L. (1997) Juvenile production of *Amphiprion bicinctus* (Pomacentridae, Teleostei) and rehabilitation of impoverished habitats. *Marine Ecology Progress Series* 151, 295–297.

Masuda, R. and Tsukamoto, K. (1998) Stock enhancement in Japan: review and perspective. *Bulletin of Marine Science* 62, 337–358.

McAllister, D.E. (1999) Is mariculture the remedy to problems of coastal communities? *South Pacific Commission Live Reef Fish Information Bulletin* 5, 47–48.

McManus J.W., Reyes, R.B. Jr and Nanola, C.L. Jr (1997) Effects of some destructive fishing methods on coral cover and potential rates of recovery. *Environmental Management* 21, 69–78.

Mous, P.J., Pet-Soede, L., Erdmann, M., Cesar, H.S.J., Sadovy, Y. and Pet, J.S. (2000) Cyanide fishing on Indonesian coral reefs for the live food fish market – what is the problem? *South Pacific Commission Live Reef Fish Information Bulletin* 7, 20–26.

Munro, J.L. and Bell, J.D. (1997) Enhancement of marine fisheries resources. *Reviews in Fisheries Science* 5, 185–222.

Naylor, R.L., Goldburg, R.J., Primavera, J., Kautsky, N., Beveridge, C.M., Clay, J., Folke, C., Lubchenco, J., Mooney, H. and Troell, M. (2000) Effects of aquaculture on world fish supplies. *Nature* 405, 1017–1024.

Ogburn, N.J. and Johannes, R.E. (1999) *Grouper Post-larvae, Fry and Fingerling Collection Devices in the Philippines*. Report prepared for The Nature Conservancy Asia/Pacific Coastal and Marine Program, May 1999, 80pp.

Ortiz, W. (1991) The *larveros*: the larval shrimp catchers. *Sea Wind* 5(2), 14–18.

Phillips, M.J., Kwei Lin, C. and Beveridge, M.C.M. (1993) Shrimp culture and the environment: lessons from the world's most rapidly expanding warm water aquaculture sector. In: *Environment and Aquaculture in Developing Countries*.

ICLARM Conference Proceedings 31. International Center for Living Aquatic Resources Management, Manila, Philippines, pp. 171–198.
Ramofafia, C., Foyle, T.P. and Bell, J.D. (1997) Growth of juvenile *Actinopyga mauritiana* (Holothuroidea) in captivity. *Aquaculture* 152, 119–128.
Rimmer, M. (1999) Reef fish aquaculture: potential, constraints and status. In: Lee, C.L. and Harvey, D.C. (eds) *Proceedings of Tropical Aquaculture in the Kimberley*, May 1999. Broome, Western Australia. Fisheries WA, pp. 47–61
Roberts, C.M. (1996) Settlement and beyond: population regulation and community structure of reef fishes. In: Polunin, N.V.C. and Roberts, C.M. (eds) *Reef Fisheries*. Fish and Fisheries Series No. 20. Chapman & Hall, New York, pp. 85–112.
Rosenberry, R. (2000) World Shrimp Farming 2000. *Shrimp News International* 13, 1–324.
Sadovy, Y. (2000) *Regional Survey for Fry/Fingerling Supply and Current Practices for Grouper Mariculture: Evaluating Current Status and Long-term Prospects for Grouper Mariculture in South East Asia*. Final report to the Collaboration APEC grouper research and development network (FWG 01/99). Asia Pacific Economic Cooperation Council, Singapore.
Sadovy, Y. and Pet, J. (1998) Wild collection of juveniles for grouper mariculture: just another capture fishery? *South Pacific Commission Live Reef Fish Information Bulletin* 4, 36–39.
Sadovy, Y.J. and Vincent, A.C.J. (2002) The trades in live reef fishes for food and aquaria: issues and impacts. In: Sale, P.F. (ed.) *Coral Reef Fishes: New Insights into their Ecology*. Academic Press, New York.
Sale, P.F. (1980) The ecology of fishes on coral reefs. *Oceanography and Marine Biology Annual Review* 18, 367–421.
Sale, P.F. and Ferrell, D. (1988) Early survivorship of juvenile coral reef fishes. *Coral Reefs* 7, 117–124.
Sims, N.A. (1993) Pearl oysters. In: Wright, A. and Hill, L. (eds) *Nearshore Marine Resources of the South Pacific, Information for Fisheries Development and Management*. Forum Fisheries Agency (FFA), Institute of Pacific Studies (IPS) Suva, pp. 409–430.
Smith, K.N. and Herrnkind, W.F. (1992) Predation on early juvenile spiny lobsters *Panulirus argus* (Latrielle): influence of size and shelter. *Journal of Experimental Marine Biology and Ecology* 157, 3–18.
Smith, P. and Rhodes, L. (1994) Plentiful supply of Kaitaia mussel spat. *Seafood New Zealand* (November) 13.
Tamaru, C.S., Cholik, J.C.M. and Fitzgerald, J.J. Jr (1995) Status of the culture of milkfish (*Chanos chanos*), striped mullet (*Mugil cephalus*) and grouper (*Epinephelus* sp.). *Reviews in Fisheries Science* 3(3), 249–273.
Tisdell, C.A. and Poirine, B. (2000) Socio-economics of pearl culture: industry changes and comparisons focussing on Australia and French Polynesia. *World Aquaculture* 31(3), 30–33ff.
Tuan, L.A. and Hambrey, J. (2000) Seed supply for grouper cage culture in Khanh Hoa, Vietnam. *Aquaculture Asia* 5(2), 39–41.
Tuan, L.A., Nho, N.T. and Hambrey, J. (2000) Status of cage mariculture in Vietnam. In: Liao, I.C. and Lin, C.K. (eds) *Cage Aquaculture in Asia: Proceedings of the First International Symposium on Cage Aquaculture in Asia*. Asian Fisheries Society, Manila, Philipppines, and World Aquaculture Society – Southeast Asian Chapter, Bangkok, Thailand, pp. 111–123.

Tzeng, W.N. (1997) Short- and long-term fluctuations in catches of elvers of the Japanese eel, *Anguilla japonica*, in Taiwan. In: Hancock, D.A., Smith, D.C., Grant, A. and Beumer, J.P. (eds) *Developing and Sustaining World Fisheries Resources, the State of Science and Management*. 2nd World Fisheries Congress, CSIRO, Brisbane, Australia, pp. 85–89.

Ventilla, R.F. (1982) The scallop industry in Japan. *Advances in Marine Biology* 20, 309–382

Walker, T. (2001) Towards rock lobster culture. *Austasia Aquaculture* 15(2), 41–42.

Watson, M., Power, R. and Munro, J.L. (2001) Use of light attracted zooplankton for rearing post-settlement coral reef fish. *Proceedings of the Gulf and Caribbean Fisheries Institute* 52, 340–351.

Whitty, J. (2001) Shoals of time, are we witnessing the extinction of the world's coral reefs? *Harpers* (January) 55–65.

Wood, E.M. (2001) *Collection of Coral Reef Fish for Aquaria: Global Trade, Conservation Issues and Management Strategies*. Marine Conservation Society, Ross-on-Wye, UK, 80pp.

Wood, P. (1999) 30 years of change – glass eel slump is nothing new. *Fish Farming International* 26(7), 44–45.

Contending with Criticism: Sensible Responses in an Age of Advocacy

18

Terrence R. Tiersch[1] and
John A. Hargreaves[2]

[1]Aquaculture Research Station, Louisiana State University Agricultural Center, Louisiana Agricultural Experiment Station, Baton Rouge, LA 70820, USA; [2]Department of Wildlife and Fisheries, Mississippi State University, Mississippi State, MS 39762, USA

Abstract

The modern environmental advocacy emerged in the 1960s in response to highly visible examples of environmental degradation such as air and water pollution, extinctions of animals and plants, and major ecosystem disruptions caused by logging, mining, dam building and mechanized agriculture. More recently, advocacy has been extended to include global concerns related to the environmental impacts of expanding human populations, widespread poverty, global climate change and unprecedented losses of biodiversity. The worldwide impact of environmental advocacy groups has increased dramatically in the recent past because of improved methods of communication, widespread media access and a large infusion of financial resources. For 15 years these groups have criticized specific sectors in aquaculture, but within the past 5 years they have begun to actively target aquaculture as a collective activity. The reasons for this shift in criticism range from legitimate identification of potential problems in aquaculture to fund-raising efforts by advocacy groups based on opinion polls of their memberships. Given these current trends, it appears that this criticism will not diminish in the future and has the potential to influence public opinion, consumer behaviour, and the activities of policymakers and regulatory agencies. Given these realities, it is incumbent upon those involved in aquaculture to gain an appreciation of the importance of responding sensibly to criticism. This includes developing an understanding of the relevant issues, communicating effectively with and through popular media, becoming familiar with the goals and tactics of advocacy groups, and increasing involvement in the formulation of policies and regulations as individuals and representatives of groups.

The Nature of Environmental Advocacy

The 21st century presents widespread access to rapid and pervasive communications media. There are many groups utilizing the media to carry their message to the public, policymakers and politicians. As with other topics of broad societal concern, environmental issues have stimulated the development and proliferation of advocacy groups during the past 40 years. Environmental issues were first brought clearly to the attention of the general public through the best-selling book *Silent Spring* (Carson, 1962), a linkage of popular media and the environmental movement that continues to this day. Environmental advocacy has shifted to a professional activity with many participants. In fact, there were more than 23,000 environmentally related non-governmental organizations (NGOs) active worldwide in 1998 (French, 2000). These organizations can be quite large. For example, the World Wide Fund for Nature has almost 5 million members in more than 90 countries, and Greenpeace claims 2.5 million members worldwide (French, 2000). These NGOs have developed powerful organizational networks employing the internet and traditional media to advance their environmental agendas by scrutiny and high-profile criticism of actions within the private and public sectors.

Within the last decade, criticism of aquaculture activities by environmental advocacy groups has escalated rapidly, evolving in format as well as in content. During the last 5 years, individuals representing these groups have begun targeting aquaculture as a collective activity, rather than directing criticism to specific sectors (Goldburg and Triplett, 1997). Prior to this, criticism was more focused on concerns with the environmental impact of coastal shrimp farms in tropical countries and, later, nearshore salmon operations in temperate waters (Naylor *et al.*, 1998).

Much of the criticism of aquaculture by NGOs began as opinion pieces in news media or as information provided by specific advocacy groups. Gradually this material began entering scientific literature as news items and recently has shifted into the arena of scientific review and technical articles, and special reports for commissions (Goldburg *et al.*, 2001) and scientific societies (Naylor *et al.*, 2001). In effect, NGOs have become clearinghouses for information critical of aquaculture. High-profile articles are now regularly seen in leading scientific journals and popular media that criticize the past and potential environmental effects of aquaculture and the use of genetically modified aquatic organisms (Table 18.1). These articles are often recycled and appear in multiple publications (Naylor *et al.*, 2000, 2001), a practice considered ethically unacceptable for publication of primary scientific data (Resnik, 1998).

Aquaculture is growing rapidly and has been associated with practices and impacts that warrant criticism. The rapid growth of aquaculture has attracted attention and has been used to suggest that greater trouble is on the horizon. Various groups have adopted attacks through popular media as a method to bring about changes in popular opinion and regulatory policy. This approach is not discouraged by the media because sensational accusations,

Table 18.1. Recent publications critical of aquaculture.

Title	Source
Marine Aquaculture in the United States	Pew Oceans Commission, 2001
Effects of aquaculture on world fish supplies	*Issues in Ecology*, 2001
Will souped up salmon sink or swim?	*Nature*, 2000
Effects of aquaculture on world fish supplies	*Nature*, 2000
Nature's subsidies to shrimp and salmon farming	*Science*, 1998
Murky Waters: Environmental Effects of Aquaculture in the United States	Environmental Defense Fund, 1997

controversy, and polarized debate are considered to be newsworthy simply for their mass appeal rather than scientific validity. As a result, policymakers and the seafood-consuming public are unable to make informed decisions based on a full set of considerations. Aquaculture producers, leaders of aquaculture organizations and research scientists should expect to continue to receive criticism and would be wise to take steps to improve their ability to respond to and benefit from it.

There are many factors that make aquaculture an easy target, especially for environmental advocacy groups. Although the diverse activities within aquaculture are not centrally organized and encompass a broad diversity of organisms, geographic locations and levels of intensity, it is often construed as a single large industry, culpable as a whole for a wide variety of problems. Critics of the aquaculture industry can make it appear large, formidable, reckless and, by implication, can blame all aquaculturists for the activities of a minority. Such an industry can be demonized easily and is without the ability to defend itself, lacking, for example, the lawyers and media consultants of large corporations. Thus, when criticized as a collective aquaculture industry, it cannot respond as such, and when participants within various aquaculture sectors make good-faith efforts to correct problems, critics outside aquaculture are often poised to claim credit for those improvements.

The current situation has resulted in an adversarial climate that often leads to polarized deadlock and failure of issue resolution. To address this, we provide some suggestions below for those involved in aquaculture to respond to criticism by advocacy groups. These suggestions focus primarily on responses and responsibilities at the level of the individual, with some mention of responses at the organization level. Such responses might take the form of one-to-one discussions with critics, debate in a public forum, press releases, scientific publications to address criticisms or interviews with the media. These points have coalesced from discussions stimulated by a presentation addressing this topic at the annual meeting of the World Aquaculture Society in Orlando, Florida (Tiersch and Hargreaves, 2001).

Suggestions for Sensible Interactions

Repond from the perspective that criticisms and solutions must be based on a comprehensive and balanced view of the total problem

The short message length emphasized in the current age of 'sound bite' media makes it difficult to present comprehensive and balanced viewpoints. Indeed, messages are often edited for maximum effect rather than greatest accuracy (Rybacki and Rybacki, 2000). This is a difficult arena for scientific discussion and the agenda of issues is often controlled by those who deliberately seek access to media or have the largest budgets.

If possible, when responding to criticism, call attention to biased assertions that lack a realistic context in order to discourage this form of attack. Consider, for example, the formulation of a written response to recent articles critical of fish meal use in aquaculture (e.g. Naylor *et al.*, 2000, 2001). An effective response should clearly point out that it would be scientifically irresponsible to advance the perception that aquaculture is a significant threat to the health of ocean fisheries without providing a comprehensive and balanced account of all the major factors affecting those fisheries. Aquaculture activities have a much smaller impact on ocean fisheries than do the larger-scale effects of widespread development and exploitation of coastal areas, poverty and lack of economic alternatives, deforestation, dam construction and other habitat alterations, pollution, international commercial fishing and associated bycatch, and a growing consumerism fuelled by the rapid expansion of the global economy. A scientific approach to this topic would identify each hazard, the magnitude of its effect, and provide an explicit assessment of the probability of occurrence (i.e. risk) of that hazard (as would be required, for example, for publication of a scientific paper on the carcinogenicity of a particular substance). Responses to criticism of this sort should insist that critics explain their selective presentation of information. For those within aquaculture, numerous references are available that provide examples of a comprehensive view of the numerous claimants to aquatic resources (Reisner, 1993; Lichatowitch, 1999; de Villiers, 2000).

Respond to criticism with clearly presented, broad-based arguments

When receiving criticism it is easy to focus on the details. Scientists intuitively respond to criticism solely with facts and data, assuming that such a response will persuade critics to accept the validity of their position. Advocacy groups commonly counter such responses with additional facts and data, and then shift to new issues that elicit additional responses. Regulators and other decision-makers attempting to sort out the facts of a debate often dismiss such situations as a case where the 'experts disagree' and therefore neutralize one another's arguments. It would be wise to remember that we all tend to disregard details or highly technical information that are outside our own sphere of interests. Do not waste opportunities for communicating effectively by burying your message in details and technical jargon.

Responding from a broader vantage point that addresses the centre of an issue rather than its periphery can be effective. For example, it would be a mistake to argue the exact area of mangrove wetlands converted to shrimp farms in Thailand while ignoring the overall global demands placed on these areas. There are many sources of information available on broad-based topics of relevance to aquaculture, such as sustainability (Hart, 1997), globalization (Friedman, 1999), and the value of natural capital and ecosystem services (Costanza *et al.*, 1997a,b). The annual and special publications by the Worldwatch Institute are especially useful in this regard (e.g. Brown *et al.*, 2001).

In responding to criticism, recognize that information can be used to achieve different ends

It is an unfortunate reality that arguments are advanced by selective presentation of facts or perceptions of reality. Similar to product advertising, specifically targeted distortions of reality (product claims) are used to 'sell' a product, in this case a world view that advances the agenda of a particular environmental advocacy group.

Bear in mind that remarkably divergent conclusions can be formed from the same data. For example, a recent book by Peter Huber (1999), entitled *Hard Green: Saving the Environment From the Environmentalists: a Conservative Manifesto*, represents a politically conservative reworking of traditionally liberal environmental concerns. Given this 'spin doctor' reality, carefully consider and anticipate the ways in which critics might misconstrue and take advantage of any statements made in response to criticism.

Refer to specific sectors rather than reinforcing the misconception of the existence of a collective aquaculture industry that is operated, regulated and culpable as a single entity

As a precedent for this, consider that aquaculture, which has been defined as 'the propagation and rearing of aquatic organisms in controlled or selected environments' (Goldburg and Triplett, 1997), must have a corollary in the propagation and rearing of terrestrial organisms in controlled or selected environments. Accordingly, we should lump the farming of turkeys, swine, cattle, cotton, rice, soybeans, tomatoes, bananas and orchids as a collective called the 'terraculture industry'. This concept of terraculture does not reflect a real-world regulatory domain or a single business entity. Similarly, the production of pearls, salmon, ornamental carp, crawfish, bait minnows and seaweeds does not comprise a single business entity. These activities constitute separate sectors and should be referred to as such. It is a large leap from the general term aquaculture to the misleading and imprecise term aquaculture industry. This argument also extends to references to the aquaculture industries of particular geographic areas. Reference to the specific aquaculture industries of Louisiana or Thailand is more accurate than combining them. This is sometimes difficult to avoid, however, given that members of small aquaculture industries are often tempted to pool their resources for specific efforts. This desire to display strength in numbers has advantages and disadvantages that should be considered before such actions are taken. Realistically, if aquaculture is to be criticized as a large unified industry, it should seek to develop a collective identity and organization to reap the benefits accorded to such an industry. This could include utilizing the services of media consultants and providing a positive recognition of aquaculture with the public, regulators and politicians.

Be familiar with the role of aquaculture in economic development, especially in developing countries

Aquaculture has many well-known values, which include generation of income and employment, protein production, and creation of foreign exchange. Less-appreciated values of aquaculture include the potential for generating economic development (Pullin *et al.*, 1993) and alleviating poverty (Hambrey *et al.*, 2001a,b). Aquaculture can serve as the nucleus or catalyst for economic development. For example, when fish farms are constructed, ancillary businesses become necessary to support the new industry, providing employment. These activities, with proper support and community acceptance (Harrison and Huntington, 2000), can be integrated with existing infrastructure to strengthen local markets and communities (Edwards, 2000), and to reduce demands on the local environment.

Know your critics, their methods and their goals

Environmental advocacy groups are well organized and well financed (Table 18.2). They are often intent on shaping political policy and public opinion rather than engaging in objective scientific debate. Such groups frequently take extreme positions in order to force movement on issues towards the middle. A common tactic is to burden the opposition with excessive paperwork and litigation, and to overload research and regulatory institutions with accusations that take minutes to formulate but years to resolve. For example, suggesting at a press conference that farmed fish contain toxic contaminants initiates a chain of actions that can waste limited resources in pursuit of a

Table 18.2. Annual income for the year 1999 for different organizations. These values were compiled from website postings in January 2001.

Organization	Income	Website
The Nature Conservancy	\$774.9 million[a]	www.nature.org
Greenpeace International	∈126 million[b]	www.greenpeace.org
World Wildlife Fund	\$111.8 million[c]	www.worldwildlife.org
National Audubon Society	\$69.2 million	www.audubon.org
Sierra Club	\$46.5 million	www.sierraclub.org
Environmental Defense (Fund)	\$31.4 million	www.edf.org
Natural Resources Defense Council	\$30.6 million	www.nrdc.org
People for the Ethical Treatment of Animals	\$17.5 million	www.peta.org
World Aquaculture Society	\$0.5 million	www.was.org

[a]Total support and revenue for 1999.
[b]Posted in euros; other values are in US dollars.
[c]Operating revenue for 2000.

definitive solution to the perceived problem. In some cases, non-violent civil disobedience or even violence are used to force change. Individuals and groups directly associated with aquaculture are typically unfamiliar or uncomfortable in responding to these tactics, yet knowledge of the methods employed by advocacy groups can facilitate the formulation of effective responses and proactive measures.

An informative five-part series of newspaper articles by Tom Knudson entitled 'Environment, Inc.' appeared recently (22–26 April 2001) in the *Sacramento Bee* newspaper (the articles can be viewed on the newspaper website at www.sacbee.com). These articles reveal the evolution of a corporate mentality in the environmental movement, sustained by aggressive fund raising and an 'ends-justify-the-means' mentality. Grassroots advocacy has yielded to an environmental defence industry that in 1999 took in an average of US\$9.6 million a day (a doubling of 1992 levels) from individual contributors, companies and foundations (Knudson, 2001). In 2000, The Nature Conservancy was listed as the ninth largest non-profit organization in the United States (from their website at www.nature.org). There is little question that advocacy groups can generate substantial revenues, well in excess of aquaculture organizations such as the World Aquaculture Society (Table 18.2). One of the strongest approaches to addressing biased or undeserved criticism is to become familiar with the financial practices of the particular advocacy group that is making the claims. Being prepared to provide accurate information concerning how donations are used for fund-raising mailshots (sometimes referred to as education), salaries of top executives, and impressive office buildings rather than assisting the environment can shift the moral authority away from disreputable groups. Financial information and evaluation of the operations of non-profit organizations in the United States are available through watchdog groups such as the National Center for Charitable Statistics (www.nccs.urban.org) and the American Institute of Philanthropy (www.charitywatch.org).

Recognize legitimate criticism

Advocacy groups can provide a valuable service by acting as an impartial watchdog of environmental issues and calling attention to legitimate concerns (e.g. Roed, 2001). With respect to aquaculture, a number of authors have addressed the potential for fish farming, the so-called blue revolution, to avoid the past mistakes of the green revolution (e.g. McGinn, 1998). It is essential that we all recognize, evaluate and minimize the potential negative consequences of our activities. Indeed, there are good business reasons for this, including improved profitability arising from increased efficiency and innovation (e.g. Porter and van der Linde, 1995; Hawken *et al.*, 1999; Bliese, 2001). Potentially negative social and environmental impacts of aquaculture can be minimized through the adoption of codes of conduct, which provide guidelines for the development of voluntary systems of management (e.g. Boyd, 1999a; Boyd *et al.*, 2001). Codes are implemented through a systematic, integrated collection of better (best) management practices (BMPs), representing creative, locally appropriate solutions for the improvement of social and environmental performance.

In addition, researchers as well as farmers need to ensure that research goals and culture practices are truly defensible, especially for issues such as the use of non-native species including black carp (Ferber, 2001) and genetically modified organisms (GMOs). It makes particular good sense to carefully evaluate the broad range of issues associated with GMOs (the so-called gene revolution) before investment of time and resources, and to be aware of the considerable controversy that surrounds this topic in the scientific community (e.g. Reichhardt, 2000) and in the popular media (e.g. Nash, 2000).

Do not shift blame to other sectors of aquaculture to deflect legitimate criticism

It is not difficult to find examples of specific mistakes that have been made as aquaculture industries have developed. It is common for critics to cite specific (often past) actions to make broad-brush indictments of aquaculture industries in general. It would be advisable when responding to criticism of this type to point out inaccurate generalizations where appropriate, and to require critics to document their claims. Regional or commodity-based aquaculture associations can play an important role in avoiding poor practices by disseminating practical information and encouraging responsible behaviour. Well-managed businesses and industries are self-correcting. Problems are identified and steps are taken to minimize or eliminate them. In the long run, this is a better way to handle criticism than diverting blame elsewhere. Deflecting criticism to other industries is a defensive posture that suggests guilt and a lack of responsible behaviour. In addition, as long as aquaculture is viewed as a single industry (as described above), all sectors will suffer from criticism of a particular sector.

Learn how the media can be used as a conduit for responses to criticism

Environmental advocacy groups are sophisticated in the use of popular media to achieve their goals. Until recently, members of the aquaculture community have generally not had much interaction with members of the media. However, given current and future criticisms, understanding the role of the media and becoming comfortable with their capabilities to communicate success stories and responses to criticisms will become increasingly important. Many environmental advocacy groups hold training sessions with the explicit intention of improving the ability of their staffs to spread their messages. Some Land Grant universities hold similar sessions for their extension agents and scientists. Although communicating through radio or television does not allow for comprehensive review of issues, producers, processors, marketers, extension agents, scientists and leaders of aquaculture organizations who learn how to communicate their message succinctly can effectively respond to criticisms by advocacy groups through those media.

An Example for Responding to Environmental Advocacy

Environmental advocates and NGOs often argue that aquaculturists should 'farm down the food web' to emphasize the culture of herbivorous fishes (Goldburg and Triplett, 1997; Naylor *et al.*, 2000, 2001). Similarly, advocates base arguments in opposition to certain forms of aquaculture on the protein equivalency or ecological footprint of those activities. These arguments can be refuted on several grounds, based on the broad principles of response described above.

These criticisms typically address the feeding of plant or animal protein to livestock or fishes instead of making that protein available for direct human consumption. First, these arguments reflect criticism of animal agriculture in general, which is a broader issue, and often are based on unrealistic requirements such as the avoidance or elimination of meat in human diets (e.g. Singer, 1975). With respect to aquaculture, these criticisms typically address the feeding of plant and animal protein to shrimp and carnivorous fishes. It is argued that herbivorous fishes would not require protein inputs in feeds and that culture of such species would reduce demand for fish meal (see

Chapter 16), protecting ocean fisheries. However, critics fail to recognize that the market for fish meal is global and that restricting its use in aquaculture feeds would only shift demand to other animal agriculture sectors. Critics also argue that the fish incorporated into aquaculture feeds would be better used for human consumption. This assertion ignores the weak market demand for direct consumption of small pelagic fishes.

While culture of food fish low on the food chain has an intuitive appeal based on principles of trophic dynamics, aquaculture is an economic activity that responds to market demands. Carnivorous fishes and shrimp are farmed because a market demand exists for those products. Without changes in consumer preferences, satisfaction of requirements of a politically correct aquaculture enterprise will be difficult. Although production of herbivorous fishes to augment local protein supplies is an appropriate goal in developing countries, it is not a reason to do away with the culture of omnivorous or carnivorous species. A blend of aquaculture production activities, ranging from herbivorous to carnivorous species cultured in extensive to intensive production systems, will be able to contribute to the satisfaction of local and global market demands for a broad range of seafood products. It is important when responding to arguments of this sort, to insist that critics provide realistic assessments of alternatives to animal agriculture, and that they explicitly disclose their views on issues such as animal rights and the consumption of meat.

How should responses be made? There is usually no need to respond directly to advocacy groups – it is more effective to make points directly to the general public or other interested parties. This is because individuals are not in a position to win an argument with advocacy groups, whose representatives are not going to change their perception of an issue based on persuasive argumentation. Thus, it would be a mistake to be lured into a debate. When addressing an audience it is important to know the message that you wish to deliver, to state it, and to come back to it. This is known as staying on message and we see it practised daily on television by politicians and spokespersons. This reasoning also works in reverse. It is important to recognize and understand the message being presented by an advocacy group. For example, they often argue based on the possibility (or perception) that a particular event or situation could occur, rather than addressing the actual probability and risk of occurrence. In this way individual concerns, such as the status of ocean fisheries, the risk of disease transmission from cultured fishes, or the potential ecological consequences of GMOs, serve as vehicles to deliver their broader message, that aquaculture threatens the well-being of the planet and its inhabitants.

Forms of Interactions with Advocacy Groups

Broadly speaking, environmental advocacy occurs at several levels, depending on the size and interactions of the organizations involved (Kirn, 2000). Large, transnational NGOs with broad-based agendas have become increasingly active in the courts, attempting to achieve objectives through litigation. Small, local environmental groups tend to respond when specific issues of local concern inspire organization and direct action from a grassroots level. As an example, a large environmental organization might advocate complete cessation of construction in mangrove areas, but a local environmental organization might advocate mangrove conservation in a particular area, while accepting development elsewhere in recognition of potential employment opportunities fostered by aquaculture development projects. Although large environmental organizations often cannot be ignored, particularly when opposition to a project is brought into court, sincere and responsive attention to the legitimate concerns of local environmental groups can go a long way towards defusing potential conflicts. Also bear in mind that the NGO community employs flexible and overlapping networks in response to various issues (French, 2000). While this allows rapid and widespread dissemination of information and planning of activities, it also sometimes blends the missions of various organizations. For example, it is not unusual to see statements made by People for the Ethical Treatment of Animals, a group formed to oppose agricultural and biomedical uses of animals, that address ecological issues, an area outside its original mission.

Responses to criticism can be made at several levels. At a personal level, responses can be made as a private citizen; in a professional capacity, responses can be made as an individual or as a member of an organization. While it is important always to have a scientific basis for arguments, it is arguably a major mistake to confine your responses only to the available data. This is especially true when advocacy is cloaked in science to present a non-scientific viewpoint, a standard practice of groups such as those that oppose teaching of Darwinian evolution (Gould, 1991; Coyne, 2001). To be effective, responses may need to be made for popular audiences outside the scientific and technical literature.

Occasionally, direct action or litigation by environmental advocacy groups in response to specific actions will lead to stalemate and polarization. In those instances, it may be necessary to implement formal methods of conflict resolution. These methods have been developed to resolve disputes within the workplace or between corporations and unions, but the techniques are applicable to disputes concerning environmental issues. The primary goal of dispute resolution is to develop a win–win solution, often through the employment of a mutually agreeable neutral party. Common dispute resolution processes include arbitration and neutral evaluation, mediated through an ombudsman or peer review. The neutral party can assist negotiations by facilitating identification and discussion of common interests and points of disagreement. Further information is available from sources such as the US Institute for Environmental Conflict Resolution (www.ecr.gov) and the Conflict Prevention and Resolution Center (www.usda.gov/cprc) of the US Department of Agriculture.

In a recent plenary address to the Aquaculture America Conference in Tampa, Florida, USA (Boyd, 1999b), it was suggested that individuals become more active in the political process. Although members of the aquaculture community are often uncomfortable in this role, representatives of environmental advocacy groups with little formal training in aquaculture have been active in this arena. Advocacy groups have effectively driven the agenda on a broad range of environmental issues. As an example, the US Environmental Protection Agency (EPA) was sued by the Natural Resources Defense Council, an environmental NGO active in the courts, for failure to enforce the 1972 Clean Water Act. As part of a consent decree that arose from the court case in the early 1990s, the EPA agreed to a timetable to review effluent limitation guidelines for industries that had not been scrutinized to that point. Publication of the *Murky Waters* report (Goldburg and Triplett, 1997) called specific attention to US aquaculture and suggested that the EPA initiate a review of effluent limitation guidelines for aquaculture. The EPA initiated rule-making activity in January 2000 that has led to institution of an aquaculture-wide questionnaire designed to disclose comprehensive financial information of aquaculture producers as an initial step in formulating regulations. This is an example of how environmental advocacy groups have affected the policy arena. There are many potential outcomes of this process, including those that may be excessively burdensome to producers and others, that would yield only modest environmental protection. Therefore, timely and effective responses to advocacy groups are necessary for rational, science-based establishment of policies and regulations.

The role of professional aquaculture organizations in providing responses to criticism by advocacy groups is not clearly defined. It has been argued by the Executive Committee of the World Aquaculture Society (WAS) that involvement should be limited to educational, scientific and technological issues as an 'apolitical organization with the mission of generating and disseminating unbiased, science-based information' (Losordo *et al.*, 2000), although it has also been pointed out that the mission of WAS includes disseminating other (non-technical) information, debating of important issues, and promoting the advancement of aquaculture (New, 2001). This is in contrast to the American Fisheries Society (AFS), which has established a position statement providing specific advocacy guidelines for its membership (available on the AFS website at www.fisheries.org). There are organizations within aquaculture with clear advocacy roles, such as the Global Aquaculture Alliance (www.gaalliance.org), which is an international, non-profit trade association dedicated to advancing environmentally responsible aquaculture. The by-laws of this organization direct it to advocate aquaculture as an answer to global food needs and to educate producers, consumers and the media, thus providing a needed voice in environmental issues. All told, it is apparent that interaction with advocacy groups is inevitable for those in aquaculture and that, at a minimum, distribution of pertinent information will be necessary. Otherwise the public, regulatory agencies and policymakers will probably receive a large portion of their information directly from environmental advocacy groups.

Conclusions

Why are advocacy groups opposed to aquaculture? Well, in addition to the points made above, including the fund-raising issues that are part of today's world (e.g. the need to stay in the headlines and on the television screen), there are deeper philosophical reasons. A core concept of the environmental movement is the precautionary principle, which basically states that it is wise to avoid unnecessary risk (Huber, 1999; McGinn, 2000). In practice this means that it would be best to restrict or prevent activities that could cause long-term or irreversible harm, even in the absence of convincing scientific data. This principle is biased towards slowing or stopping the development of new activities, and shifts the burden of proof from environmental advocates to practitioners such that new activities, like aquaculture, must show that they will not be a problem in the future. This is in contrast to the situation for established industries – detractors must prove that the established industry presents a problem. Of course, newer industries also lack the financial and political resources of groups such as logging, mining and petroleum extraction interests and large chemical corporations. It is easier to restrict or stop aquaculture projects, despite their much smaller environmental risk, than it is to attempt to control more damaging established activities. Thus, opposing aquaculture development is viewed by advocacy groups as applying an ounce of prevention now instead of the pound of cure that would be required later.

There are some negative outcomes that could arise from the intensified efforts of NGOs to publicize ecological distress and to criticize corporations and agricultural industries, including those of aquaculture. First, the general public can become desensitized to environmental issues owing to overexposure. Aggressive fund-raising by direct mail may already be causing this to happen, producing so-called compassion fatigue. Second, biased and unfounded attacks by environmental advocacy groups will eventually be recognized as such. When this happens and is linked to revelations concerning high-powered fund-raising efforts – which annually produce enough paper to encircle the Earth more than twice (Knudson, 2001) – an emergent corporate mentality, and sky-high salaries for top executives (more than US$200,000 for the top ten groups), the public could react negatively to advocacy groups. The negative reaction would be intensified if the public realized that higher prices in the supermarket for seafood and other commodities were caused by the actions of NGOs. Third, if the dire warnings of NGOs are not followed by global ecological collapse and extinction of wolves, whales and pandas, the general public could come to view NGOs as the watchdogs who cried wolf. This would be a serious blow to their credibility and that of the environmental movement in general. Indeed, conservative groups have begun to attack environmentalists on these grounds (see Huber, 1999). Because the public generally makes no distinction between environmentalists and ecologists, there could be a loss of credibility for environmental scientists as well. Fourth, the activities of NGOs and the possible outcomes described above could alienate colleagues of environmental advocates within the scientific community, who would otherwise support a responsible, science-based environmental agenda.

Thus, alone or in combination, desensitization of the public, corporate hypocrisy, loss of credibility and alienation of sympathetic members of the scientific community could weaken the ability of NGOs to protect the environment. Indeed, by diminishing the economic opportunities afforded by aquaculture worldwide, advocacy groups could indirectly cause environmental damage in economically disadvantaged areas because of a lack of alternatives for local people. Ironically, these actions by advocacy groups would compromise the sustainability of their own environmental defence industry and ultimately the sustainability of the natural resources they profess to protect.

Finally, it is important to remember in any interaction that issues can quickly become polarized, leaving factions on either side of a broad divide. Once polarized, it is difficult for the factions to re-establish common goals and to work together. This is especially true for issues with strong emotional connections or with heated media coverage. Accordingly, we should strive to act professionally while maintaining a factual, comprehensive and balanced view of issues. Realistically, the best approach to dealing with advocacy groups is to devote effort in gaining a strong personal understanding of the relevant issues, and to be proactive in addressing problems and communicating solutions. This has been emphasized, for example, for interactions with leaders of the animal rights movement to avoid embarrassment of ill-prepared scientists and the scientific community (Macrina, 2000). It is ill-advised to allow advocacy groups to set the agenda and to control the debate on aquaculture-related issues. There are many additional sources of information concerning broad issues beyond those cited in this chapter. An increased awareness of social, economic, ecological and political issues will allow those involved in aquaculture to be proactive and avoid taking a defensive, reactionary position. Indeed, it is likely that aquaculturists and environmental advocates share values at the heart of most issues, and it is the tactics used in addressing the inappropriate actions of a minority within aquaculture and environmental advocacy that drive groups apart.

Addendum

The relationship between aquaculture development and environmental advocacy groups has been affected by recent events. First, the pendulum of moral high ground in environmental issues has swung rapidly to the right in the past months. A number of recent publications (including those cited above) have confronted advocacy groups with compelling evidence of misrepresentation or distortion of facts to create a false impression of impending ecological collapse to fuel advancement of their environmental agenda. This backlash has provided the ammunition for a political battle. A book by Bjorn Lomborg (2001) is being heralded by the media and opponents of advocacy groups as a definitive work identifying the fallacies put forth by environmentalists, complete with more than 2900 notes and sources. It is clear that advocacy groups will need to regroup and do a better job of using creditable facts and figures. Thus, the backlash has created a demand for accountability.

Second, the declaration of a war on terrorism in the aftermath of the attacks of 11 September has reorganized national priorities and diminished the importance of environmental issues and advocacy groups. Although the issues that stimulated environmental advocacy remain, the context of operation and activity prioritization have changed. Downturns in the US and world economy, and shifts in charitable contributions to disaster relief funds are likely to reduce donations to advocacy groups, thereby forcing the groups to reconsider their agendas and approaches. It would be appropriate to continue to evaluate case studies of responding to environmental advocacy by discussing sensible responses to claims (e.g. such as damage to ocean fisheries by inclusion of fish meal in aquaculture feeds) and to remain informed of advocacy group activities to shape aquaculture regulations, policies and perceptions.

Acknowledgements

We thank C. Tucker, G. Lockwood, G. Chamberlain, L. D'Abramo, M. New, M. Broussard, T. Losordo, J. Tomasso, P. Sorgeloos, P. Edwards, C. Browdy, C. Boyd, D. Scarfe, R. Romaire, J. Avault, R. Reigh, G. Lutz, R. McClain, A. Kelly, P. Lang, A. Nickens, E. Herbst, G. Poleo, K. Riley and L. Parr for assistance in developing this manuscript. Approved for publication by the Director of the Louisiana Agricultural Experiment Station as manuscript number 01-66-0554.

References

Bliese, J.R. (2001) *The Greening of Conservative America.* Westview Press, Boulder, Colorado.

Boyd, C.E. (1999a) *Codes of Practice for Responsible Shrimp Farming.* Global Aquaculture Alliance, St Louis, Missouri, 42pp.

Boyd, C.E. (1999b) Aquaculture sustainability and environmental issues. *World Aquaculture* 30, 10–13, 71–72.

Boyd, C.E., Hargreaves, J.A. and Clay, J.W. (2001) Codes of conduct for marine shrimp aquaculture. In: Browdy, C.L. and Jory, D.E. (eds) *The New Wave. Proceedings of the Special Session on Sustainable Shrimp Culture.* The World Aquaculture Society, Baton Rouge, Louisiana, pp. 238–256.

Brown, L., Flavin, C. and French, H. (2001) *State of the World 2001*. The Worldwatch Institute, WW Norton and Company, New York.

Carson, R. (1962) *Silent Spring*. Houghton Mifflin, New York, 368pp.

Costanza, R., d'Arge, R., de Groot, R., Farber, S., Grasso, M., Hannon, B., Limburg, K., Naeem, S., O'Neill, R., Paruelo, J., Raskin, R., Sutton, P. and van den Belt, M. (1997a) The value of the world's ecosystem services and natural capital. *Nature* 387, 253–260.

Costanza, R., Cumberland, J., Daly, H., Goodland, R. and Norgaard, R. (1997b) *An Introduction to Ecological Economics*. St Lucie Press, Boca Raton, Florida.

Coyne, J.A. (2001) Creationism by stealth. *Nature* 410, 745–746.

Edwards, P. (2000) Aquaculture, poverty impacts and livelihoods. Natural Resource Perspectives, Number 56, June 2000. Overseas Development Institute, London, 4pp.

Ferber, D. (2001) Will the black carp be the next zebra mussel? *Science* 292, 203.

French, H. (2000) Partnerships for the planet. In: *Vanishing Borders: Protecting the Planet in the Age of Globalization*. WW Norton and Company, New York, pp. 163–176.

Friedman, T.L. (1999) *The Lexus and the Olive Tree*. Farrar, Straus and Giroux, New York.

Goldburg, R. and Triplett, T. (1997) *Murky Waters: Environmental Effects of Aquaculture in the United States*. Environmental Defense Fund, Washington, DC, 195pp.

Goldburg, R.J., Elliot, M.S. and Naylor, R.L. (2001) *Marine Aquaculture in the United States*. Pew Oceans Commission, Arlington, Virginia, 33pp.

Gould, S.J. (1991) Justice Scalia's misunderstanding. In: *Bully for Brontosaurus: Reflections in Natural History*. WW Norton and Company, New York, pp. 448–460.

Hambrey, J., Beveridge, M. and McAndrew, K. (2001a) Aquaculture and poverty alleviation I. Cage culture in freshwater in Bangladesh. *World Aquaculture* 32(1), 50–51ff.

Hambrey, J., Tuan, L.A. and Thuong, T.K. (2001b) Aquaculture and poverty alleviation. II. Cage culture in coastal waters of Vietnam. *World Aquaculture* 32(2), 34–36ff.

Harrison, L.E. and Huntington, S.P. (eds) (2000) *Culture Matters: How Values Shape Human Progress*. Basic Books, New York, 348pp.

Hart, S. (1997) Beyond greening: strategies for a sustainable world. *Harvard Business Review* January–February, 66–76.

Hawken, P., Lovins, A. and Lovins, L.H. (1999) *Natural Capitalism: Creating the Next Industrial Revolution*. Little, Brown and Company, New York, 396pp.

Huber, P. (1999) *Hard Green: Saving the Environment from the Environmentalists: a Conservative Manifesto*. Basic Books, New York.

Kirn, W. (2000) The new radicals. *Time* 156, 42–46.

Knudson, T. (2001) Environment, Inc. *Sacramento Bee* 22–26 April.

Lichatowitch, J. (1999) *Salmon Without Rivers*. Island Press, Washington, DC, 317pp.

Lomborg, B. (2001) *The Skeptical Environmentalist: Measuring the Real State of the World*. Cambridge University Press, Cambridge.

Losordo T., Sorgeloos, P., Tidwell, J. and Browdy, C. (2000) President's column. *World Aquaculture* 31, 3.

Macrina, F. (2000) Use of animals in biomedical experimentation. In: *Scientific Integrity: An Introductory Text with Cases*. ASM Press. Washington, DC, pp. 101–130.

McGinn, A.P. (1998) Blue revolution: the promises and pitfalls of fish farming. *Worldwatch* March–April, 11–19.

McGinn, A.P. (2000) Why poison ourselves? A precautionary approach to synthetic chemicals. In: *Worldwatch Paper 153*, Worldwatch Institute, Washington, DC, pp. 17–18.

Nash, J.M. (2000) Grains of hope. *Time* 156, 38–46.

Naylor R., Goldburg, R., Mooney, H., Beveridge, M., Clay, J., Folke, C., Kautsky, N., Lubchenco, J., Primavera, J. and Williams, M. (1998) Nature's subsidies to shrimp and salmon farming. *Science* 282, 883–884.

Naylor R., Goldburg, R., Primavera, J., Kautsky, N., Beveridge, M., Clay, J., Folke, C., Lubchenco, J., Mooney, H. and Troell, M. (2000) Effects of aquaculture on world fish supplies. *Nature* 405, 1017–1024.

Naylor R., Goldburg, R., Primavera, J., Kautsky, N., Beveridge, M., Clay, J., Folke, C., Lubchenco, J., Mooney, H. and Troell, M. (2001) Effects of aquaculture on world fish supplies. *Issues in Ecology* 8, 1–13.

New, M. (2001) Letter to the editor. *World Aquaculture* 32(1), 4.

Porter, M.E. and van der Linde, C. (1995) Green and competitive: ending the stalemate. *Harvard Business Review* September–October, 120–134.

Pullin R S.V., Rosenthal, H. and Maclean, J.L. (eds) (1993) *Environment and Aquaculture in Developing Countries*. International Center for Aquatic Living Resources, Management Conference Proceedings 31, Manila, Philippines, 359pp.

Reichhardt, T. (2000) Will souped up salmon sink or swim? *Nature* 406, 10–12.

Reisner, M. (1993) *Cadillac Desert: The American West and its Disappearing Water*. Penguin Books, New York, 582pp.

Resnik, D. (1998) Standards of ethical conduct in science. In: *The Ethics of Science: An Introduction*. Routledge, New York, pp. 53–73.

Roed, H (ed.) (2001) *The Status of Wild Atlantic Salmon: a River by River Assessment*. World Wildlife Fund. Full text is available at: http://www.wwf-uk.org/news/pdfs/atlanticsalmon.pdf

Rybacki, K.C. and Rybacki, D.J. (2000) What is argumentation? In: *Advocacy and Opposition: An Introduction to Argumentation*. Allyn and Bacon, Boston, Massachusetts, pp. 1–16.

Singer, P. (1975) *Animal Liberation: a New Ethics for Our Treatment of Animals*. Avon Books, New York.

Tiersch T.R. and Hargreaves, J.A. (2001) Sensible interactions with advocacy groups. Annual meeting of the World Aquaculture Society, 22 January 2001, Orlando, Florida.

Villiers, M. de (2000) *Water: the Fate of Our Most Precious Resource*. Houghton Mifflin, New York.

Index

Figures in **bold** indicate major references.
Figures in *italic* refer to diagrams, photographs and tables.

abalone 65
acadja-enclos 117
acid volatile sulphides (AVS) 167
acoustic deterrence devices (ADDs) 50, 51–52, 56
acoustic harassment devices (AHDs) 50, 51–52, *52*, 56
Admiralty Bay, Marlborough Sounds, South Island, New Zealand *48*
advection 275, 276
advocacy groups 19, 356, 358, 363
  income 361–362, *361*, 367
  interactions with 364–367
  tactics 360–361, 366
Africa 11, 149, 245
African sabellid worm 65
agriculture 101, 256–257
Alaska, USA 6, 72
  attitudes to aquaculture 73, 75
  permit system 212–213
  salmon farming 75, 99
algae 278, 294, 308
  fouling of marine cages 114
  macroalgae
    DO concentration 290–291, 294
    farms 96, *288*
    genetic improvement 248, 254
    nutrient content 279, 284–286, *285*, *286*
  as nutrient removers 95–96, 100, 285
    polyculture 99
algal blooms 66, 95, 177, 178, 268, 275
alwives 99
Amazon basin 46
American Fisheries Society 1, 254, 366
American Institute of Philanthropy 362
Americoat 675 168–169
ammonia 166, 179, 282, *307*
ammonium cycle 287
ammonium fluxes 288
anaerobic environments 123
anaerobic microbes 94

androgenesis 243
aneuploidy 244
angling *see* recreational fishing
animal nutrition 313–314
anoxia 95
antibiotics 110, *113*, 188
antifoulants 168–169
apparent effects threshold (AET) (zinc) 167
Approved Shellfish Harvest Classification 170
aquaculture 225
  changes in 221–222
  coastal *see* coastal aquaculture
  crawfish–rice 132
  ecological 107–108
  extractive aquaculture 15, 100–101, 265, 294
  fed 15, 265, 294
  freshwater 119
  land-based aquaculture 46
  marine *see* marine aquaculture industry
  offshore *see* offshore aquaculture
  polyculture *see* polyculture
  top ten species 247, *247*
Aquaculture 2001 xiii, 1, 235
Aquaculture America Conference 366
aquaculture cage classification scheme 109–110, *111*
aquaculture production 225, 229
  growth in xiii–xiv, 222, 315–316
  species, production weight *247*
  systems 23, 255
    fish cage culture 277, 278, 280, 288, 291–292
aquaculture technologies 26
aquaculture, criticisms of 248, 355–369
  environmental advocacy 356–357, 363–364
  response to criticism 358–363
aquarium fish 342, 345
Aquatic Lands Leases (ALLs) 166, 185
Arizona 209
Arroyo Aquaculture Association (AAA), Texas 298, *300*, *301*, 305–307, *307*
Arroyo Colorado, USA 66, 66–67
Aspen Corporation 29, 37
Atlantic salmon *see* salmon, Atlantic
Atlas of North American Freshwater Fishes 245
Australia 49, 50, 146, 212, 332, 335, 339, 344
aversive conditioning 4, 53–54, 55, 56
azamethiphos 175
azoic zones 12

Bacon, Francis 85
bacteria 94
  colonization of marine cages 113–115
  risks to human health 171
    faecal coliform bacteria 170
  salmon disease 213
balanced ecosystem management 91–94
  fish 98–100
  human community 100–102
  marine plants 95–96
  microbes 94–95
  polyculture *see* polyculture
  shellfish 97–98
  see also sustainable marine aquaculture
Ballard Locks, Seattle, Washington, USA 49
ballast water 206–207, 211
Bangladesh 122, 125
Bangkok Principles (FAO) 237
barriers (to predators) 53, 54, 55–56
bass 13, 209, 243, 251
bays 266, 293
bees 148
benthic community 276
  analysis of 175–176, 185, 186
    TOC screening 187
  effects of organic wastes 12, 27, 97, 180, 181–182
    management of 189–190
    measurement of 164
  impacts on 47–48, 190
    'halo' region 110, *112*
    neutral impact zone (NIZ) 163
    sediment impact zone (SIZ) 187, 188

best management practices (BMPs) 163, 168, 188, 362
Bintuni Bay, Indonesia 121
biochemical conditioning 113, 115
biochemical oxygen demand (BOD) 123, 162
biodeposition 274, 293
biodiversity 94, 95, 146, 181
biofouling organisms 112, 113, 114–16, 163
  see also wastes
biological feed conversion ratio (BFCR) 162
biological remediation (of sediments) 183–185
biosecurity 213, 214
biotechnology 239, 240
birds 10–11, 49, 132, 147, 170
bivalves 97–98, 100, 276, 329, 332
brackish water 146
British Columbia Draft Waste Management Policy 164
broodstocks 228–229, 243, 246, 336, 341
Brundtland Commission 108
Brunei 210
bubble curtains 53
bycatch 341

cages *see* marine cages
Calcutta, India 125
calicide 174
California, USA 13, 65, 209
Canada 46, 74, 322
  ADDs 52
  Bay of Fundy 180–181
  British Columbia 48, 161, 163
    Atlantic salmon 207, 208
    blue mussels 214
    clams 210
    copper benchmarks 169
    finfish aquaculture 48, 49, 52
    performance-based waste management policy (WMP) 189–190
    salmon farm sediments 165, 180, 184–185
    sewage outfalls 164
  Nova Scotia 31
  production losses 50
  Vancouver Island 52
Canadian Department of Fisheries and Oceans 207
Capitella capitata 12, 173
capture fishing 71–73, 100
  aquaculture's impact on seafood prices 73–74
  impact on natural populations 223–224
  improvements due to aquaculture 109
  loss of fishing grounds 74–75
  mangrove ecosystems 121–122
  offshore aquaculture 75–76
  stock enhancement 75
carbon
  from feed 162, 164
  measurement of 176
    TOC 183, 185, 186, 187–188, *188*, 190
  modelling salmon farm wastes 186–187
carbonaceous biochemical oxygen demand (CBOD) 307, *307*
carp 234, 246–247, 315–316, 318, 319
carrying capacity (of ecosystem) 94, 100, 108, 264–265
catfish 17–18, 67, 74, 247, 249, 251, 255
Cedar Key, Florida, USA 72, 73
charcoal 148
chemical remediation (of sediments) 182
chemicals 28, 239
Chesapeake Bay, USA 66, 95, 97, 98, 100, 131, 209
Chile 75, 190, 210, 213, 314, 315, 322, 323
China 15, 91, 95, 125, 334, 342
  Jiaozhou Bay Coupled Model Experiment *see* ecosystem modelling
  Xincun Bay Modelling Experiment *see* ecosystem modelling
chlorophyll 270–271, 272–273

chromosome manipulation 243–245, 252
cichlids 245–246
clams
farming 6, 72, 74, 210
non-indigenous 210, 212
prices 73
Ruditapes philippinarum 268
water filtration 97
Xincun Bay 277
Clean Water Act, 1972 (USA) 366
cloning 244, 253
coastal aquaculture 92–93
criticisms of 234–235
mangroves 145–146, 148–149
stock enhancement 80–81, 84, 85
wild-caught juveniles 329–335
Coastal Conservation Association (CCA) 68
coastal development 124–125
coastal ecosystems
management 26, 38
of nutrient levels 96, 100, 264–265
shoreline protection 122, 147
water column 94
wetlands 132
coastal ocean circulation model 288
forcing functions
tidal oscillations 268, 270, *271*, 275
time-variable inputs of rivers 268
wind driven features 268–269, 270
Coastal Zone Management Act (USA) 35
cobia 101
Cobscook Bay, Maine, USA 72, 75–76, 285
cockle beds 148
cod 73–74, 75, 80, 99
codes of conduct 35–36, 235–237, 362
GAA 153
genetics 14–15
siting 24–25
species introductions 211
Cold War 240
collection of juveniles 329–335, *330–331*
coloration (fish) 250
Columbia 125, 210
Columbia River 223
commercial fishing 6–7, 72
communities
coastal 25–26, 38, 39, 149
involvement 38–39
oyster gardening 131
sustainable 72
compensation depth 177
Conflict Prevention and Resolution Center, US Department of Agriculture 365
conflict resolution 68–70
Congo 149
contract services 72
Cook Islands 332
cooperatives 7, 77
copper 168–170
coral reef fishes 18, 98, 101, 329, 337, 342–347, *344*, *346*
coupled biological model 269–270
nutrient simulation 269–270, *272*
plankton measurement 270, *272*, *273*
stock density 270, 271–272
crest nets 344–345, 345–347
Criteria for Surface Water Quality Classifications 291
crustaceans 100, 181–182
Cuba 254
culling 54
currents 177–178
cypermethrin 174–175

Dagu River, China 266, 267, 268
dams 223, 239
denitrification 93, 94
Denmark 314, 315
deterrents (of marine mammals) 50
dichlorvos *see* cypermethrin
diets, farmed salmon 161, 162, 166
diffusion 275
discharge canals 301–302, *304*
disease resistance 249
diseases 13, 28, 64–65, 68
bacteria 170, 171
cage fouling 116

control of 188, 190
epidemics 239
molluscs 212
non-indigenous species 211–213, 214
shrimps 278
viruses 170, 211–212
wild juveniles 336
dissolved inorganic nitrogen (DIN) 167
dissolved oxygen (DO) 280–281, *281*
cycle 286–287
effect of salmon culture 179–180
low levels 95, 180, 284
modelling 286, 289–291, *289*, *290*, *291*, *292*, 294
dogs 242
dolphins 47, 48, *48*, 49, 51, 53
domestic stocks 247–248
domestication 225, 228, 229, 234, 237–238, 246, 308
double-crested cormorants 132
dugongs 47

earthworks 154
eco-labelling 345
ecological aquaculture 107–108
ecological concerns 27–29, 64
benthic community *see* benthic community
genetic changes *see* genetic changes
water quality *see* water
ecological engineering 119, *120*
ecological footprint 108–109, 236
ecological functions 93
economic feed conversion ratio (EFCR) 162
ecosystem modelling 15–16, 263–266
Jiaozhou Bay Coupled Model Experiment 266–271, *267*, *269*, *271*, *272*, *273*
effects of aquaculture 271–274
nutrient uptake and regeneration 274–276
river discharge 266, *268*
Xincun Bay Modelling Experiment 276–277, *276*, 291–293
aquaculture industry characteristics 277–278, *278*
ecosystem study and model 279–287
determination of major sinks and sources 287–288
model results 289–291, *289*, *290*
ecosystems
benthic *see* benthic community
coral reef 341, 342–347, *344*, *346*
effects of using wild-caught juveniles 341–342
management
balanced *see* balanced ecosystem management
sustainable *see* sustainable marine aquaculture
mangroves *see* mangroves
pelagic 106, 116–117
phosphorus limited 267, 270
restoration
oyster reef and shellfish 130–131
salt marshes 128–130
seagrass 94, 96, 127–128, *129*, 131
stressed, symptoms of *107*
Ecuador 151, 211, 333, 341
eelgrass (*Zostera marina*) 128, *129*
eels 334–335, 338, 340
effluent management 16, 297, *298*
discharge rates 306–307, *307*
feed management 305, *305*, 308–309
harvest techniques 308, *308*
using algae 95–96, 99, 100
water reuse 297–305, *299*, *300*, *301*, *302*, *306*, *307*, 308
see also sewage; wastes
effluents, pulp mill 164
Egypt 210
El Niño 312, 314, 315, 317
emamectin benzoate 174
endangered species 50
England, UK 212
entanglement 49, 55–56
'Environment Inc.' 361

- environmental advocacy 146, 356–357, 363–367
- Environmental Defense Fund 68, 72
- environmental effects 27
  - genetic resources 222–223
  - human activity 92, 93, 98
  - management of
    - modelling salmon farm wastes 186–187
    - monitoring experiences 185–186
    - risk management 187–190
  - netpen waste *see* netpen waste
  - sediment analysis 175–176
- environmental sustainability 237
- epidemics *see* diseases
- epiphytes 96
- erosion 28
- escapement 213, 224–225, 245, 253
  - Atlantic salmon 208
  - fish, site fidelity 117–118
  - shrimps 67–68
- Esposito, Beniamino 77
- Establishment (introduced species) 213–214, 214–216
- estuaries 93, 94, 96, 98
- ethical and stewardship practices
  - criteria for assessing 236
  - evaluation of genetic practices 245–248
  - genetics in aquaculture 240–241
  - historical lessons from crop genetics 237–240
- ethical codes *see* codes of conduct
- European Union 323
- eutrophication 95, 264–265, 280
  - WASP5 water quality model 286
- Excis 175
- exclusion (of marine mammals) 53
- extension professionals 73, 77
- extinctions 215, 223
- extractive aquaculture 15, 100–101, 265, 294

- faecal coliform bacteria 170
- FAO (United Nations Food and Agriculture Organization) xiii, 3, 10, 22–23, 24, 92
  - Code of Conduct for Responsible Fisheries 35, 108, 235, 236
  - State of World Fisheries and Aquaculture 80
- farming, mechanized 239
- farms, small 67, 239–240, 251, 253
- fed aquaculture 15, 265, 294
- feed conversion ratios (FCRs) 162–163
- feedback technologies 161–162
- feeding methods 161, 162, 305, 308
- feeds
  - animal 313–314, 316
  - fish 17–18, 96
    - composition 161, 162, 167, 313
    - management 305, *305*, 308–309
    - production 316, *316*
    - protein sources 312, 314, 317, 318, 319–321
    - waste 161–162
  - fish meal *see* fish meal
  - pelleted 312, 315, 345
- fertilizers 239
- field surveys 279
- filter feeders 97–98, 100, 123, 130, 270
- financial investment 24, 26, 30, 32, 39, 81, 85
- finfish aquaculture 47–49
  - conflicts with marine mammals 50, 54–56
  - dissolved nutrient levels 96
  - nested cage arrangement *110*
  - opposition to 73
  - stock enhancement 75
- fish
  - coloration 250
  - evolution 98
  - faeces 162–163
  - as food source xiii–xiv, 100, 363–364
    - demand for 215, 223
    - prices 73–74
    - yield 101
  - game 63, 67, 101
  - industrial 314–315, 322
  - net cleaning 116
  - nutritional requirements 318

role in balanced ecosystems 99–100
benthic 117
coral reef species 98, 101, 329
forage species 98
pelagic species 116–117
polyculture 95–96, 99–100
solid organic wastes 161–163
year classes 190
fish aggregating devices (FADs) 112, 116–119, *118*
fish cage culture *see* marine cages
fish farms 67, 77, 360
construction of 153–155
fallow periods 183–184, 190
sediments *see* sediments
small 152
wastes *see* wastes
fish meal 17–18, 31, 311–313, 363–364
alternative protein sources 18, 319–321, *320*
aquaculture production trends 315–316
challenges for the future 323–324
factors affecting abundance 323
history of use 313–314
production trends 17, 311, 314–315, *316*
sustainable supplies 321–322
use in feeds 64, 317–319, *317*, *318*, *319*, *320*
fish oil 313, 314–315
fish processing 73–74, 312
fish stocks xii
collapse 322
decline in 80, 92, 95, 97
enhancement of *see* stock enhancement
restoration 339
strength 321–322
fisheries
balanced ecosystem approach 93
and mangroves 126–127, 148
yields 92, 93
fishermen
business opportunities 72, 76
cooperatives 77
inshore 72, 76
lobster 72, 74–75, 76
surveys of 71, 72, 73, 76
fishing 6–7, 72
capture *see* capture fishing
destructive 342
recreational *see* recreational fishing
shrimp *see* shrimp farming
fishing grounds, loss of 74–75
Flexgard XI 169
Florida, USA 6, 19, 209
clam aquaculture 72
mangrove forests 120–121, 122, 148
sport fish 101
flounder 73, 74, 75, 80, 96
food conversion efficiency (FCE) 249, 277
food safety 324
food security 234, 240
food web 276
forage fish 98
forest canopy 122
fouling *see* biofouling organisms
French Polynesia 332, 337, 344, 345
freshwater 67, 122, 223
fuelwood 148

genetic changes 14, 221–222, 224
responsible aquaculture 14–15, 236–237
risk of *see* genetic risk
sources and magnitude
aquaculture 224–225
capture fisheries 223–224
environmental changes 222–223
genetic conservation 227–229
genetic diversity 222, 224, 225, 236
decline in 241–242, 245
genetic drift 228
genetic improvement programmes 238, 249, 256
genetic resources
management 224–225
natural environment 222
ownership 238, 241, 242

genetic risk 27, 225–227, *226*, *227*, 228
impact on native populations 28, 65–66
genetic technologies
application of 248–254
genetic modification 241
chromosome manipulation 243–245, 252
gynogenesis and androgenesis 243, 251
hybridization 243, 251–252
inbred lines 242, 250–251
selective breeding 242, 248–250, 251, 256
genetic monitoring 240, 241–242, 245, 246
genetic variability 225–227, *226*, *227*, 228, 242
genetics, responsible use of 14–15, 233–234, 240–241
ethical and stewardship practices 14–15, 236, 237–241, 245–248
objectives 236–237
Georgia, USA 210
germplasm 'banking' 229
GESAMP 175–176
Global Aquaculture Alliance (GAA) 153–154, 211, 366
global production, aquaculture species *247*
GMOs (genetically modified organisms) 13, 215, 362
goldfish 234
governments 24, 30–31
information and education 39–40
policy-making 36–37, 68–69, 152
regulations *see* regulations
role of 32, 33–34, 35–36
Gracilaria parvispora (alga) 99
grain production 240
GRAS classification 238
gravity cages 109–110
Great Barrier Reef 346
Great Lakes, USA 206, 207
Green Revolution 238, 239, 240
Greenpeace 356
groupers 101, 334, 338, 341
growout
juveniles 336–337
long-line system *115*
sites 117, *118*
Gulf of Maine 50
Gulf of Mexico 6, 73, 95, 101, 212
Gulf of St Lawrence 207
gynogenesis 243, 251

habitats 9–11
birds 10–11, 132, 147
destruction of 47, 62–63, 338
fish 106
mangroves 122–123
marine cages 109–113, *110*, *111*
shellfish 130–131
haddock 73, 80
harassment (of marine mammals) 4, 51–52
Hard Green: Saving the Environment From the Environmentalists: a Conservative Manifesto (Huber) 359
Harlingen Shrimp Farms, Texas *302*, *304*
harvesting 223, 308, *308*
hatcheries 80–82, 329, 334, 335–336
Hawaii, USA 86, 99, 122, 209, 212
heavy metals 123, 167–170
herring 49
heterosis 242, 243, 251
Hong Kong 342
hormonal shifting 244
human communities
diet 363–364
disadvantages of aquaculture 235
environmental effects 92, 93, 98, **100–102**
eutrophication 264–265
on genetic resources 222–223
wastes 108, 123
impact of aquaculture 246, 255
health risks 170, 171, 254, 324
population growth 124, 215–216
hybridization 243, 245, 251, 252
hydrogen sulphide gas production 178–179

hypothesis testing 84, 85
hypoxia 95

Iceland 314, 315, 322
IN/IP ratio 282–283, *283*
inbreeding 238, 240, **242**, 243, 248, 250–251
India 211, 341
indicator taxa 186
Indonesia 121, 126, 150, 211, 212, 332, 334
inland waterways 46
inorganic wastes 166–170
inshore fishermen 72, 76
International Marinelife Alliance 342
intertidal zone 276, 277
introductions *see* non-indigenous species
invertebrates, succession 184
Ireland 211
Israel 210, 212
Italy 77
ivermectin 172

jacks 121
Japan 7, 74, 95, 322, 332, 334, 335
Jiaozhou Bay Coupled Model Experiment *see* ecosystem modelling
Joint Subcommittee on Aquaculture (JSA) (USA) 32, 37
juveniles
  after settlement 334–335
  large 335
  limitations to supply 340
  pre-settlement and settlement stages 329, 332–334, 341, 344
  propagation 335

Kappaphycus sp. 277, 284, *285*, 294
kelp 151
killer whales 50, 51, 52, *52*
Knudson, Tom 361
Korea 211, 334

Lacey Act, USA 210
Laguna Madre, Texas, USA 67
Lake Erie 97
Laminaria sp. 95
Land Grant universities 363
Latin America 11
leaf fall 121, 147
leases, marine areas 7, 30
legislation 29–30, 32–33, 35–36, 37, 67
light availability 177
light traps 344, 345–347
lithium chloride (LiCl) 53
live food market 342, *343*, 346
live reef fish trade 342–347
livestock production 148, 313–314
lobsters 72, 74–75, 76, 175
  American 209
  spiny (*P. ornatus*) 333, 338–339, 345
Louisiana, USA 101, 130, 132

mackerel 80, 99, 322
macroalgae *see* algae
Madeira Archipelago, Portugal 210
Maine, USA 74, 75, 95–96
maize breeding 238–240
Malaysia 148, 334
management of marine aquaculture
  active-adaptive 85–86
  assessing stocking impact 82–84, *83*
  domestic stocks 246–248
  environmental effects of waste *see* netpen waste
  modelling salmon farm wastes 186–187
  polarized fisheries management 80–81
  risk management 187–190
  site selection 25
  sustainability *see* sustainable marine aquaculture
  wild stocks 245–246
manatees 47
mangrove controversy 149–153, *150*, *151*, *152*
mangrove protection 153–155
mangrove restoration aquaculture 126–127

mangroves 11, 119–120
coastal aquaculture 145–146
ecosystems 109, 121, **146–147**, *147*
habitats 122–123
shoreline protection 122
value to fisheries 121–122
importance of 121
attributes 148
functions 147
social issues 149
uses 148–149
loss of 28, 47, 147, **149–153**, *150*, *151*
reasons for destruction 149–150
aquaculture development 125–126
expanding urban areas 124–125
uncontrolled population growth 124
shrimp pond wastes 123–124
soil 152, *152*
value of 121
Marco Polo 313
Marina Di Camerota, Italy 77
marine agronomy 109
marine aquaculture facilities 105–109
habitats and ecosystems 109
biofouling community 113–116
fish aggregating devices (FADs) 116–119, *118*
marine cages *see* marine cages
marine aquaculture industry
conflict with recreational fishing 5–6, 61–70
growth and development 46, 61
challenges to 22–23
impact on natural populations 50, 221–230
nearshore 75–76
offshore 7, 72, 75–76
predation losses 46, 50
sustainability *see* sustainable marine aquaculture
marine cages 277, 278, 280, 288
cage mooring systems 117
classification scheme 109, *111*
gravity cage 109–110, 116
colonization of 113–114
ecological engineering 119, *120*
fouling of 110, 112, 113, 114–116
nested arrangement *110*, 112–113, 116
oysters 131
pollution 291–292
see also netpen waste
marine chronic water quality criteria 168, 169, 179
Marine Mammal Protection Act (USA) 49, 50
marine mammals 4, 45–46, 170
aversive conditioning 4
exclusion 4, 53
removal and population control 4, 53–54
effect on gross production 46, 50
finfish predation *see* finfish
mitigation measures for conflicts 50–51
harassment 4, 51–52
shellfish predation *see* shellfish
see also wild populations
marine plants 95–96
marketing 31, 39, 239
Marlborough Sounds, South Island, New Zealand 46, 47, *48*
Maryland, USA 131
Massachusetts, USA 72, 73, 74
media criticisms 356–357, *357*
menhaden 98, 99, 313
metal toxicity 167
Miami, Florida, USA 120
microbes 12, 94, 96
microsatellite marker technology 248, 249
milkfish 150, 334, 340
mink 4
Mobile Bay, Alabama, USA 131
molluscs 65, 181–182, 212
moral codes 235
mullet 85, 99, 121, 210
multicellular fouling 113, 114, 115
Murky Waters: Environmental Effects of Aquaculture in the United States 72, 366
mussels 46, 47, 49, 108
blue *115*, 214, 329, 332

debris 163
green 210, 332, 340
long-line growout system *115*
water filtration 97
zebra (*Dreissena* sp.) 206

Narragansett Bay, Rhode Island, USA 130
National Academy of Sciences (NAS), USA 313
National Research Council (NRC) 25, 31
Committee on Animal Nutrition (CAN) 313–314
National Aquaculture Act, USA 32, 35, 36, 37
National Aquaculture Development Plan (USA) 36, 37
National Aquaculture Information Center, USA 32
National Centre for Charitable Statistics 362
National Invasive Species Council, USA 210–211
National Marine Fisheries Service (NMFS), USA 14, 36, 81, 92, 93, 207, 209
National Oceanic and Atmospheric Administration (NOAA), USA 35, 93
National Science and Technology Council (NSTC), USA
Committee on Health, Safety and Food Research 32
National Shellfish Sanitation Programme (NSSP), USA 170
National Shellfisheries Association 1
natural populations *see* wild populations
Natural Resources Defense Council 366
Nature Conservancy 361, 342
nearshore fisheries 121–122
nearshore marine aquaculture 75–76
nested cage arrangement 110
netpen waste 159–160
inorganic wastes
dissolved nitrogen and phosphorus 166–167
heavy metals 167–170
pathogenic organisms 171–172
effects of therapeutic compounds 172–175
faecal coliform bacteria 170
solid organic wastes
biofouling organisms 163
fish carcasses 163
fish faeces 162–163
measurement of 164–165
waste feed 161–162
see also sediments; marine cages
netpens 49, 75, 92, 213
nets
cleaning 115, 168, 169
fouling of 114–115
fyke 341
lantern 270
neutral impact zone (NIZ) 163
New England, USA 6, 7, 71, 74, 76
New Hampshire, USA 71
New Jersey, USA 73, 74
New York, USA 74
New Zealand 46, 47, *48*, 49, 50, 339, 340
Ninety Mile Beach, Marlborough Sounds 332
nitrate cycle 287
nitrates 281, *282*
fluxes 288
nitrogen
inorganic 282–283, *283*
Jiaozhou Bay 266–268
mangrove sediments 123
removal of 93, 95, 96, 100
solid form 164
water column loading 92, 99, 162, 177–178
salmon farms 166–167
noise pollution 51–52
non-governmental organizations (NGOs) 356, 365, 366, 367–368
non-indigenous species 12–13, 28, 67, 205–207
ballast water 206–207, 211
case against introductions
diseases 211–213, 214
establishment 213–214, 214–216
definition 205–206

non-indigenous species *continued*
examples of introductions
American lobsters 209
Atlantic salmon 207–208
freshwater shrimp 209–210
green mussel 210
Manila clams 210
mullet 210
oysters 209, 210, 247
scallops 210
striped bass 209
tilapia 206, 210, 245
regulations 210–211
nori 95–96, 248
North Sea 94
Northern Aquaculture 73
Northern Ireland, UK 208
Norway 7, 75, 93, 162, 188, 190, 251
fish meal 313, 314, 322
NPDES permit standards 187–189
NPZ model 274–275
nursery grounds 147, 148
nutrient content
macroalgae 284–285, *285*
seagrass 285–286, *286*
nutrient fluxes 275–276, 288
nutrient pump 95
nutrient sinks 279
nutrients
levels 93
balance 264
distribution 118–119, 274, 281–283, *282*, *283*
Jiaozhou Bay 268
removal of 99–100
algal 95–96, 99, 285, 292–293
bacterial 94
mangrove forests 123–124
uptake and regeneration 274–276
Nutritional Requirements of Domestic Animals 314

offshore (open ocean) aquaculture 7, 72, 75–76
oil and gas platforms 10
organic wastes 97, 161–165
Orlando, Florida xiii, 1, 19
ornamental market 18–19, 342, *343*
otters 4, 46, 49, 50
outcrossing 238, 239, 250
overfishing 92, 97, 223–224, 343
oxygen, dissolved *see* dissolved oxygen (DO)
oyster reef restoration 130–131
oysters 27, 46, 47
spat 247, 329, 332
cupped *227*
ecological role 97, 98, 109
non-indigenous 209, 210, 212–213
Pacific 13, 247
pearl 277, 278, 332, 337
rock 252

Pacific Northwest 4, 46
extinctions 223
salmon farming
introductions 207
netpen waste *see* netpen waste
phytoplankton 177
removal of pinnipeds 53–54
sea lice 172
Pacific Ocean 93
Pacific threadfin 85
paper production 148
parasites 64–65
patents 239, 242, 255
pathogenic organisms 171–175
PCBs (polychlorinated biphenyls) 18
pearl industry 332, 337
People for the Ethical Treatment of Animals 365
Perkin-Elmer CHN elemental analyser 279
permits, fishing 71
Peru 9, 93, 314, 315, 322, 323
pesticides 123, 239
Philippines, The 125, 148, 150, 332, 334, 340, 341
phosphorus
concentrations 282
inorganic 282–283, *283*
Jiaozhou Bay 266–267, 270
mangrove sediments 123
removal of 93, 96

solid form 164
water column loading 99, 162, 177–178
salmon netpen farms 166–167
photosynthesis 93, 95
physiochemical changes 180–181
physiochemical end-points 175, 184–185, 186
phytoplankton *see* plankton
Picasso triggerfish 345
pinnipeds 47, 48, 49, 51, 53–54, 55
plankton 95, 96, 97, 99
measurement of 270
non-indigenous 206–207
phytoplankton 274
consumption by shellfish 274–275
fluxes *273*, 275–276
nutrient consumption 282
production 177–178, 265, 272–273, *273*
Jiaozhou Bay 267–268
salinity range 280–281
zooplankton 274
planning for marine aquaculture
balanced ecosystems 93–94
national 32, 33–34, 37–38
polarized fisheries management 80–81
policies 25, 37, 69
components of 30–31
USA 32–33, 35, 36–40
pollack 80
pollution 46, 291–292, 324
polychaetes 181
polychlorinated biphenyls (PCBs) 323
polyculture 9, 15, 265–266, 292–293
salmon farming 95–96, 99, 119, 285
shrimp farming 99
polyploidy 243–244, 252
pools 122, 148
population control (of marine mammals) 53–54
population growth 124
Porphyra sp. 95–96
porpoises 47, 52
Portugal 210
poultry 313, 314, 317
'Preadoption of Open Ocean Aquaculture by Northern New England's Inshore Commercial Fishermen' 76
precautionary principle 367
predators 49
marine mammals *see* marine mammals
principal components analysis (PCA) 183
production oriented stocking 80–81
productivity, increasing 337–340
professional aquaculture organizations 366–367
property rights 25, 33–34, 38
protein sources
fishmeal 312, 314, 317, 318, 322
plant-derived 314, 317, 320–321
rendered products 321
Providence River, USA 97
public health 170, 171
public perceptions (of aquaculture) 34–35, 50, 146, 357
Puerto Rico 209
Puget Sound estuary protocols (PSEP) 188

Qingdao, China 264, 266
quantitative trait locus selection (QTL) 249
quarantine 212

rabbitfish 341
Ranong, Thailand 121
recirculation systems *see* water recirculation
recombinant DNA technology 244–245, 254
record-keeping 228–229
recreational fishing 5–6, 61–64
conflict resolution 68–70
conflicts with aquaculture 64–68, 101, 338
recycling waste products 99, 100
red alga *see* nori
red drum 63–64, 75
red swamp crawfish 132

redox potential 175, 176, 185, 186
reefs
artificial 117, 119
coral 18, 98, 101, 342–347, *344*, *346*
regulations 69–70
genetic modifications 238
non-indigenous species 210, 212–213
NPDES permit standards 187–189
shellfish sanitation 170
Texas 67, 68
removal (of marine mammals) 53–54, 55, 56
reproduction 246, 252
control of 229
success (farmed salmon) 208, 228–229
research agenda 33, 34, 39
resource management 35, 63–64
responsible aquaculture
forms of responsibility 235–236
genetics *see* genetics, responsible use of
irresponsible aquaculture 234–235
mangrove protection 153–155
sustainable marine aquaculture *see* sustainable marine aquaculture
Responsible Marine Aquaculture symposium 1
Rhode Island, USA 74, 130
rice 132, 148
Rio Grande Valley, Texas, USA 67
risk management
British Columbia, Canada 189–190
NPDES permit standards 187–189
rivers 97, 266, *268*
Robertson, Robert 72, 76
RoLande, Ray 73
Russia 46

Sacramento Bee 361
salmon
Atlantic 13, 161, 163, 184, 251
domesticated 247, 248
escapement 208, 215
introductions 207, 210
reproduction 208, 246
transgenic 253–254
chinook 168–169, 171
coho 213
copper levels 168–169
diseases 213
masu 207
Pacific 207, 208, 210, 215, 223, *227*
selective breeding goals 249
sockeye 166
urea excretion rate 166
salmon farming 6
antibiotic use *113*
control of predators 49, 52
diet 161, 318
ecological footprint 108–109
effect on dissolved oxygen (DO) 179–180
effect on prices 73
escapes 208
management of environmental effects 185–190
algal removal of nutrients 95–96, 99
modelling farm wastes 186–187
netpen waste *see* netpen waste
non-indigenous species 13, 207–208, 213
opposition to 74–75
PCBs 323
physiochemical changes in sediment 180–181
stock enhancement 75
salmonids 80, *120*, 223
Salmosan 175
salt marsh restoration 128–130
salt tolerance 146
sardines 322
Saudi Arabia 210
scallop rafts 270
scallops 210, 268, 270, 271–274, 293, 332, 341
Scotland, UK 164, 172–173, 174, 208
Scottish Environmental Protection Agency (SEPA) 173, 174, 175
sea cages *see* marine cages
sea cucumbers 99

sea lice 172
sea lions 4, 49, 50, 53, 54
sea pens 49
seabream 77, 250, 251
seafood xii–xiv, 100, 308
contaminants 324
demand for 215–216, 223, 312
prices 73–74
waste processing 312, 322, 323
yield 101
seagrasses 94, 96, 277–278
nutrient content 279, 284–286, *285*, *286*
restoration aquaculture 109, 127, 128, *129*, 131
seal bombs 51, 52
seals 4, 46, 47, 49, 52, 53, 54, 55, *55*
seaweed *see* algae
Second International Symposium on Tilapia in Aquaculture 250
sediment impact zone (SIZ) 187, 188
sediment oxygen demand (SOD) 282, 286–287, 288
sedimentation 122, 147, 165
sediments 97
allowable concentrations
therapeutic compounds 173–175
zinc 167–168
assessment 283–284, *284*
biological changes 176
dissolved oxygen 179–180
effects of wastes on benthic infauna 181–182
hydrogen sulphide gas production 178–179
number of animals 188
physiochemical changes near salmon farms 180–181
water column changes 177–178
BOD 162
management standards 187–189
mangrove 123
monitoring environmental effects 175–176, 185, 186
recovery and remediation
biological remediation 183–185
chemical remediation 182
Xincun Bay 284, *284*
see also netpen waste; water
seed planting (mangroves) 126–127
seed production 238, 239
seedstock 252
selective breeding 242, 248–250, 251, 341
settlement, juveniles 329, 332–335, 338–340
sewage 63, 92
alternative treatments
mangroves 123, 154
nori culture 96
shellfisheries 98
effect on benthic infauna 164
harvesting of shellfish 170
see also effluent management; wastes
sex manipulation 253
sex ratio 244
Sgiegie Market, Tokyo, Japan 74
Shannon's diversity index 183
shellfish
habitats 130–131
oysters *see* oysters
role in balanced ecosystem 97–98
sanitation of 170
scallops 210, 268, 270, 271–274, 293
spat collection 114
shellfish aquaculture 72
conflicts with marine mammals 46–47, *48*
opposition to 73, 74–75
small businesses 77
shoreline protection 122, 147
shrimp farming 6, 46
code of conduct 211
conflict with recreational fishing 5–6, 61, 62–63, 66–70
effect on prices 73
effluent management 16, 297, *298*
discharge rates 306–307, *307*
feed management 305, *305*, 308–309
harvest techniques 308, *308*
pond wastes 123–124, 152–153

shrimp farming *continued*
effluent management *continued*
using algae 99
water reuse 297–305, *299*, *300*, *301*, *302*, *306*, *307*, 308
environmental impacts 11, 66–68, 125–126
estimated land use 151
hybridization 251–252
management of wild stocks 245–246
mangrove forests
loss of 149–153, *150*, *151*
protection of 153–155
marine mammals 47
Middle East 210
non-indigenous 209–210, 211
disease transmission 211–212
production rates 305, 307, *307*
vagile postlarvae 332–333
Xincun Bay 277, 278
shrimp feed blowers 305
shrimps
freshwater 13, 209–210, 247
Marine 247, 248
mortality 173, 175
Pacific white 67, 246, 308
tiger (*Penaeus monodon*) 13, 245, 246, 248, 341
white 68
Sierra Club 68
Silent Spring 356
silvofisheries 127
Sino-US Living Marine Resources Panel 264
site fidelity 117–118, *118*
siting (aquaculture facilities) 24–25
ecosystem preservation 106
extractive aquaculture 100–101
finfish aquaculture 101
shrimp ponds 125–126
to avoid predators 54
snails, marine 65
snappers 101, 121
socio-cultural considerations 25–26
sodium cyanide solution 342
soils 122, 123, 152, 154
solid organic wastes
biofouling organisms 163
fish carcasses 163
fish faeces 162–163
measurement of 164–165
waste feed 161–162
Solomon Islands 344–345, 346
sound discomfort 51
South Atlantic 101
South Carolina 209
Southeast Asia 124, 125, 146, 147, 149, 212
spat 114, 252, 329, 332
spawning 208
species extinction 101–102
species richness *see* biodiversity
Sprague, Sonny 74
Sri Lanka 122, 210
St Lawrence River 206–207
stewardship 235–236, 237, 245, 246
stock enhancement 7–9, 75, 79–80
genetic technologies *see* genetic technologies
opposition to 81
pros and cons 81–82
scientific basis 82–84
uncertainties in 83, *83*
wild juveniles 336
stock enhancement programmes 207, 341
accountability 84–87
Striplin Environmental Associates 183
'strong inference' hypothesis testing 85, 86
successional series 184
sulphides 167, 168, 176, 180–181, 182, 186
surveys (of fishermen) 71, 72, 73, 76
suspended rake culture 268, *269*
sustainable marine aquaculture 2–3, 21–23, 106–107, **108–109**
complexities of industry 23–24
definition 10, 235–236
environmental and ecological concerns 27–29, 236, 237
institutional and legal environment 29–30
policy framework 30–33, 33–34, 35
siting 24–25

socio-cultural considerations 25–26, 72, 236
challenges to 34–40
see also balanced ecosystem management
Swan's Island, Maine, USA 74
Sweden 108, 164

tagging technology 81, 341
Taiwan 67, 211, 334, 338, 340, 342
Tango-Lowy, Torene 76
Tasmania, Australia 339
taste aversion 53
Taura syndrome virus (TSV) 5, 68, 211–212
technology transfer 39–40
teflubenzuron *see* calicide
telemetry research 117–118
temperate marine environments 177
teraploids 244
Texas
finfish stock enhancement 75
shrimp farming 209, 212, *300*
annual production 298, *299*, *303*
conflicts with recreational fishing 5, 61, 62–63, 66–70
discharge averages 309
environmental impacts 66–67
stocking density 301
water reuse 16, 298–305, *299*, *301*, *302*
sport fish 101
Texas Department of Agriculture 67
Texas Natural Resources Conservation Commission (TNRCC) 67
Texas Parks and Wildlife Department (TPWD) 66, 67, 68
Texas Shrimp Association 68, 73
Thailand 121, 125, 211
loss of mangrove forests 149–150
water recirculation on shrimp farm *299*
therapeutic compounds 172–175
tidal mixing 275
tides
amplitude 266, 279
current 266, *271*, 279–280
DO concentration 289–291, *289*, *290*
model of 268, 270
salinity 289
temperature 289
tilapia 67, 205, 206, 210, 243, 244
interspecific crosses 251
non-native habitats 245
red 250
sex manipulation 253
transgenic 254
tinies 333–334, 341
total organic carbon (TOC) 170, 181
total suspended solids (TSS) 306, *307*
total volatile solids (TVS) 165, 167, 174, 182, 183, 186, 190
trait selection
aquaculture 249–250
maize 238, 240
transgenics 239, 244–245, 253–254
translocation 206, 213
'trialability' 26
triploids 229, 243–244, 252
trout
rainbow 247, 248
selective breeding goals 249
steelhead trout 49, 117, 171
whirling disease 65
tuna 49, 335, 340
turbidity 66–67

unicellular fouling 113, 114, 115
United States Department of Agriculture Plant and Materials Center (PMC) 130
United States Department of Agriculture Soil Conservation Service 120
United States Emergency Wetlands Restoration Act 128
United States of America (USA)
animal feeds 313–314
commercial fishing 6–7
fish stock enhancement 7–9, 75, **79–87**
government policy 32–33
marine mammal protection 4
overfishing 92

United States of America (USA) *continued*
oyster reef restoration 130–131
production losses 50
recreational fishing 5, 64–70
regulation of aquaculture 29–30
shrimp farming 5, 13, 16
wetlands 128–130
aquaculture ponds 132
eelgrass 128
mangroves 120–121, 124
University of Alaska Marine Advisory Programme 73
University of California, Santa Barbara, USA 65
University of New Hampshire Open Ocean Aquaculture Demonstration Project 77
University of Rhode Island, USA 25, 128
urban expansion 124–125
urea 166
US Bureau of Commercial Fisheries *see* National Marine Fisheries Service (NMFS)
US Environmental Protection Agency (EPA) 166, 168, 304, 366
US Exclusive Economic Zone (EEZ) 29, 36, 37
US Food and Drug Administration (FDA) 238
US Institute for Environmental Conflict Resolution 365
US Joint Subcommittee on Aquaculture 29
US National Marine Fisheries Service 14

vaccination 188
vagile postlarvae 332–334
Vibrio vulnificus (bacteria) 171
Vietnam 125, 333, 334, 335, 338, 339
Virginia Institute of Marine Science, USA 131
viruses 170, 211–212, 244
volatile suspended solids (VSS) 308

Wales, UK 212
walrus 46
Washington State, USA 166, 167, 171, 172
Atlantic salmon escapes 208, 215
clams 210
invertebrate analysis 185
Puget Sound 179, 180, 184, 207
sediment management standards 187–189
wastes
effects on benthic infauna 181–182
inorganic 166–170
modelling 186–187
pathogenic organisms 170–172
performance-based waste management policy (WMP) 189–190
recycling 99, 100
solid organic 161–165
treatment 108, 154
microbial degradation 12
pelagic and benthic layer 106
utilization of
fishmeal 312, 322, 323
polyculture 119, *120*
see also biofouling organisms; effluent management; sewage
water
brackish 146
filtration
by bivalves 97–98
by mangroves 147, 148–149
flow in marine cages 116
fresh 67, 122, 223
public use of 23–24
quality 23, 27, 28, **63**, 247, 278
assessment 283–284, *284*, 286
bacteria levels 170
coastal 94, 147, 148–149
copper levels 168, 169
eutrophication 95, 264–265, 280, 286
models 286–287, 288, 292
parameters *307*
and shrimp farms 66–67
undissociated sulphide 179
Xincun Bay 283–284, *284*
salinity 270, 280, *281*, 289, 305

temperature 270, 289
see also sediments
water column changes 12
denitrification 94
nitrate dissimilation 94
nitrogen and phosphorus loading 92, 93, 99, 177–178, 288
water column oxygen demand (WOD) 286–287
water recirculation
biosecurity 213, 214
nori culture 96
shrimp farming 5, 16, **297–305**, *299*, *300*, *301*, *302*, *304*, 308
West Africa 117
wetlands 119–121
aquaculture ponds 5, 10–11, 132
loss of 128–129
Texas 66, 308–309
tidal 109, 127
waste management 123–124
whales 49, 50, 51, 52, *52*
whirling disease 65
wild populations
adverse effect of stocking 82
bacterial infection 171–172
FADs 117, 119
genetic impact of aquaculture 14, 27, 215, 221–230, 252, 253
harvest of 101–102
decline in 6, 73–74
management of 245–246
reducing impacts on 336
see also marine mammals
wild-caught juveniles 327–329
advantages of
bypassing propagation 335–336
increasing productivity 337–340
reducing impacts on wild stocks 336
socioeconomic benefits 336–337
collection of *330–331*
juveniles well after settlement 334–335
large juveniles 335
pre-settlement and settlement stages 329–334
disadvantages of
effects on ecosystem 341–342
lack of selective breeding 341
limitation to supply and quality 340
postlarval coral reef fishes 342–347, *344*, *346*
Wilson, Woodrow (President, USA) 313
winds 268–269
World Aquaculture Society (WAS) xiv–xv, 1, 19, 357, 361, 366
World Resources Institute 149
World Wide Fund for Nature 356
Worldwatch Institute 359
WSSV (white spot syndrome virus) 212

Xincun Bay Modelling Experiment *see* ecosystem modelling

Yangtze River, China 46
year-classes 322
Yellow Sea, China 264, 266
yellowtail 74, 335
YHV (yellow head virus) 212

zebra mussel 206
zinc 167–168
zooplankton *see* plankton

Printed and bound by CPI Group (UK) Ltd, Croydon, CR0 4YY
16/09/2026
14963432-0001